Soil Erosion at Multiple Scales

PRINCIPLES AND METHODS FOR ASSESSING CAUSES AND IMPACTS

Soil Erosion at Multiple Scales

Principles and Methods for Assessing Causes and Impacts

Edited by

F.W.T. Penning de Vries

International Board for Soil Research and Management
Thailand

F. Agus

Centre for Soils and Agroclimate Research
Indonesia

and

J. Kerr

International Food Policy Research Institute
Washington DC

CABI *Publishing*

in association with the

International Board for Soil Research and Management (IBSRAM)

CABI Publishing
Wallingford
Oxon OX10 8DE
UK

Tel: +44 (0)1491 832111
Fax: +44 (0)1491 833508
Email: cabi@cabi.org

CABI Publishing
10 E 40th Street, Suite 3203
New York, NY 10016
USA

Tel: +1 212 726 6490
Fax: +1 212 686 7993
Email: cabi-nao@cabi.org

A catalogue record for this book is available from the British Library, London, UK

Library of Congress Cataloging-in-Publication Data
Soil erosion at multiple scales : principles and methods for assessing causes and impacts / edited by F.W.T. Penning de Vries, F. Agus, and J. Kerr.
p. cm.
Based on papers from a workshop held in Indonesia Nov. 1997.
Includes index.
ISBN 0-85199-290-0 (alk. paper)
1. Soil erosion—Congresses. 2. Soil erosion—Simulation methods—Congresses. I. Penning de Vries, F.W.T. II. Agus, F. III. Kerr, J.M.
S622.2.S655 1998
631.4′5—dc21 98-22770
CIP

ISBN 0 85199 290 0

Published in association with:
IBSRAM (International Board for Soil Research and Management)
PO Box 9-109
Jatujak
Bangkok 10900
Thailand

This volume is published as IBSRAM Proceedings no. 18 and is an output of 'The Soil, Water, and Nutrient Management (SWNM) Programme' of the Consultative Group on International Agricultural Research (CGIAR).

Typeset in Garamond by Columns Design Ltd, Reading
Printed and bound in the UK by Biddles Ltd, Guildford and King's Lynn

Contents

Contributors

A. Abdurachman, Center for Soil and Agroclimate Research, Jln. Juanda 98, Bogor 16123, Indonesia

Fahmuddin Agus, Center for Soil and Agroclimate Research, Jln. Juanda 98, Bogor, 16123, Indonesia

E. Amézquita, Centro Internacional de Agricultura Tropical (CIAT), A.A. 6713, Cali, Colombia

J. Ashby, Centro Internacional de Agricultura Tropical (CIAT), A.A. 6713, Cali, Colombia

Belinda Barnes, School of Mathematical Sciences, The Australian National University, Canberra, ACT 0200, Australia

E. Barrios, Centro Internacional de Agricultura Tropical (CIAT), A.A. 6713, Cali, Colombia

Sara G. Beavis, Centre for Resource and Environmental Studies and Centre for Integrated Catchment Assessment and Management, The Australian National University, Canberra, ACT 0200, Australia

J. Beltran, Centro Internacional de Agricultura Tropical (CIAT), A.A. 6713, Cali, Colombia

Jean Pierre Bricquet, Laboratoire d'hydrologie, ORSTOM, BP 5045, 34032 Montpellier cedex, France

C.A.A. Ciesiolka, Department of Natural Resources, 203 Tor Street, Toowoomba, Queensland 4350, Australia

Jacques Claude, Laboratoire d'hydrologie, ORSTOM, BP 5045, 34032 Montpellier cedex, France

Robert Collins, Institute of Hydrology, Crowmarsh Gifford, Wallingford, Oxon OX10 8BB, UK

Kep J. Coughlan, ACIAR, GPO Box 1571, Canberra, ACT 2601, Australia

R.A. Cramb, Department of Agriculture, University of Queensland, Brisbane, Australia

Eric T. Craswell, International Board for Soil Research and Management, PO Box 9-109, Jatujak, Bangkok 10900, Thailand

Adam Dickinson, HR Wallingford, Howbery Park, Wallingford, Oxon OX10 8BA, UK

Claude R. Dietrich, Faculty of Environmental Sciences, Griffith University, Nathan, Queensland 4111, Australia

Thomas Enters, 16 Jalan Tan Jit Seng, 11200 Penang, Malaysia

Jason P. Evans, Centre for Resource and Environmental Studies and Centre for Integrated Catchment Assessment and Management, The Australian National University, Canberra, ACT 0200, Australia

Dennis P. Garrity, International Centre for Research in Agroforestry, Southeast Asian Regional Research Programme, PO Box 161, Bogor 16001, Indonesia

A. Ngaloken Gintings, Center for Forestry and Natural Conservation Research and Development, Jln. Gunung Batu No. 5, Bogor, Indonesia

Arturo A. Gomez, Department of Agronomy, University of the Philippines, Los Baños, College, Laguna, the Philippines

Timothy R. Green, CSIRO Land and Water, Floreat Park, Western Australia 6014, Australia

Peter Grist, Centre for Resource and Environmental Studies, ANU Canberra, 0200, Australia

Ghulam M. Hashim, Malaysian Agricultural Research and Development Institute (MARDI), PO Box 12301, Kuala Lumpur 50774, Malaysia

Bernd Huwe, Institute of Soil Science and Geography, University of Bayreuth, 95440 Bayreuth, Germany

Anthony J. Jakeman, Centre for Resource and Environmental Studies and Centre for Integrated Catchment Assessment and Management, The Australian National University, Canberra, ACT 0200, Australia

Richard Johnson, Institute of Hydrology, Crowmarsh Gifford, Wallingford, Oxon OX10 8BB, UK

Klaus Kaiser, Institute of Soil Science and Geography, University of Bayreuth, 95440 Bayreuth, Germany

John Kerr, International Food Policy Research Institute, 1200 17th Street NW, Washington, DC 20036–3006, USA

E.K. Knapp, Centro Internacional de Agricultura Tropical (CIAT), A.A. 6713, Cali, Colombia

U. Kurnia, Center for Agroclimate Research, Jln. Juanda 98, Bogor 16123, Indonesia

Padma Lal, ACIAR, GPO Box 1571, Canberra, ACT 2601, Australia

Rattan Lal, School of Natural Resources, The Ohio State University, Columbus, Ohio, USA

M.A. Lucila Lapar, Social Sciences Division, International Rice Research Institute, PO Box 933, 1099 Manila, the Philippines

Johannes Lehmann, Institute of Soil Science and Geography, University of Bayreuth, 95440 Bayreuth, Germany

Betha Luisana, International Centre for Research in Agroforestry (ICRAF), PO Box 161, Bogor, Indonesia

Ellen L. McCallie, Cornell University, Ithaca, USA

Ken Menz, ACI Centre for Resource and Environmental Studies, ANU Canberra, 0200, Australia and ACIAR, GPO Box 1571, Canberra, ACT 2601, Australia

Agustin Mercado Jr, International Centre for Research in Agroforestry, Southeast Asian Regional Research Programme, PO Box 161, Bogor 16001, Indonesia

K. Müller-Sämann, Centro Internacional de Agricultura Tropical (CIAT), A.A. 6713, Cali, Colombia

R.A. Nelson, Forestry Technical Services (FORTECH), Canberra, Australia

Chalinee Niamskul, International Board for Soil Research and Management, PO Box 9-109, Jatujak, Bangkok 10900, Thailand

Sushil Pandey, Social Sciences Division, International Rice Research Institute, PO Box 933, 1099 Manila, the Philippines

Frits Penning de Vries, International Board for Soil Research and Management, PO Box 9-109, Jatujak, Bangkok 10900, Thailand

I.M. Rao, Centro Internacional de Agricultura Tropical (CIAT), A.A. 6713, Cali, Colombia

H. Ravnborg, Centro Internacional de Agricultura Tropical (CIAT), A.A. 6713, Cali, Colombia

Calvin Rose, Faculty of Environmental Sciences, Griffith University, Nathan, Queensland 4111, Australia

J.I. Sanz, Centro Internacional de Agricultura Tropical (CIAT), A.A. 6713, Cali, Colombia

Marco Stark, International Centre for Research in Agroforestry, Southeast Asian Regional Research Programme, PO Box 161, Bogor 16001, Indonesia

David E. Swete Kelly, Maroochy Research Station, Queensland Horticultural Institute, Nambour, Queensland, Australia

J. Keith Syers, Department of Agricultural and Environmental Science, University of Newcastle, Newcastle upon Tyne NE1 7RU, UK

R. Thomas, Centro Internacional de Agricultura Tropical (CIAT), A.A. 6713, Cali, Colombia

G. Trébuil, Agronomy, Plant Physiology and Agroecology Division, IRRI, Los Baños, Laguna, the Philippines

Francis Turkelboom, Department of Land Management, Catholic University of Leuven, K. Mercierlaan 92, 3001 Leuven, Belgium

P. van der Poel, National Watershed Management and Conservation Project, Jln. Taman Makam Pahlawan Kalibata No. 10, Jakarta, Indonesia

Meine van Noordwijk, International Centre for Research in Agroforestry (ICRAF), PO Box 161, Bogor, Indonesia

Mirjam van Roode, International Centre for Research in Agroforestry (ICRAF), Nairobi, Kenya

Wolfgang Wilcke, Institute of Soil Science and Geography, University of Bayreuth, 95440 Bayreuth, Germany

Bofu Yu, Faculty of Environmental Sciences, Griffith University, Nathan, Queensland 4111, Australia

Wolfgang Zech, Institute of Soil Science and Geography, University of Bayreuth, 95440 Bayreuth, Germany

Li Zhang, Centre for Resource and Environmental Studies and Centre for Integrated Catchment Assessment and Management, The Australian National University, Canberra, ACT 0200, Australia

Foreword

Soil erosion by water is the most important land degradation problem in the humid tropics. To the estimated 500 million poor people farming tropical steep lands, soil erosion may be a less immediate concern than producing enough food. Nevertheless, the accumulative effects of erosion on soil productivity are soon reflected in declining crop yields. Off-site impacts on hydrology and sedimentation are also soon felt by communities living at lower levels in the landscape. The problem is complex because it results from human actions influenced strongly by key parameters of the ecological, biophysical, sociological, economic, and policy environment.

Despite the seriousness of the erosion problem, our knowledge base on the causes and impacts of erosion is fragmented across disciplines, and across scales. Small erosion plot studies have been the mainstay of the soil science discipline, although the results do not extrapolate easily to the catchment level. Participatory research at the farm level is being advocated widely, but the government policy at the national level may have an overriding influence on farmer decisions.

The concept of this workshop derived from a degree of frustration felt by staff of some participating institutions that inadequate attention had been paid to the need to draw together the latest thinking in soil erosion research across a range of disciplines. The workshop set out to address this gap. The international organizing committee that was formed included scientists from Ohio State University, AARD, ACIAR, CIAT, IBSRAM, ICRAF, IFPRI, and IRRI. The meeting was organized under the auspices of the global Soil, Water, and Nutrient Management programme of the CGIAR, and the IRRI-led ecoregional programme for the humid and subhumid tropics and subtropics of Asia. The Centre for Soils and Agroclimatic Research, of the Agency for Agricultural Research and Development in Indonesia, agreed to host the meeting. We thank the staff of CSAR who worked hard to make the meeting a success. I also acknowledge the generous support of the donors, including ACIAR, the Department for International Development, of the UK, and USAID.

The papers published from the workshop cover a wide range of disciplines and levels or scales of research. Papers in the biophysical arena discuss research methods ranging from the intricacies of measuring runoff rates and nutrient fluxes, to the uncertainties of measuring streamflows integrated across large river basins. The socio-economic dimensions include such difficult problems as understanding farmer decisions about soil conservation and improving the economic analysis of soil erosion impact. Most importantly, some papers and the working group discussions report excellent progress in integrating the social, economic, and biophysical dimensions of the erosion problem in a holistic analysis and in identifying bottlenecks requiring further research.

This publication is the main tangible output of the meeting, but the major impacts will flow through resulting improvements in the efficiency and effectiveness of national and international research to tackle soil erosion. We hope for significant impact, albeit indirectly, on the ultimate beneficiaries who include both the poor farmers in upper catchments, and the rural and urban populations that rely on those catchments.

Eric T. Craswell
Director General
IBSRAM

1

A Framework for the Economic Assessment of Soil Erosion and Soil Conservation

T. Enters
16 Jalan Tan Jit Seng, 11200 Penang, Malaysia

This paper presents a framework and some broad guidelines for assessing the costs of soil erosion. The approach suggested is cost–benefit analysis (CBA), which links erosion and sedimentation to changes in productivity and benefits forgone. It concentrates on the comparison between soil erosion and soil conservation and stresses the importance of socio-economic parameters in assessing damage.

It proposes six principles to be clarified before applying CBA. They include: the assessment objective, the need for a multidisciplinary approach, the divergence in perspective, the hierarchical scale of the analysis, the actual methodological approach and the benchmark land use against which to measure erosion.

A brief discussion of problem identification and quantification is followed by crucial valuation issues concentrating on discount rates, time horizons and labour inputs. Regarding the on-site impacts, the approach is based on changes in productivity. Off-site costs of sedimentation are estimated according to forgone benefits. To scale-up results to higher hierarchical levels, landscape units based on potential yields, erosion risks, estimated yields for alternative practices and net production benefits are suggested.

In conclusion the paper stresses the importance of the comparative approach and emphasizes that CBAs provide meaningful results only if both bio-physical and socio-economic processes are understood.

Introduction

Soil erosion is widely believed to be a major threat to sustainable crop production, if not the long-term viability of agriculture. Soil erosion is especially problematic in tropical countries because of high rainfall intensities and generally less fertile soils. It is a threat to those countries where agricultural production is crucial to development and the majority of the rural people base their livelihood strategies on the primary sector. Unfortunately, many rural residents have been pushed to the margins of agricultural production, such as shallower and poorer soils, the sloping lands and the forest frontiers. In these situations agriculture can be the main cause of soil erosion, watershed deterioration and land degradation, although the impacts of other activities such as road construction or logging operations should not be downplayed. Cultivating upland soils often leads to a reduction in natural soil fertility and crop productivity, undermining future income generation.

The impacts of erosion are potentially far reaching, not only for the agricultural sector. Farming can inflict off-site costs on lowland economies through the processes of erosion and sedimentation and changes in hydrological patterns and water quality. Ultimately, the continuous exploitation of soil resources can reduce future economic growth considerably (Alfsen *et al.*, 1997). Hence, the magnitudes of immediate and future costs of erosion are questions that not only individual farm households should be interested in. Policy makers require answers to questions related to the cost and benefit structure of erosion to prioritize problems and design incentive structures that make soil conservation measures more attractive. An important task is thus to appraise the actual extent and impact of soil degradation and to evaluate its economic significance. This is only possible if an appropriate assessment framework and tools are available that allow for the identification, quantification and valuation of the impacts of erosion.

There are many different ways to derive and use values as inputs to the decision-making process. The diversity in methodological approaches and underlying assumptions, the tendency to ignore complications, as well as the focus on on-site costs and financial aspects, make it very difficult to compare valuation results and to assess their usefulness. Recent studies and reviews of economic analysis of soil erosion and conservation indicate that different approaches to valuation can lead to inconsistent reporting of outcomes. The following framework outlines a standardized methodology. As an assessment framework it describes principles on the appropriate use of economic analysis and broad guidelines for the practical application of cost–benefit analysis (CBA), a well recognized and established means of project and policy evaluation. It links erosion to changes in productivity and benefits forgone for on- and off-site impacts. It concentrates on the comparison between soil erosion and soil conservation, as the hypothetical and unattainable 'no-erosion' benchmark is deemed inadequate for costing erosion.

Appropriateness and Limitations

Numerous methods exist for evaluating projects, although not all lend themselves to assessing the cost of soil erosion. Bojö (1986, p. 5) suggests general criteria for a 'good' method. They should be:

- inexpensive – compared with the utility of the results;
- comprehensive – covering all relevant and important aspects;

- understandable – thus open to scrutiny and criticism;
- based on democratic values – considering the individual's own opinions as the prime basis for social evaluation; and
- producing succinct results relevant to decision-making.

Although not necessarily inexpensive, CBA is comprehensive, benefits from open discussions about its application, and produces relevant results. Furthermore, it allows for evaluations from different perspectives. As a financial analysis it is conducted from the vantage point of the individual firm, in our case the farm household. As an economic analysis the vantage point is that of the whole society.

If the worth of a project or an activity is expressed in the net present value (NPV), it is possible to aggregate results to higher levels. Thus one can start with individual fields and then move up slowly to the farm and subsequently to even higher levels such as communities, watersheds or whole economies. This can be done separately for on- and off-site costs. However, many issues concerning agricultural production and soil erosion are very location specific (Magrath and Arens, 1989; Pagiola, 1993; Clark, 1996; Pender and Kerr, 1996, Grepperud, 1997).

CBA explicitly accounts for environment–economy interactions. Soil *per se* does not contribute to well-being directly, but serves as an essential input in crop production. The result of using less soil or less fertile soil in the production process is that yield declines, which is the on-site cost of soil erosion to the farm household. While society can be affected indirectly by on-site costs, its immediate concerns are the off-site costs of erosion. Clark (1996) lists 16 off-site effects of which the most important are sedimentation effects on reservoirs, a decrease in the availability of irrigation water, increased dredging or siltation, in-stream problems of water quality and quantity, the degradation of potable water, accelerated runoff leading to localized flooding, reduced hydrological cycling and recharge of groundwater. The costs are not internalized and affect downstream landowners and water users in various ways. They are therefore termed externalities and can be positive or negative, albeit economists usually discuss only negative externalities and their impacts. The off-site costs and benefits are not reflected in farmer decision-making, but they are an integral part of the economic impacts of soil erosion (Barbier, 1996).

CBA is premised on the notion that the values to be assigned to any impacts should be those of the affected individuals, not the values held by economists or government departments (Arrow *et al.*, 1997). It lends itself to developing into an interactive, participatory process whereby those people affected by decisions assist in the definition of potential alternatives and their likely impacts. At the same time stakeholders involved in the process learn more about potential impacts. Ultimately, CBA can only benefit from a participatory approach (see Nelson *et al.*, 1996a). When pursued with integrity, implicit judgements are made explicit, subject to further analysis and discussion.

Unfortunately, CBA also has several disadvantages, which are common to many other evaluation methods. It has received considerable criticism in the literature for various reasons (Enters, 1992). Most critiques have been refuted by economists as irrelevant because they are rather directed at traditional economics and not CBA *per se*. Critics and advocates of CBA are still debating valuation as a decision-making aid. Its objectivity (Tacconi, 1995) and value-neutrality (Söderbaum, 1994) are questioned. This is not the place to provide an in-depth analysis of the points raised by economists, ecologists and philosophers over the last ten years. It should suffice to say that CBA is not a means to an end, but a coherent method for organizing and presenting information expressed in monetary values, which allows for direct comparison among alternative options such as soil eroding and soil conserving land-use practices.

A key limitation or problem for critics and advocates alike – an issue that practitioners frequently skip over when applying economics to environmental problems such as soil erosion – is the high degree of interconnectedness within and between ecosystems (Gilbert and Janssen, 1997) and the lack of knowledge about the true nature of economy–environment interactions (Winpenny, 1991; Hanley, 1992; Aylward *et al.*, 1995; Chomitz and Kumari, 1996). These interactions are so complex that 'obviously, much empirical work is needed in this area' (Lutz and Munasinghe, 1994, p. 39). Although the main purpose of assessing the economics of soil erosion should not be 'to provide fine-tuned numbers, but to indicate orders of magnitude' (Lutz and Munasinghe, 1994, p. 44). Enters (1998b) has shown that the effects of changing assumptions can be quite dramatic on the final result of a CBA and that limited knowledge may mean that the assessment is not done correctly. In fact, CBA can be damaging and dangerous (Chichilnisky, 1997).

There is no doubt that doing it correctly can provide much useful information for decision-making (Gregersen *et al.*, 1995). However, this does not mean that the valuation theory and methodologies have to be refined further. Rather, actual applications need to be rigorous, data inputs cannot be based solely on 'guesstimates' and more effort has to be spent on identifying problems and quantifying effects than on the actual translation of impacts into monetary values.

In summary, CBA is an appropriate tool for assessing the impacts of soil erosion but is potentially limited by data paucity and our limited knowledge of environment–economy interactions.

Principles in CBA

Before conducting a CBA, a number of principles have to be understood and agreed upon. They include issues such as its scope, multidisciplinarity, the vantage point of the analysis, its hierarchical scale, the actual methodological approach and the benchmark against which changes over time are measured.

Assessment objective

The objectives of the assessment need to be clearly spelled out. This appears to be rather obvious, but, it is rare that analysts differentiate properly between assessing the impacts of erosion, the impacts of farming systems and the impacts of different land uses. If erosion is the focus of a study, then the hydrological responses to land use changes are of only little importance. Vice versa, if the hydrological impacts are to be included, then the study investigates land use changes or land degradation, and not erosion *per se*, although erosion remains a very important part of the analysis. A change from agroforestry to grass may illustrate the point. In terms of erosion control, grass is probably more effective. On the other hand it may generate more runoff. Thus, the structure of benefits and costs is rather different.

Farmers are most interested in the private returns to alternative crops and land use management options. They are only indirectly interested in the costs of erosion inasmuch as it affects crop productivity. Two cropping systems can produce the same amount of erosion but very different returns to land and labour. Ultimately, the assessment needs to be directed at the productivity of a system with a predominant view on the sustainability of inputs and outputs.

Private versus social analysis

It is important to shape one's analysis into a form that is recognizable by the target audience and addresses their interests. From the onset it needs to be clear which vantage point the analysis takes, the perspective of the farmer or the perspective of the society – which is not to say that farmers are not part of the society. In other words, it has to be made explicit whether the focus is a private CBA (financial analysis) or social CBA (economic analysis)[1].

The economic analysis in which the allocation of productive factors is assessed from the society's perspective differs from the financial analysis in three ways. First, distortions induced by regulations, imperfections in the market and overvalued currencies give rise to deficiencies in the use of existing price data (market prices) as the basis for evaluating inputs and outputs. These deficiencies require a re-evaluation of market prices and the adjustment for direct transfer payments, issues which are discussed in detail by Squire and van der Tak (1975) and Gittinger (1982). Second, discount rates have to be adjusted as a farmer's private rate of discount will be higher than the social rate of discount (Barbier, 1996). Third, externalities are incorporated in the analysis.

Multidisciplinarity

The breadth of issues to be dealt with when assessing the impacts of soil erosion requires a multidisciplinary approach. It posits an in-depth understanding of the effects of erosion on the physical and chemical properties of soil, the relationship between soil erosion and crop yields, farmer decision-making and agriculture, geomorphology and the hydrological responses to land cover changes, as well as economic systems such as labour markets and credit availability. It is not within the scope of this chapter to provide detailed

[1]Editors' footnote: In this paper as well as throughout the following chapters the terms 'financial and economic' and 'private and social' are used as synonyms.

information on all these issues. It concentrates on the main issues concerning the practical application of CBA. To be practical, the multidisciplinary aspect cannot be emphasized enough.

Hierarchical scale

The most logical place to start with is the cropping system and the farm as well as those locations where off-site impacts are expected to be problematic, such as at a reservoir. It is possible to aggregate values for individual locations later to higher hierarchical levels. However, at the beginning of the analysis there should be an agreement about its geographical scope to facilitate the subsequent stratification of the total area into land units, which may differ in their resistance to erosion and resilience to change as well as socioeconomic variables.

Approaches for cost and benefit calculations

Three valuation techniques are commonly used to assess the on-site costs of soil erosion as well as the cost and benefits of soil conservation:

- hedonic pricing (property valuation);
- replacement costs; and
- change of productivity.

The application of any of the three approaches is usually flawed and may lead to overestimation, on site, of costs, if not carefully applied. The change of productivity approach treats the investment costs as sunk costs and only looks at the immediate relevant cost items, whereas the replacement costs take changes of the capital stock into account. Another problem is that each approach assumes that the comparison is between a situation 'with' and 'without' erosion, as if it was possible to eliminate or at least reduce erosion without any extra cost (Barbier, 1996). The advantages and drawbacks of each method have been discussed recently (Grohs, 1994; Barbier, 1996; Clark, 1996; Enters, 1998a).

At this point, the change of productivity approach is suggested. Following the approach, the on-site erosion damage equals the value of the lost crop production at market prices (Grohs, 1994). As a physical measurement it relies on crop yields under different practices. Crop yields are then multiplied by the unit price of the crops.

The change in productivity approach does not rely on proxy measures, is logical, and is straightforward to apply. The use of the latest simulation models allows for making yield forecasts under a variety of conditions, although there are some inherent problems, such as isolating the impacts of technological progress over time as well as soil depletion that occurs even without erosion. The correct measure of yield damage from erosion is the present value of incremental net returns (PVINR), which calculates differences in long-term crop yields applying erosive and soil conserving techniques, respectively, against differences in various inputs (Eaton, 1996). Particular attention is paid to labour input, as it is the major cost of conservation (de Graaff, 1993) and the need to perform certain tasks during a season of low labour availability may be a significant cost factor.

The use of the PVINR does not only allow the comparison of erosive with conserving practices for the same crop but also enables the analyst to switch crops at the beginning of the analysis or after a certain time. Thus it is possible to compare, for example, corn monocropping with mixed cropping, corn-based agroforestry systems, fruit orchards, wood-producing plantations or even natural forests. While not neglecting erosion, this approach emphasizes the significance of comparing cropping systems, which is, from the farmers' point of view, extremely important.

For off-site effects the change of productivity approach is suggested too. Conceptually, the approach is identical to the quantification of on-site effects. Focusing on sedimentation, it evaluates the income forgone from not being able to irrigate fields or produce hydroelectricity caused by the reduction in dam yield, increased operating and maintenance costs of irrigation schemes or ports, and the higher costs for removing sediments through dredging and replacing damaged equipment such as turbines. It can also deal with other damages, such as to fisheries or tourism (see Hodgson and Dixon, 1988).

In a careful analysis of the situation and expected changes over time, it is possible to consider most costs and benefits that can be quantified. However, there are also no quantifiable variables and sociocultural impacts that cannot be readily translated into monetary values. In such cases, contingent valuation (CV) is a technique that can be applied for incorporating the willingness to pay (WTP) for environmental benefits, in general, or soil conservation and catchment management, in particular, into a CBA[2]. It is a useful tool for assessing economic impacts where environment–economy interactions are difficult to predict.

Benchmark land use or practice

A meaningful cost assessment is only possible in a comparative approach. Erosive cropping practices can be compared with a variety of land uses that may have very different ecological characteristics. As early as possible in the CBA a decision needs to be made regarding one or several benchmark land uses. The following land uses will produce diverse results in a comparative study:

- conservation forests;
- production forests (available for timber harvesting – destructive or reduced impact logging);
- shifting cultivation;
- tree plantations (for wood and non-wood forest products) and fruit orchards;

[2] For a recent case study of contingent valuation and integrated catchment management see Cameron (1997).

- soil-conserving permanent cropping;
- agroforestry systems; and
- grassland (sustainably managed or degraded).

The discussion of the six issues above completes the more general introduction to the framework. There has to be an agreement on all of the issues before conducting the assessment itself. If not, results may be meaningless and/or comparisons and aggregation to higher hierarchical levels may not be possible.

The main steps of CBA

The basic methodology of CBA is straightforward and easy to comprehend. It consists of three main steps:

- identification of all components relevant for the analysis, i.e. setting the scope;
- quantification of physical variables, i.e. environmental impacts and resources; and
- valuation, i.e. translation into monetary terms in the formal project analysis.

The discussion below concentrates on the third step, although it is in the first two steps in which the crucial groundwork is laid to provide a meaningful economic assessment. If problems are discarded or misrepresented, or if environmental impacts are measured incorrectly, cost and benefit estimates will be, at best, misleading.

Problem identification

The identification of potential environmental impacts as well as relevant costs and benefits is not an area where economists have a comparative advantage (Angelsen and Sumaila, 1995). Their contribution in the scoping exercise may be in the selection of specific effects that need to be studied in more detail. The identification of costs and benefits is a truly multidisciplinary exercise, particularly if on-site as well as off-site impacts are to be quantified and assessed. It is crucial to note the inputs and outputs for each land use practice to be investigated, to understand the farming system as well as the landscape, for identifying relevant off-site costs. Particular attention has to be paid to the lowlands, where positive or negative externalities are translated into on- and off-site costs. In fact, it could be argued that the starting point for any investigation should be the potentially affected lowland population and its activities. Ultimately, if no problems can be identified in the lowlands, there is no point in attempting to proceed to the next step of the analysis for off-site effects.

The result of problem identification is a checklist of variables that need to be quantified in the next step. It should include:

- cost items such as capital, material, labour and land;
- relative changes in yields due to erosion;

- potential impacts of sedimentation;
- potential impacts of hydrological patterns (if focus is on land use changes and not on erosion); and
- population and lowland economy.

Quantification of physical variables

The most important on-site effect of soil erosion is a decline in yields. The on-site impacts of erosion on productivity are exceptionally complex, and both soil and plant specific (Stocking, 1996). Soil conserving practices will show clearly different yield patterns over time compared with erosive practices. This may concern crops, fodder grass, fuelwood, fruits and timber, for example. They may be used collectively in the concept of 'yield' (Bojö, 1986) or output.

The change of productivity approach requires that the relationship of erosion and yields is quantified over time, an issue that in the past did not receive sufficient attention :

> Our review of applied CBA studies has revealed that the estimation of yield differences is often a weak spot in the analysis. Scant evidence and ad-hoc assumptions often underlie conclusions whose certainty tend to increase as we approach the executive summary.
>
> (Bojö, 1986, p. 26)

Today, the use of models allows for predicting yield changes due to erosion under a variety of conditions and estimates have become more reliable (Nelson *et al.*, 1996b; Enters, 1998a). Simulation models or dynamic growth models describe the basic processes of crop growth mathematically, which allows in principle the prediction of yields at any location in the landscape, although models need calibration. An important consideration in choosing a particular model is the degree of institutional back-up. If none is available, it may be better to conduct formal or informal expert studies, or to rely on existing data if available. Extra care should also be taken to separate the effects of soil depletion from erosion.

The most critical question is not the cost of soil erosion *per se* but whether the long-term benefits of reduced erosion and soil degradation make the current costs of abatement, namely additional inputs, worth bearing (Pagiola, 1994; Ekbom, 1995). Any reduction in erosion rates has to enter the analysis as an investment cost. Soil erosion and erosion–yield decline estimates are only a partial input in any cost calculations. Data are also needed on secondary products, such as those cultivated in buffer strips. Most critical is the exact quantification of labour inputs for individual activities. It is very important to categorize labour inputs by season, as demand and supply may follow different patterns. Other inputs also need to be quantified.

The quantification of off-site impacts is the final task before valuation can begin. It is probably the most daunting task and is beset with sometimes insurmountable problems (Enters, 1998a).

The evidence of on-site runoff generation suggests that land use changes

– that is, from forest to agriculture – result in soil compaction, the collapse of macropores and the decline of soil organic matter (Turkelboom *et al.*, 1991). These factors lead to decreased surface infiltration rates, thereby increasing runoff (Tangtham, 1991). However, this does not mean that increases in infiltration rates accompanying the adoption of soil-conserving land management significantly reduce the incidence of downstream flooding (Gilmour *et al.*, 1987). This is particularly the case for large-scale floods. Analysts may want to quantify the damage of local floods, but only if they can show that there has been an increase in flood severity, which is quite distinct from an increase in flood damage. Land subsidence and urbanization lead to increases in damages, even though the incidence of flooding may decrease. Also, as introduced earlier, if the focus of the assessment is erosion, it is advisable to exclude the impacts of changing water yields and regimes from the analysis.

The quantification of externalities and their impacts requires a careful investigation of actual and potential effects in the lowlands, that is, off site. Social surveys of stratified population samples provide detailed accounts of:

- population structure;
- livelihood strategies and wealth of downstream residents;
- sociocultural aspects;
- economic activities;
- prices for various products and services;
- demand for goods and services; and
- future trends for the above four categories.

In developed economies, the main negative externality of soil erosion is non-point water pollution, which affects water supplies for residential and industrial purposes, results in water nutrient enrichment and reduces the biological diversity of water bodies as well as the recreational and amenity values of water resources. In developing countries on the other hand, the main impact is probably, albeit not exclusively, the sedimentation of hydroelectric, flood control and irrigation facilities.

In most situations, the analyst will focus on the impacts of sedimentation. The majority of recent CBAs rely on assumptions about complex environment–economy interactions for which the knowledge base is still rather weak. This is particularly the case when sediment rates are quantified. It is not the intention here to review the various problems associated with measuring changes and making the linkage between agriculture, erosion and sedimentation that affects lowland facilities (see Enters, 1998a, for a recent analysis). The following issues need more attention and substantial research efforts:

- sediment sources, particularly the contribution of roads to sedimentation rates;
- spatial distribution of sediment and delivery ratios;
- time horizons; and
- reliability and objectivity of measurements.

This is not to say that it is impossible to estimate the costs of sedimentation, but it remains unclear what the relationship between agricultural land use, erosion and sedimentation is, as long as the four issues above do not receive more serious attention.

The off-site costs are measured normally in terms of the net present value of forgone net benefits from any loss of downstream economic activity (Barbier, 1996) or additional operating and maintenance costs, such as dredging costs for canals, reservoirs or port facilities. Examples are production cuts through the loss of irrigation water and/or reduced hydroelectricity production as well as diminishing harvests of aquatic wildlife or reduced tourism revenue. It is therefore crucial to estimate reductions in dam yields and water quality and to assess their impacts on present and future uses.

Additional costs and benefits may have been identified earlier in the CBA, such as the reduction in flood control capacity, the degradation of potable water and/or damage to coral reefs. In each case damages need to be quantified so that they can be translated into monetary terms in the following step.

Valuing on-site costs and benefits

The previous sections have provided a list of costs and benefits to consider as well as some insights into how best to capture important cost and benefit aspects based on the change of productivity approach. Once all important physical variables have been quantified, price tags can be put on inputs and outputs.

It is very important to keep in mind the differences between the private and social perspectives. The private CBA (financial analysis) forms the basis for the economic analysis and market prices are used for all inputs and outputs. A CBA of alternative agricultural practices with and without, or with reduced, erosion is futile if it does not consider a farm household's circumstances and different input requirements. Farming systems are highly heterogeneous. Amongst other factors, they differ in the degree of mechanization, cash or subsistence crop orientation, the area under perennial crops, and the involvement in livestock rearing. Furthermore, there are differences in social, ethnic and household characteristics as well as in accessibility and marketing opportunities. All these factors influence the costs that a farm household faces in agricultural production.

It is not possible to discuss all aspects of conducting a CBA from the farmers' point of view in this chapter. Instead the following discussion will focus on three critical valuation issues. They are:

- the discount rate;
- the time horizon; and
- valuing labour.

Choosing the 'right' discount rate

The calculation of NPVs requires the determination of an appropriate discount rate, both as a private and/or social rate. A 'weak spot' in many CBAs is the rationale for the choice of discount rate (Bojö, 1986; Ekbom, 1995). The selected interest rate obviously influences the results of the CBA and needs careful consideration. According to Gittinger (1982, p. 314):

> ... for financial analysis, the discount rate or cut-off rate is usually the marginal cost of money to the farm or firm for which the analysis is being done. This often will be the rate at which the enterprise is able to borrow money.

Frequently, borrowing and saving rates are not available because such facilities are not accessible to small farmers (Hoekstra, 1987; Moll, 1989). Where equity capital is used, it is possible to determine the rate of return a farmer normally expects from existing long-term farm enterprises such as livestock or orchards.

According to Hoekstra (1985), discount rates differ among farmers. They are based on several factors regarding the farmers' current status, their outlook, the farmers' attitudes towards risk and uncertainty and the length of waiting time before consumption. Applying the same discount rate to all farmers should be avoided.

Schreier (1989) suggests that small farmers base their investment decision on their marginal time preference rate (MTPR). He acknowledges that it is problematic to discover the appropriate MTPR but indicates that it is usually greater than a government-subsidized interest rate. This is reiterated by Barbier (1996, p. 4) who stresses that 'private individuals employ higher discount rates, on average, than society as a whole'. Since the social rate of discount is usually assumed to be in the order of 10% – Bojö (1986) suggests between 2 and 12% – the private rate of discount would be higher, that is, between 10 and 15%. Betters (1988) recommends using a range of discount rates in financial analysis because of the difficulty in specifying appropriate rates. This can be done in a sensitivity analysis, but probably private discount rates should not be lower than 5% and not much higher than 25%.

Choice of time horizon

The NPV is not only very sensitive to discount rates but also to time horizons. Here again the analyst is faced with making assumptions. Dixon and Fallon (1989) argue that many poor subsistence farmers may have a time horizon that only goes to the next season because of pressing current needs. This short-sightedness, they continue, explains exploitative patterns of land use and the farmers' unwillingness, incapability or disinterest in investing in resource conservation. Barbier (1987) points out that poor people often have no choice but to opt for immediate economic benefits at the expense of the long-term sustainability of their livelihoods.

While there is no doubt that many upland farmers make decisions to fulfil their short-term needs, it would be wrong to assume that their decision-making is generally guided by short-term thinking. If we accept the villagers' short-sightedness, a CBA of growing perennials becomes meaningless. Using a time horizon of, for example, 5 years would result predominantly in costs and only negligible benefits.

A minimum criterion should be that the time horizon conforms with the production characteristics of the species under investigation (Enters, 1992). In addition, the 'right' time horizon depends partially on the discount rate applied. The higher the discount rate, the lower will be the NPV of soil conserving practices that predominantly have higher returns in the future.

It is probably best to apply a time horizon of between 15 and 25 years. If the purpose of the financial analysis is to examine the returns to investments, it is also instructive to calculate the number of years to break even. In this case the choice of time horizon is irrelevant.

Valuing labour

In their review of six studies that estimated the costs of erosion, Fox and Dickson (1988, p. 27) conclude:

> Farmers have been reluctant to adopt conservation tillage, not because of a lack of information, not because of 'perceptions' and not because 'old habits die hard', but because they would lose money, both in the short and in the long run.

They also find (p. 1) that 'soil erosion costs farmers dearly' but that soil conservation costs them even more. The reason for this discrepancy is that, as Fox and Dickson (1988) criticize, in many assessments technologically unattainable standards are used as a basis for measuring erosion costs. Many CBAs are based on the assumption that sustainable crop production is possible without any additional costs. There is no or only minimal value ascribed to labour, and its greater use may even be counted as a benefit (Wiggins and Palma, 1980).

Investment costs and labour requirements in soil conservation are usually high (de Graaff and Nibbering, 1996). Given the considerable proportion of family labour in both crop production and soil conservation and the fact that assumptions on labour input and wage rates drive the results of any economic assessment, it is absolutely crucial to obtain an accurate value of the opportunity cost of labour (Ellis-Jones and Sims, 1995). Unfortunately, the importance of labour inputs and costs (establishment as well as maintenance) are often underestimated.

Identifying the appropriate opportunity cost of labour is not easy. It depends on the nature of the activity to be performed, on the characteristics of the labourer (age, wealth and gender), the season (growing season or slack season) and the availability of non-farm and off-farm employment. Most studies are characterized by a strong farming bias even though household income generation is increasingly influenced by off-farm employment opportunities. This development severely restricts labour availability at certain times of the year and raises the opportunity cost for labour especially of younger farmers who favour non-farm employment opportunities over agriculture (Parnwell and Taylor, 1996; Rigg, 1997).

How to value labour is an issue, debated as the choice of discount rate and time horizon. While it has been suggested that under certain circumstances the opportunity cost of labour can be zero (Stocking and Abel, 1993), it is doubtful whether zero return to labour is acceptable even in conditions of high unemployment. What is clear is that labour costs vary enormously according to circumstances that need considerable attention in extrapolating results.

Renaud (1997) uses a cost of 50 baht day^{-1} which corresponds with the agricultural salary paid in the region. Nelson *et al.* (1996b, p. 19) report that a 'farm labourer would expect to be paid 120 pesos day^{-1} and receive meals valued at 40 pesos'. However, none of the interviewed farmers was working for these daily wages, neither could they afford to hire in at this cost. Since off-farm labour was only very sporadic, the authors used beside 160 a daily rate of 35 pesos. The second estimate not only decreased the absolute cumulative NPV of all three alternative farming methods (open-field, fallow, hedgerows), it also made the soil conserving land-use practices relatively more attractive.

The correct valuation of labour is critical for assessing alternative courses of action, which highlights again the importance of a participatory approach. Farm household members are in the best position to provide information on seasonal wage rates and the conditions under which they would give up leisure. In an ex-ante analysis they can also assist in quantifying labour inputs. They are the best source for costs on various inputs including equipment, seeds, pesticides and fertilizer as well as the farm-gate price from produce.

Social analysis

The first step in the social analysis is to correct market prices, eliminate transfer payments and adjust the opportunity cost of labour and capital, namely the discount rate. Gittinger (1982) and Bojö (1986) provide an overview of methods for correcting distortions. The issue will not be treated here. Instead the discussion concentrates on the valuation of off-site impacts.

Valuing off-site costs and benefits

The costs of unexpected and excessive sedimentation of reservoirs can be calculated as the income forgone from not being able to irrigate fields caused by the reduction in dam yield, increased operation and maintenance costs of irrigation schemes, and the higher operation and maintenance costs for removing sediments through dredging and replacing damaged equipment such as turbines. The estimation in itself is straightforward, although decisions have to be made regarding the economic activity most affected if water can be used for numerous mutually exclusive economic activities such as the production of electricity or agriculture. If only agriculture is affected, then the forgone benefits are the incremental benefits between a hectare of irrigated crop production and a hectare of dryland farming. If water is used for industrial or residential purposes then costs are higher. Getting the price right is extremely important, because an error in prices can radically change the results (Chichilnisky, 1997).

The costs for replacing equipment should be readily available as well as the cost for dredging canals and harbours, the prices of fish, tourism revenue or the expenditures for constructing additional flood prevention structures.

Based on social surveys that provided the baseline information of the lowland conditions, social impact assessments provide insights into those aspects that cannot be quantified and are required for contingent valuation (CV) during which participants of lowland communities are asked about their WTP for upstream soil and water conservation or better water quality. Cost–benefit analysis permits collective action that can harm individuals. Therefore, the distributional and equity effects of downstream impacts are assessed in an additional step based on information concerning population distribution, livelihood strategies and wealth.

Scaling-up from Field to National Levels

A CBA of the on-site impacts of soil erosion can be conducted at various hierarchical scales. The discussion above focused predominantly on cropping systems or the field and the farm household as the decision-making unit. What happens at one scale usually influences the outcome of other scales. For example the demand for off-farm employment may have repercussions for what is happening on the farm. These interactions should be considered when moving from one scale to the next.

Izac and Swift (1994) defined five hierarchical levels for measuring agricultural sustainability in small-scale farming in sub-Saharan Africa (p. 109):

- cropping system
- farming system
- village/catchment system
- regional system
- supra-regional system

According to Izac and Swift (1994) the lowest level in the hierarchy, the cropping system, is the smallest spatial unit, or the scale 'at which specific biological processes such as nutrient uptake or plant competition are regulated and may be studied (p. 109)'. This is where the baseline information about the natural system is derived from and erosion–yield relationships are analysed. Decisions regarding crop choices and management practices are made at the farm. Here numerous interactions occur with the next higher levels, especially the catchment system. This third level is important for considering the flow of nutrient and sediments and is therefore relevant for quantifying the off-site impacts of soil erosion.

The on-site costs of erosion can be aggregated to any higher level from the lowest spatial unit upwards. An important level is most likely the national one, though it is desirable to aggregate by individual catchments, mountain ranges or administrative units in an intermediate step to identify 'hot spots' or priority areas for soil conservation investments.

The main objective of a national level study is to 'put a value on a natural resource that is "used up" through agricultural production by applying a 'national resource accounting' approach to quantify the cost of soil degradation to the economy' (Grohs, 1994, p. 25).

The valuation techniques discussed above can be used at any level, although it is not feasible to estimate the cost of erosion for each individual field or farm in a catchment. Therefore, landscape units for which the same or similar impacts can be expected must be defined. Hamilton and Pearce (1991) suggest geology and terrain, soils and hydrological data as basic information for any assessment. A slightly different approach would be to store for each unit information on:

- potential yields;
- erosion risks (see Batjes, 1996), translated into loss of topsoil;
- actual yields for alternative practices; and
- net production benefits.

The most straightforward approach is then to add up the annual costs of soil erosion for each unit and thereby derive the total cost due to erosion nationwide (Bishop, 1995). Once spatial data are stored in a geographical information system, data manipulation and analysis become very simple.

The simple approach described above has a number of problems that are not immediately apparent when only aggregate figures are provided (Bishop, 1995). If, originally, soil loss was measured on the smallest spatial unit, then the aggregation will also only deal with soil loss. Adding up all the soil losses basically means that deposition is ruled out. All the eroded material is treated as a negative externality entering the nearest stream or lake. As a result, total cost to the national economy is overestimated as the redistribution of soil and its potentially productivity-enhancing deposition are not accounted for.

Once it is possible to include depositions in the spatial analysis, upscaling will produce more useful results, particularly if parameters that delineate landscape or mapping units are not only of a biophysical nature. The cost of erosion depends on biophysical as well as economic processes. Net production benefits do not only depend on soil type and climatic data but also on such variables as crop prices, off-farm employment opportunities, discount rates, credit availability and farmers' perspectives. Each of these variables affects the cost of agricultural production and erosion in the same way as do slope, soil quality, rainfall and erosivity. Thus, they also need to be mapped.

Upscaling off-site impacts and costs is performed in accordance to upscaling on-site impacts and costs, although great care is needed to avoid double-counting. The same material can only be trapped or dredged once. Also, what may be counted in one micro- or mesocatchment as a negative externality, may turn out to be a benefit contributing to income generation in another location. A basic requirement is thus knowledge of soil movement within farms and catchments in quantitative terms with a particular focus on the effects of depositions on present and future crop productivity.

Conclusions

Any assessment of the cost of soil erosion is beset with problems. Even a very carefully conducted CBA will leave room for criticism and further refinement. Having said that, certainly much can be done to get the orders of magnitudes right. First of all, however, we have to accept that the cost of erosion as a measure is meaningless if it is not compared with the cost of soil conservation. If we follow the comparative approach, we will find that in a number of cases farmers behave rationally when they let erosion continue. Whether, as a society, we behave just as rationally can only be answered by extending the CBA to the off-site impacts of soil erosion.

In conducting a CBA, rationale should be provided for using particular variables, parameters, prices and rates. Figures should not be selected randomly. The choice of any number should always be based on rational decision-making, as we expect farmers also to be rational decision-makers.

In the past, data collection has focused on the uplands and what is happening to soil and water under exploitative cropping regimes or increasing land degradation. For on-site assessments this is the sensible approach. However, for off-site impacts the focus of any analysis is not the soil but economic activities as well as sociocultural practices in the lowlands. Hence the lowland conditions, the livelihood strategies of its population and the processes that lead to a transformation in economic activities have to be investigated with the same diligence as soil erosion and sediment transport processes.

Finally, the audience for any CBA should be constantly in the back of the mind of the analyst. CBA provides a logical framework for the systematic collection, interpretation and presentation of information from the perspective of trade-offs in decision-making. The final results, hopefully, are presented to an audience composed of non-economists. There is no need to impress them with endless tables and repetitious sensitivity analysis. This is not to say that the report should consist only of one monetary value as the bottom line. Rather it should take its reader in a comprehensible way through the various steps of a CBA and allow the layman to understand why certain decisions were made during the course of the analysis. Ultimately the reader, whether a farmer, project planner or government official, should be able to perform his/her own sensitivity analysis to examine what happens to the bottom line.

References

Alfsen, K.H., Bye, T., Glomsrød, S. and Wiig, H. (1997) Soil degradation and economic development in Ghana. *Environment and Development Economics* 2, 119–143.

Angelsen, A. and Sumaila, U.R. (1995) *Hard Methods for Soft Policies. Environmental and Social Cost–Benefit Analysis. Working Paper 1.* Chr. Michelsen Institute, Bergen, Norway.

Arrow, K.J., Cropper, M.L., Eads, G.C., Hahn, R.W., Lave, L.B., Noll, R.G., Portney, P.R., Russel, M., Schalensee, R., Smith, V.K. and Stavins, R.N. (1997) Is there a role for cost–benefit analysis in environmental, health, and safety regulation? *Environment and Development Economics* 2, 196–201.

Aylward, B., Echeverría, J. and Barbier, E.B. (1995) *Economic incentives for watershed protection: A report on an ongoing study of Arenal, Costa Rica. CREED Working Paper Series No. 3.* Vrije Universiteit Amsterdam, and International Institute for Environment and Development, London.

Barbier, E.B. (1987) The concept of sustainable economic development. *Environmental Conservation* 14, 101–110.

Barbier, E.B. (1996) *The Economics of Soil Erosion: Theory, Methodology and Examples. EEPSEA Special Paper.* Economy and Environment Program for Southeast Asia, Singapore.

Batjes, N.H. (1996) *A Qualitative Assessment of Water Erosion Risk using 1:5 M SOTER Data – an Application for Northern Argentina, South-East Brazil and Uruguay. Report 96/04.* International Soil Reference and Information Centre, Wageningen.

Betters, D.R. (1988) Planning optimal economic strategies for agroforestry systems. *Agroforestry Systems* 7, 17–31.

Bishop, J. (1995) *The Economics of Soil Degradation: an Illustration of the Change in Productivity Approach to Valuation in Mali and Malawi. LEEC Paper DP 95–02.* International Institute for Environment and Development, London.

Bojö, J. (1986) *A Review of Cost–Benefit Studies of Soil and Water Conservation Projects. Report No. 3.* Soil and Water Conservation and Land Utilization Programme, Maseru.

Cameron, J.I. (1997) Applying socio-ecological economics: a case of contingent valuation and integrated catchment management. *Ecological Economics* 23, 155–165.

Chichilnisky, G. (1997) The costs and benefits of benefit–cost analysis. *Environment and Development Economics* 2, 202–206.

Chomitz, K.M. and Kumari, K. (1996) *The Domestic Benefits of Tropical Forests. Policy Research Working Paper 1601.* Policy Research Department, The World Bank, Washington, DC.

Clark, R. (1996) *Methodologies for the Economic Analysis of Soil Erosion and Conservation. CSERGE Working Paper GEC 96–13.* University of East Anglia, Norwich.

de Graff, J. (1993) *Soil Conservation and Sustainable Land Use – An Economic Approach.* Royal Tropical Institute, Amsterdam.

de Graff, J. and Nibbering, J.W. (1996) Indicators for monitoring and evaluating watershed development. *Paper presented at the 9th Conference of the International Soil Conservation Organisation, Bonn, 26–30 August.*

Dixon, J.A. and Fallon, L.A. (1989) The concept of sustainability: origins, extensions, and usefulness for policy. *Society and Natural Resources* 2, 73–84.

Eaton, D. (1996) *The Economics of Soil Erosion: a Model of Farm Decision-making. Discussion Paper 96–01.* Environmental Economics Programme, International Institute for Environment and Development, London.

Ekbom, A. (1995) *The Economics of Soil Conservation.* Unit for Environmental Economics, Department of Economics, Gothenburg University, Göteborg.

Ellis-Jones, J. and Sims, B. (1995) An appraisal of soil conservation technologies on hillside farms in Honduras, Mexico and Nicaragua, 1995. *Project Appraisal* 10(2), 125–134.

Enters, T. (1992) Land degradation and resource conservation in the highlands of Northern Thailand: the limits to economic evaluation. PhD dissertation. The Australian National University, Canberra.

Enters, T. (1998a) Methods for the economic assessment of soil erosion and soil conservation. IBSRAM Monograph. *Issues in Sustainable Land Management,* Vol. 2. International Board for Soil Research and Management, Bangkok, 60 pp.

Enters, T. (1998b) The economics of forest conversion in Northern Thailand: challenging assumptions. In: Roper, C.S. and Park, A. (eds) *The Living Forest: Non-Market Benefits of Forestry.* The Stationery Office, London (in press).

Fox, G. and Dickson, E.J. (1988) *What's Economic about the Economic Costs of Soil Erosion to Canadian Farmers? Discussion Paper DP88/3.* Department of Agricultural Economics and Business, University of Guelph, Guelph, Ontario.

Gilbert, A.J. and Janssen, R. (1997) *The Use of Environmental Functions to Evaluate Management Strategies for the Pagbilao Mangrove Forest. CREED Working Paper*

Series No. 15, Vrije Universiteit, Amsterdam and International Institute for Environment and Development, London.

Gilmour, D.A., Bonell, M. and Cassells, D.S. (1987) The effects of forestation on soil hydraulic properties in the Middle Hills of Nepal: a preliminary assessment. *Mountain Research and Development* 7, 239–249.

Gittinger, J.P. (1982) *Economic Analysis of Agricultural Projects. EDI Series in Economic Development.* Johns Hopkins University Press, Baltimore.

Gregersen, H.M., Arnold, J.E.M., Lundgren, A.L. and Contreras-Hermosilla, A. (1995) *Valuing Forests: Context, Issues and Guidelines. FAO Forestry Paper 127.* Food and Agriculture Organization of the United Nations, Rome.

Grepperud, S. (1997) *The Impact of Policy on Farm Conservation Incentives in Developing Countries: What can be Learned from Theory? Documents 97/2.* Statistics Norway, Research Department, Oslo.

Grohs, F. (1994) *Economics of Soil Degradation, Erosion and Conservation: a Case Study of Zimbabwe. Arbeiten zur Agrarwirtschaft in Entwicklungsländern.* Wissenschaftsverlag Vauk Kiel KG, Kiel.

Hamilton, L.S. and Pearce, A.J. (1991) Biophysical aspects in watershed management. In: Easter, K.W., Dixon, J.A. and Hufschmidt, M.M. (eds) *Watershed Resources Management – Studies from Asia and the Pacific.* ISEAS, Singapore and East–West Center, Honolulu, pp. 33–52.

Hanley, N. (1992) Are there environmental limits to cost–benefit analysis? *Environmental and Resource Economics* 2(1), 33–59.

Hodgson, G. and Dixon, J.A. (1988) *Logging Versus Fisheries and Tourism in Palawan. Occasional Paper No. 7.* East-West Environment and Policy Center, Honolulu, Hawaii.

Hoekstra, D.A. (1985) Choosing the discount rate for analyzing agroforestry systems/technologies from a private economic viewpoint. *Forest Ecology and Management* 10, 177–183.

Hoekstra, D.A. (1987) Economics of agroforestry. *Agroforestry Systems* 5, 293–300.

Izac, A.-M.N. and Swift, M.J. (1994) On agricultural sustainability and its measurement in small-scale farming in Sub-Saharan Africa. *Ecological Economics* 11, 105–125.

Lutz, E. and Munasinghe, M. (1994) Integration of environmental concerns into economic analyses of projects and policies in an operational context. *Ecological Economics* 10, 37–46.

Magrath, W. and Arens, P. (1989) *The Costs of Soil Erosion on Java: a Natural Resource Accounting Approach. Environment Department Working Paper No. 18.* The World Bank, Washington, DC.

Moll, H.A.J. (1989) *Farmers and Finance: Experience with Institutional Savings and Credit in West Java.* Agricultural University, Wageningen.

Nelson, R.A., Dimes, J.P., Silburn, D.M. and Carberry, P.S. (1996a) *Erosion/Productivity Modelling of Maize Farming in the Philippine Uplands. Part I: A Simple Description of the Agricultural Production Systems Simulator. SEARCA-UQ Uplands Research Project. Working Paper No. 12.* Los Baños, The Philippines.

Nelson, R.A., Cramb, R.A., Menz, K. and Mamicpic, M.A. (1996b) *Bioeconomic Modelling of Alternative Forms of Hedgerow Intercropping in the Philippine Uplands using SCUAF. Imperata Project Paper 1996/9.* CRES, The Australian National University, Canberra.

Pagiola, S. (1993) Soil conservation and the sustainability of agricultural production. PhD dissertation, Stanford University, California.

Pagiola, S. (1994) Soil conservation in a semi-arid region of Kenya: rates of return and adoption by farmers. In: Napier, T.L., Camboni, S.M. and El-Swaify, S.A. (eds)

Adopting conservation on the farm. Soil and Water Conservation Society, Ankeny, pp. 171–188.

Parnwell, M.J.G. and Taylor, D.M. (1996) Environmental degradation, non-timber forest products and Iban communities in Sarawak. In: Parnwell, M.J.G. and Bryant, R.L. (eds) *Environmental Change in South-East Asia.* Routledge, London, pp. 269–300

Pender, J. and Kerr, J. (1996) *Determinants of Farmers Indigenous Soil and Water Conservation Investments in India's Semi-arid Tropics. EPTD Discussion Paper No. 17.* International Food Policy Research Center, Washington, DC.

Renaud, F. (1997) Financial cost–benefit analysis of soil conservation practices in Northern Thailand. *Mountain Research and Development* 17, 11–18.

Rigg, J. (1997) *Southeast Asia – The Human Landscape of Modernisation and Development.* Routledge, London.

Schreier, D.F. (1989) Agricultural project investment analysis. In: Tweeten, L. (ed.) *Agricultural Project Analysis. Tools for Economic Development.* Westview Press/IT Publications, Boulder, pp. 238–277.

Söderbaum, P. (1994) Towards a microeconomics for ecological sustainability. *Journal of Interdisciplinary Economics* 5, 197–220.

Squire, L. and van der Tak, H.G. (1975) *Economic analysis of projects.* A World Bank Research Publication. The Johns Hopkins University Press, Baltimore.

Stocking, M. (1996) Soil erosion – breaking new ground. In: Leach, M. and Mearns, R. (eds) *The Lie of the Land.* Viliers Publication, London, pp. 140–154.

Stocking, M. and Abel, N. (1993) Labour cost: critical element in soil conservation. In: Tato, K. and Hurni, H. (eds) *Soil Conservation for Survival.* Soil and Water Conservation Society, Ankeny, Iowa, pp. 206–218.

Tacconi, L. (1995) Rethinking the economic analysis of forests: theory and practice. *Forest Ecology and Management* 73, 229–238.

Tangtham, N. (1991) Erosion study and control in Thailand. *Paper presented in the Workshop on Soil Erosion and Debris Flow Control organized by the Indonesian Institute of Science under UNDP-UNESCO-IRTCES-LIPI Regional Training Programme on Erosion and Sedimentation for Asia, 5–8 November 1991, Yogyakarta.*

Turkelboom, F., Ongprasert, S., van Look, K. and Vlassak, K. (1991) Soil dynamics under traditional and alternative cropping systems in the highlands of Northern Thailand. *Text of a poster presented at The International Workshop on Evaluation for Sustainable Land Management in the Developing World. Workshop held 15–21 September, Chiang Rai.*

Wiggins, S.L. and Palma, O.G. (1980) *Acelhuate River Catchment Management Project, El Salvador. Cost–benefit Analysis of Soil Conservation. Project Report 104.* Land Resources Development Centre, Overseas Development Administration, Surbiton.

Winpenny, J. (1991) Environmental values and their implications for development. *Development Policy Review* 9(4), 381–390.

2

The Economics of Soil Degradation: From National Policy to Farmers' Fields

J. Kerr
International Food Policy Research Institute, 1200 17th Street NW, Washington, DC 20036-3006, USA

Socio-economic factors operating at the level of the national economy, the farm, the household, the watershed and the village all play a strong role in shaping farmers' incentives to manage soil resources in ways that lead to erosion or conservation. Three main points are made in this chapter. First, if prices do not reflect resource scarcity because of market failure, farmers will have insufficient incentive to invest unless an appropriate economic policy is introduced to correct the market failure. Special soil conservation projects and improved technologies stand little chance of promoting adoption under these conditions. Second, household-level socio-economic and cultural conditions will play a strong role in determining a household's optimal level of conservation investment and the optimal type of practice. Third, when erosion problems cross farm boundary problems, lack of collective action among farmers can constrain conservation investments even if generally favourable economic conditions and policies are in place to promote conservation. Because conservation practices must fit into the household- and watershed-level socio-economic system, the conservation practice that stands the best chance of success may not be the one that is most effective from a technical standpoint.

Introduction

Knowledge about soil erosion and conservation is growing rapidly. The bio-physical causes and effects of erosion are increasingly well understood and methods to protect soil resources are well known. However, erosion remains widespread and adoption of conservation practices is limited. This is in part because underlying the immediate biophysical causes of soil erosion are socio-economic conditions that dictate whether land users encourage or discourage erosion. In this way, socio-economic forces constrain technical

 Soil Erosion at Multiple Scales
(eds F.W.T. Penning de Vries, F. Agus and J. Kerr)

solutions. As a result, understanding socio-economic factors, and designing policies that harness them in a way that promotes conservation behaviour, will play an important role in promoting soil conservation.

Farmers have multiple objectives in managing their resources and also various constraints that limit their choices. These objectives and constraints determine both the extent to which farmers are interested in investing in conservation and the kinds of conservation practices they are likely to adopt. Objectives and constraints can be quite complex, involving multiple decision-makers and multiple, overlapping concerns, some of which have no obvious connection to soil management decisions. They operate at many levels, including the national economy, the plot, the household, the community and the watershed. Economic policies can influence farmers' objectives up to a point, but in some cases it is important to design policies and programmes to suit farmers' objectives, rather than try to change them. In this chapter, policies are discussed at multiple levels, covering issues of technology design, programme management and the national economy. Most of the examples are drawn from India, based on the author's previous experience there.

Economic Forces, 'Optimal' Land Degradation and Conservation

Estimates of the extent and consequences of soil erosion and other forms of land degradation throughout the world vary widely. The Global Assessment of Soil Degradation (GLASOD) study of land degradation is widely cited as having relatively reliable figures (Oldeman *et al.*, 1990). It estimated that about 22% of the world's 8.7 billion ha of land area has been degraded since the Second World War, and that severe degradation is claiming an estimated 5–10 million ha of crop lands annually. Studies in Mali, Mexico, Costa Rica, Indonesia and Malawi calculated that soil erosion leads to losses ranging in value from 0.5 to 1.5% of gross domestic product (GDP) (World Bank, 1992, p. 56).

Despite these problems, much of the world agricultural development appears to be sustainable and technology and institutions continue to develop and adapt to the environmental stress of growth. The widely publicized case of natural resource management in Kenya's Machakos District (Tiffen *et al.*, 1994) illustrates the success that can occur. In the 1930s, Machakos' hilly lands were marked by erosion, deforestation, low productivity and pasture degradation. People were poor and the population was considered greater than the land could support. By 1990, however, incomes were higher, the resource base had recovered, and the real value of agricultural output per capita had risen by 300%, even though the population had quintupled and agriculture had spread to even more marginal lands. According to the study's authors, good roads offered access to the large Nairobi market for agricultural products, increasing the returns to agriculture, and off-farm income provided resources for land improvements such as terraces, trees and hedges, and organic matter applications. Local institutions also developed as rules emerged for managing communal resources.

Evidence from around the world suggests that farmers are conscious of the causes and consequences of natural resource degradation. Farmer surveys consistently show awareness of erosion and nutrient loss, deforestation, and range decline. A growing body of literature and field experience shows that the Machakos experience, where people have developed effective resource management practices in the face of rapid population growth without external assistance, is not unique (Gill 1991; Cleaver and Schreiber, 1994; Reij *et al.*, 1996).

Induced innovation

Induced innovation theory helps explain the paradox that soil erosion is widespread in some places but conservation prevails in others. The theory holds that over time, technological innovations and institutional changes take place to economize on scarce resources and utilize abundant ones. In other words, innovations are induced by economic forces, so that farming systems evolve differently in different places.

Induced innovation also suggests that degradation may sometimes be self-correcting as resource scarcity induces development of more sustainable land use, as in path I in Fig. 2.1 (Scherr and Hazell, 1994). Consider the case of land that can be managed as forest, pasture, or extensive or intensive agriculture. The first development stage (A) is a bush–fallow system in which there are no land productivity investments, but fallow periods are long and

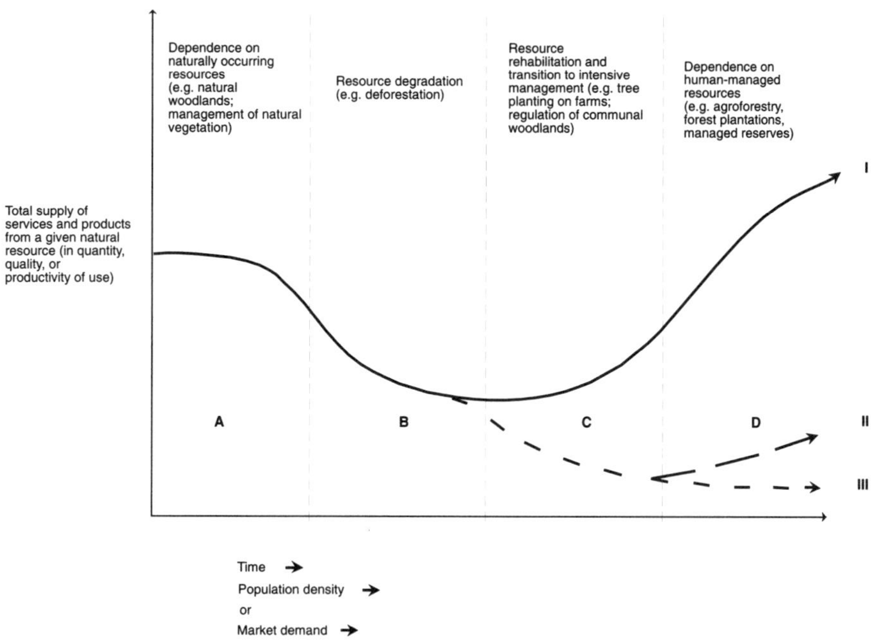

Fig. 2.1. Induced innovation in natural resource management.

pressure is low enough to avoid degradation. There is no incentive to develop additional conservation methods. Productivity losses begin in stage B as the growing population and market pressure lead to reduced fallows. Degradation continues while the returns to investing in land improvement are not worth the costs. Degradation under these circumstances can be considered economically optimal, since it is not profitable to arrest it under current conditions, and since it is not irreversible. At some point, land becomes scarce, and its productivity so low, that investments in conservation and improvement become profitable. Here, in stage C, a transition to more intensive agriculture occurs; farmers invest to plant trees, build terraces and apply fertilizer and organic matter, and regulations are imposed on communal forest, pasture and water resources. In a sustainable, intensive agricultural system (stage D), ultimately land productivity may be higher than under the original forest system. An example is the intensive rice growing areas of South-East Asia, where resource scarcity required the development of elaborate terraces, irrigation systems and nutrient management systems that enabled continuous cultivation without land degradation.

Besides the optimistic scenario of path I, Fig. 2.1 also illustrates two other natural resource management scenarios, paths II and III. In path II, resource-improving investments are made only following degradation so severe that the system can never recover high productivity (except perhaps at uneconomical costs). In path III, resource-conserving investments are never made and productivity dwindles to virtually nothing. Paths II and III reflect the fact that in some places, natural resource degradation continues even though people both recognize the problem and have the technical know-how to stop it. Favourable economic conditions and sound policies are needed to encourage sustainable natural resource management practices in path I.

Economic Conditions, Market Failures and Policy

The health of the rural economy has mixed implications for soil erosion, depending in large part on the stage of agricultural intensification. When land is abundant and agriculture is less intensive, rising prices for agricultural commodities will lead to expansion of cultivated area, disturbing natural vegetation and leading to increased erosion. This situation may be analogous to that in stage B of Fig. 2.1. In a more intensive system with little or no room for expansion, a healthy economy raises the incentives to invest in land improvements that increase yields. In stage C or D of Fig. 2.1, growth of the rural economy is certainly required to encourage investments that cause natural resource management to follow path I rather than paths II or III.

As a result, investments and policies that support economic growth will favour conservation, on the whole. Investments in rural infrastructure, such as roads that help farmers transport their products to market, will raise agricultural profits and encourage better land management.

Economic forces also affect farmers' preferences among different types of conservation measures. For example, biomass-based approaches such as retaining stubble in the field or adding manure or straw as an amendment to cover the soil have been shown to be effective, low cost approaches. However, experience shows that such approaches stand little chance of adoption in economies where biomass is in high demand for alternative uses. In dry areas of India, for example, straw has such high value as animal fodder, and manure such high value as fuel, that households are most unlikely to devote these resources to soil conservation. A different approach is likely to be required under these circumstances.

Market failures

Free markets have many attractive features, but they do not always provide adequate incentives for sustainable natural resource management. Various characteristics of natural resources, including soil, make them particularly vulnerable to market failure. A market failure occurs when market prices encourage people to produce, consume or invest in ways that are economically optimal for the individual but not economically optimal for society as a whole. In the language of economics, under these circumstances the private (or financial) costs and returns of an activity do not equal the social (or economic) costs and returns (Enters, Chapter 1, this volume). This situation is referred to as a market failure. Under a market failure, prices of natural resources do not accurately reflect their scarcity, making them poor indicators of the need to conserve.

Market failures typically lead to increased soil erosion and reduced soil conservation for two reasons. First, people may undertake some erosion-causing economic activity in which they gain all the benefits but do not bear all the costs. Second, people may face a situation in which they would bear all the costs of investing in soil conservation, but they would not gain all the benefits. Both of these cases can be illustrated by the case of a farmer in a watershed that drains to a reservoir that provides hydroelectric and irrigation benefits. Suppose that the farmer is considering clearing an area of steep land for cultivation, which would deposit silt into the reservoir. This situation would represent a negative externality, or an unintended cost that affects someone other than the one who produces it and that does not enter the market. Externalities can be beneficial, for example if eroded soil is deposited on someone else's farm, or harmful, such as if it is deposited behind a hydroelectric dam.

Figure 2.2 illustrates the situation where the farmer gains all the benefits of cultivating the land, but costs are also imposed on the reservoir. The marginal benefits from clearing and cultivating the land are represented by *D*. In this example, these benefits (i.e. food production) are about the same from both the private and social perspective. The marginal private cost to the farmer, labelled *MPC*, represents the value of the farmer's labour and any other costs he faces, including on-site degradation resulting from erosion. The marginal external cost of clearing and cultivation, labelled *MEC*, is the cost associated with siltation of the reservoir, which affects the reservoir but not the upstream farmer. The *MPC* and *MEC* curves slope upward because both production costs and erosion costs increase as more area is cleared and cultivated.

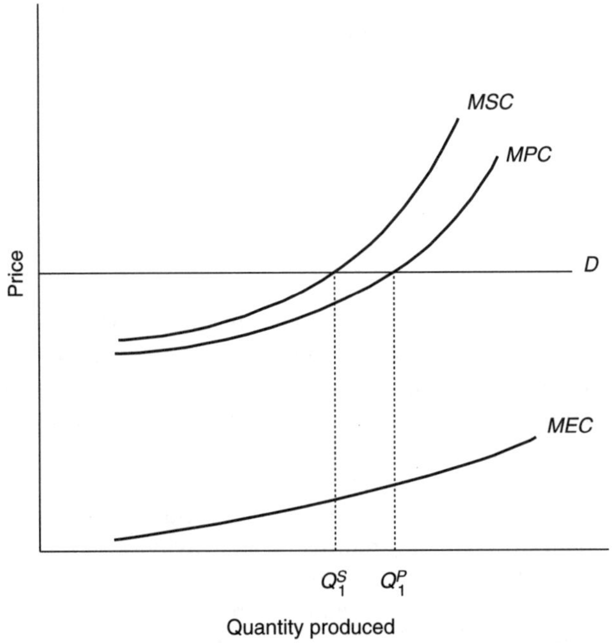

Fig. 2.2. Optimal private and social utilization of resources – social costs exceed private costs. (Labels are defined in the text.)

The marginal social cost, *MSC*, is the sum of the marginal private cost and the marginal external cost. Because private costs are lower, actual area cultivated (Q_1^s) will exceed socially optimal cultivation (Q_1^p), leading to excessive erosion.

Figure 2.3 illustrates the situation if the same farmer in the same watershed chose to terrace his land in order to reduce erosion. The private and social costs of the investment are assumed to be identical in this example, but they would be borne entirely by the farmer, while the benefits would be shared between the farmer and the reservoir. The cost of investment is represented by *MPC*. The marginal private benefits of this investment accruing to the farmer are indicated by D^p; for the purposes of this example they may be considered as constant for each increment of investment. The downstream benefits to the reservoir, meanwhile, are equal to *MEB*. The marginal social benefits are equal to D^s, or the sum of D^p and *MEB*. Under these circumstances, the farmer is likely to invest at the level of $Q^p{}_2$, which is less than the optimal level of investment $Q^s{}_2$.

In both of these cases, the price mechanism creates incentives for the farmer to address his own private costs and benefits but ignore the external costs and benefits to society. This is the essence of a market failure.

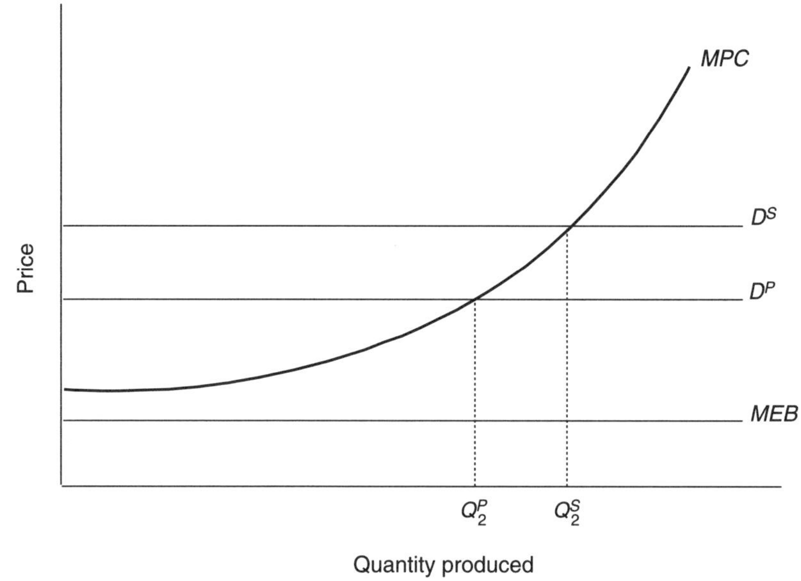

Fig. 2.3. Optimal private and social investment in resource improvement when social returns exceed private returns. (Labels are defined in the text.)

Consequently, the problem for policy makers is to introduce ways to equate private and social costs and benefits. In the context of Fig. 2.2, the objective of a policy intervention is to raise the marginal private cost (*MPC*) of exploiting a natural resource so that it is equal to the marginal social cost (*MSC*). Analogously, in Fig. 2.3 a policy intervention would seek to increase the returns to private investment so that D^p equals D^s.

Specific market failures and policy responses to promote soil conservation

Two important reasons behind the prevalence of market failures in natural resource management are the very long time period over which processes affecting their productivity take place, and the poor specification of property rights governing their control. These general characteristics lead to several more specific problems, such as lack of information, risk and uncertainty, insecure tenure, credit constraints, externalities and collective action. These are discussed in more detail below.

Lack of information

Processes that take a long time, such as gradual sheet erosion, may be difficult to perceive. If a farmer does not understand the long-term yield implications of ongoing soil erosion, investment to conserve the soil is less likely to be forthcoming. Clearly, the appropriate policy intervention to confront lack of information is simply to generate and disseminate more information. Educating farmers in general, and providing them with more information about erosion and conservation in particular, can help them make their own informed decisions about what kinds of investments to make.

While farmers undoubtedly lack detailed information on some long-term processes related to soil erosion, growing evidence suggests that farmers do perceive erosion and understand its associated yield losses (Ashby, 1985; Kerr and Pender, 1997; and others). Other factors, particularly market failures, may bear greater responsibility for lack of investment.

Risk and uncertainty

Even if a farmer had complete information about the long-term effects of soil erosion on production, the optimal level of investment would be subject to question due to the element of risk. The appropriate soil conservation measures depend on the intensity of storms, a factor that is uncertain. Therefore the farmer's investment carries a certain degree of risk. For example, if he builds a bund and it is too small, he risks that the bund will be breached in a large storm, actually contributing to erosion rather than protecting against it. If the bund is too large, he risks having devoted his time and money to an unnecessarily large investment.

Policy intervention for risk and uncertainty is a difficult task. In principle, economists would recommend insurance markets to cover insurable risk. For example, farmers could be covered by climate insurance that pays them if a storm of a certain intensity does not occur. However, this kind of arrangement would be expensive to administer and its effectiveness is unknown. (Experience with crop insurance has been disastrous in several developing countries, with claims far exceeding premiums (Hazell *et al.*, 1986). Another approach is to develop and encourage different kinds of conservation practices that bear less risk. However, not all risk can be overcome by technical solutions; more research and experimentation is needed to determine how to reduce risk in soil conservation investments.

Insecure tenure

Farmers are unlikely to invest in soil conservation if they have a short time horizon, meaning that they do not expect to be affected by long-term productivity losses due to soil erosion. The main reason why they would be likely to have a short time horizon is insecure tenure, which can take several forms. For example, farmers who encroach on public land and are likely to be evicted have insecure tenure. On the other hand, if they encroach on public land but know they will not be evicted, then their tenure is secure. Likewise, the effect on investment of holding land under lease varies by location, depending on the nature of rental contracts. In many Indian states, land tenure laws that discourage leasing mean that leases are rarely more than one or two seasons; the longer a lease contract continues, the greater the owner's risk that the tenant will be able to claim ownership rights to the land. As a result, tenants have insufficient incentive to invest in soil conservation since they cannot recoup all the benefits during the brief period that they hold the land.

Policy intervention to overcome the problem of insecure tenure first requires gaining a clear understanding of how existing tenure systems work and their impact on incentives to manage resources. After that, it is still necessary to proceed with caution to replace tenure systems that do not provide security but retain those that do. Research in different countries has shown that a tenure arrangement that appears insecure in one context may provide adequate security in another, so it is impossible to make blanket recommendations regarding tenure reform (Feder *et al.*, 1988; Atwood, 1990; Migot-Adholla *et al.*, 1991).

Credit constraints

Investors may require credit to make a profitable investment with only long-term pay-offs. However, in many developing countries, rural credit markets are weak, and credit for soil conservation may be particularly difficult to obtain. In India, for example, institutional credit (from banks) is rationed and informal credit (from village moneylenders) is expensive. Credit from institutional lenders is available for land improvement investments related to irrigation, but not rainfed land. Special credit arrangements tied to soil conservation improvement programmes have tried to make credit available to farmers, but these efforts have had limited effectiveness because farmers rejected the specific conservation practices eligible for credit.

The obvious policy response to credit constraints is to make institutional credit more widely available and less restricted. Even at relatively high rates of interest, institutional credit could have important soil conservation benefits. Moreover, strengthening rural credit markets would have generally favourable effects on the rural economy, stimulating productive investments of all kinds.

Externalities

Externalities have already been discussed above. The appropriate policy intervention to correct externalities depends on the particular situation. If an externality is localized, say between two neighbouring farmers, solving it should be possible through intervention by a local-level authority. For example, if one farmer's mismanagement leads to formation of a gully that threatens to extend to a neighbour's land, the local authority could help establish or enforce rules regarding the rights and duties of each farmer as pertains to the case at hand.

In the case of a more complicated externality, where the effect is felt far away or dispersed among numerous people, it would be difficult to come up with simple rules for people to follow, and even more difficult to enforce them. Negotiating a solution would simply be too much trouble and too expensive. In this case, a government subsidy for conservation investment might make more sense. In principle, the subsidy would be designed to raise the farmer's private net returns to investment so that it equals the social net returns. Note, however, that subsidies can introduce other problems which will be discussed below.

Collective action problems

In many cases, investment to correct soil erosion requires collective action by more than one person. First and simplest, research in India showed that when two adjacent farmers are in dispute about the location of the boundary between their farms, they are less likely to make an investment to correct an erosion problem that affects both farms (Kerr and Sanghi, 1992). The reason for this is straightforward. Boundary-based bunds are the most common soil conservation measure in the area under study. However, if the boundary is in dispute, construction is difficult because it is not clear where the boundary actually lies. One farmer or the other surely will object to any given placement of the bund. As a result, the erosion problem goes untreated. As in the case of externalities, the boundary dispute could be solved by a local-level authority, paving the way for investment in soil conservation.

A second kind of problem arises when a large gully crosses several farms, and stabilizing it requires simultaneous action by all the affected parties. (When a gully is large enough, it must be treated in several places to ensure that runoff water cannot gain enough momentum to breach any given structure.) Typically, the more farmers whose action is required, the more difficult it is to make the investment, because the chances are greater that at least one party will be constrained from investing for one reason or another. Solving this kind of problem is difficult, but in principle it could be done through the coordinating efforts of some kind of local-level organization or authority. It is also a good area for intervention by special soil and water conservation programmes.

A third kind of problem is on land that is not owned individually. This may refer to land owned by the state, or by the community as a group, or by no one in particular. Similarly, the land may be nominally owned by an individual, a group, or the state, but if the ownership right is not enforced then it might as well be non-existent. Resources managed by an identifiable group are referred to here as common property, whereas resources not managed by anyone are referred to as open access, in which no property rights apply. There is a large literature on the distinctions between different property rights regimes and the circumstances under which the open access situation is likely to reign, but it is too big an issue to address here. The main points to emphasize here are as follows. First, under a common property regime, the decision to invest in soil conservation is complicated by the challenge of convincing all the users to invest jointly. This is conceptually equivalent to the case described above of the gully that runs through several properties. The second point is that under conditions of open access, investment is almost certain not to be forthcoming. The reason is obvious: the cost of investment would be borne by those who invest, but they would not be able to capture the benefits.

Solving open access and common property problems is a complex task. As discussed above in the section on insecure tenure, the success of different property rights arrangements varies by situation and location, so it is not possible to make a blanket recommendation. Location-specific research and experimentation are needed to find out what kinds of property rights arrangements work or do not work.

Some caveats on proceeding with policy intervention

The discussion in the previous section makes it clear that policy makers have a range of tools to work with, such as granting subsidies, levying taxes, assigning and specifying property rights, improving credit and insurance markets, and many others. Direct government investment also can be used to provide goods and services that the market does not provide.

The first question to address when considering policy interventions to promote soil conservation is whether intervention is justified. The key issue is whether there is a market failure that causes investment to be 'too low'. As discussed above in the context of Fig. 2.1, conservation may be low in some cases simply because erosion does not have much of an effect on productivity, so a conservation investment would incur costs but not produce returns. If this is the case, the situation is better left alone until conditions change and erosion reaches the stage where it threatens to impose significant productivity losses. Private investors might undertake such investment on their own, without government intervention.

On the other hand, as discussed in the previous section, there are many cases where a conservation investment would make good economic sense, but it is inhibited by some kind of market failure. In this case it is important to determine what kind of market failure is contributing to resource degradation. In short, the first question to answer is whether some policy intervention is needed, and the second is what kind of policy.

In general, policy interventions should be designed to target the specific market failure in question. A policy not specifically targeted to the problem at hand is likely to be ineffective and may even make it worse. Policies that compound existing market failures or create new ones are quite common. This situation occurs when an inappropriate policy tool is used to address a problem, or when a policy is targeted to a problem's symptoms rather than its cause. A good example is the effort to reduce deforestation in India and Thailand. Policy makers in these countries, alarmed at the loss of tree cover, introduced laws against cutting and marketing trees from both public and private land. This legislation had the opposite effect of that intended on private land: farmers, fearing that they would not be able to sell trees, reduced the number of trees they planted. Thus tree production, already low due to uncertainty and credit constraints, fell further because this policy reduced the private returns to growing trees even further. In this case, policy makers confused the cause of a problem with its symptoms. The problem was a scarcity of wood, and the symptom was that farmers were cutting trees because they fetched a high price. The cause of the problem, however, was insufficient incentive to plant trees on private land because of uncontrolled logging on public land, particularly in open access forests. Clearly, a better policy would have sought to encourage farmers to plant trees and help them gain secure marketing outlets while controlling logging in public forests (Chambers *et al.*, 1989)

Sometimes policies targeted to other objectives distort markets for natural resources. In Sudan, for example, subsidies for fuel and land encouraged extensive cultivation by the mechanized agricultural sector. The objective was to raise food production in underutilized areas, but the result was overcultivation of marginal areas that led to degradation. This is because subsidized tractors made it feasible to cultivate previously unexploited areas that were prone to erosion and other forms of soil degradation. Moreover, land acquisition laws made it easier for farmers to claim ownership of a plot of land if they cleared it of trees. This was an outdated policy designed to settle property claims in more fertile areas. When applied to marginal areas, it hastened the degradation process (Nelson, 1990). Similar examples abound of unintended side-effects of misdirected policies.

Another case of misdirected policies concerns subsidies. As mentioned above, subsidies can serve as a policy tool to encourage soil conservation, but many countries have had disappointing experience with subsidies. A common approach is to link conservation objectives to employment programmes, for example in which people are paid to plant vegetation and build bunds. In this case two separate objectives are addressed by the same policy. The problem is that this approach can encourage farmers to 'accept' practices that they do not want and have no intention of maintaining beyond the life of a project. This outcome is especially likely if underinvestment in soil conservation is the result of some other market failure that the subsidy fails to address (Kerr *et al.*, 1996).

Household and Plot Level Factors Affecting Investment in Soil Conservation

Market conditions and economic policies interact with conditions at the level of the plot and farm household to determine farmers' conservation investments. This section focuses more sharply on plot and household factors to show how they can shape farmers' incentives to invest in conservation measures.

Farm households everywhere have a wide variety of interests and objectives. Soil conservation will always be just one of many objectives, and it is unlikely ever to be any given farmer's single most pressing interest. As a result, making soil conservation attractive to farmers will require making it fit into the overall household economy.

This line of thinking is exemplified by the case of farmers' adoption of soil and water conservation practices in semi-arid India. As is the case in low productivity regions in other developing countries, farmers in this area tend to follow a portfolio approach to managing risk and earning a livelihood (Walker, 1989). Their farming systems are diverse, with numerous crops and even crop varieties to meet different needs. For example, farmers may plant one crop or crop variety that tolerates low rainfall alongside another that is better suited to high rainfall, with a third that provides greater resistance in the event of a severe pest attack. Most natural resources have a variety of purposes for these farmers. Cow manure, for example, is used as an agricultural input, a fuel, and a raw material for making floors and walls in low-income housing. Off-farm income sources, including migration, may also be important components of overall livelihood strategies.

Not surprisingly, this complexity of farming systems creates difficulties for the introduction of new technologies to promote soil conservation. Farmers are keenly aware that new agricultural technologies and natural resource management practices usually incur opportunity costs by competing with one or more of the many components of their livelihood strategies. This effectively reduces the net benefits of adopting new technologies and management practices, thus increasing the challenges facing agricultural researchers and project managers. An important implication for cost–benefit analysis is that great care is required to estimate the true net returns to adopting a particular technology or management practice.

Kerr and Sanghi (1992) examined the reasons why farmers did not appear to accept recommended soil conservation practices. They found that farmers have developed a wide variety of indigenous soil and water conservation practices, subject to the multiple objectives and constraints they face. These practices vary with the soil and water management problems associated with different agroclimatic conditions, such as rainfall zone, slope and soil type.

Some of farmers' criteria for selecting soil conservation practices correspond to those of scientists, but others do not. Farmers' objectives that determine their choice of conservation practice include minimizing soil erosion, demarcating field and ownership boundaries, producing fodder for animals and other items of economic importance, protecting against trespassers and stray animals, creating new fields or microenvironments, making field operations convenient, and facilitating land partitioning for inheritance. Farmers thus face trade-offs in considering 'improved' soil and water conservation (SWC) techniques that conserve soil more effectively but do not meet their other objectives. As a result, the best soil conservation practice from a farmer's perspective is not necessarily the one that conserves the most soil. By extension, the way to conserve as much soil as possible in the aggregate may be to promote practices that are second best from a technical perspective but first best from the perspective of a farmer's multiple objectives. This way adoption will be most widespread. The Indian experience shows that introducing the technically optimal practice may not conserve any soil at all, because farmers widely fail to adopt or maintain it.

Boundary-based bunds are an example of a successful second-best approach to soil conservation. Farmers in the Indian semi-arid tropics understand the technically superior contour-based approach to soil conservation; they use it to build terraces on sloping rainfed lands and to irrigate paddy in tank-irrigated areas in lowlands. It is common to find the same farmer practising contour farming on tank-irrigated paddy land but boundary–based erosion control measures on rainfed land. Farmers agree that contour farming is more effective in conserving soil and water than a boundary-based system, but they feel that adhering to the contour will not allow them to achieve their other objectives discussed above.

In deep black soil areas, many farmers build large earthen and stone bunds on the lower boundary of their plot, trapping soil that erodes from the upper end. A stone drain at the lower corner of the plot encourages safe runoff. In some cases farmers actively encourage soil movement within the plot, while in others they wait for it to move at its own pace. Soil movement from the upper end of the plot toward the lower end creates natural terracing that distributes soil moisture evenly. Most soils where this system is practised are several metres deep, so soil movement within the plot does not leave some areas too shallow to be productive. Other physical properties of deep black soils also favour this soil management system. It may be second best from a technical perspective, but many farmers adopt it and it works.

On the other hand, soil erosion remains a problem on unprotected plots where farmers do not adopt SWC measures. Research that combines indigenous and scientific knowledge is needed to develop new systems that both accommodate local constraints and conserve soil effectively in order to increase adoption. Some non-government organizations (NGOs) in South India are already making this approach pay off by designing soil conservation programmes in collaboration with farmers (Kerr, 1991). After learning farmers' perceptions of soil erosion problems and understanding indigenous practices, the NGOs demonstrate alternatives based on scientific recommendations. From there, the farmers and the NGOs jointly design soil conservation interventions that satisfy farmers' multiple objectives while incorporating some technical improvements. This approach requires that the NGO must have both technical competence and an open mind to see things from farmers' point of view.

Watershed and Community Level Forces Affecting Conservation Investments

Adoption of soil conservation practices becomes more complicated when soil conservation is promoted on a watershed basis. This is because a watershed is far more complex than an individual farm household because it involves multiple decision-makers who are not affected equally by the introduction of watershed development technology. For example, a typical microwatershed in semi-arid India contains the best farmland in the lower reaches, much of it irrigated by groundwater or surface water from a tank or pond. The wealthiest farmers typically hold land in this part of the micro-watershed. In the middle reaches, the agricultural plots may be less fertile and of greater slope, and they are less likely to be irrigated. The upper reaches may contain steep, uncultivable land suitable only for pasture or forestry. These areas are utilized mainly by the poorest people in the village, who graze their herds and gather products from common pastures and forests because they have no land of their own.

When a watershed project is introduced, soil conservation measures are installed throughout the microwatershed, and restrictions are placed on access to the upper reaches. Eliminating grazing and felling of trees in the upper reaches increases infiltration and percolation of water, which increases the supply of water available for irrigation in the lower reaches. As a result, the people who utilize the upper watershed bear the brunt of the costs of watershed development, which mainly benefits the wealthiest people in the lower watershed.

A similar situation arises with the large contour bunds traditionally introduced in Indian government-sponsored watershed projects. Contour bunds transcend farm boundaries, and their costs and benefits are distributed unevenly among those affected by them. On the research station or in large-scale agriculture, such uneven distribution is acceptable as long as overall productivity rises sufficiently. In smallholder Indian agriculture, however, it means that some people gain from contour bunds and others do not. Those situated where central waterways are constructed lose, for example, as do those with contour bunds bisecting their small fields.

Not surprisingly, those who are made worse off by a watershed project commonly refuse to go along with it, actively undermining the project's efforts. Herders, for example, may refuse to abide by grazing bans, trespassing on the common lands if they are able to. Similarly, farmers have been known to plough over bunds introduced on their fields. If one section of a contour bund is not maintained, the entire structure becomes useless, since water flows through the breach and none is conserved. In general, watershed technologies are likely to fail if they divide benefits unevenly but require near-universal cooperation to make them work. In this case, equity becomes a prerequisite to efficiency (Sanghi, 1989).

This situation has obvious implications for cost–benefit analysis of watershed projects. In particular, analysis at the microwatershed level that suggests positive aggregate profits are not sufficient to justify investment, because the analysis will ignore the fact that the 'losers' might not go along with the system, thus making the whole arrangement unworkable. If so, the aggregate cost–benefit analysis is irrelevant. This suggests the need to augment the watershed-level cost–benefit analysis with additional analysis to identify incentives for each group of users within the system.

The uneven distribution of benefits and costs of some otherwise promising soil conservation techniques suggests the need to organize communities to develop mechanisms for compensating those who lose so that they will go along for the greater good. Experience in India and elsewhere suggests that this approach can work. An example comes from MYRADA, an NGO in southern India that is actively involved in community-based dryland watershed development (Mascarenhas *et al.*, 1991). In all of its rural development projects, MYRADA organizes people around specific small-group objectives to work toward one common purpose. In the context of watersheds, the complexity of building a consensus among everyone affected is easily seen by considering the different groups involved:

- those who live in the watershed and have land in the watershed;
- those who live in the watershed but have land out of the watershed;
- those who live out of the watershed but have land in the watershed;
- those with land in the lower reaches of the watershed;
- those with land in the upper reaches of the watershed; and
- the landless.

To devise a feasible approach to community organization for watershed management, MYRADA must reconcile the features of its successful small-group approach to the large size and heterogeneity of the watershed groups. The first step is to work in 'miniwatersheds' of no more than a few hundred hectares and a hundred farmers. Second, MYRADA helps form small subgroups of farmers based on homogeneity of location, socio-economic conditions or interests. The miniwatershed group serves as an informal apex body for the various subgroups within the miniwatershed. This preserves the participatory and socially functional character of the smaller, homogeneous subgroups while also retaining advantages of scale in planning watershed works and interacting with government agencies, banks and input suppliers. Subgroups also provide a vehicle for airing complaints and settling disputes among people in the watershed.

Experimental research is needed to test different technical approaches and institutional arrangements for collective action to manage natural resources. NGOs are in a good position to undertake such experimentation because they have the flexibility to try a variety of approaches. Researchers can team up with project managers to help devise new approaches and systematically document how resource users respond to different arrangements. Often NGOs do not document their tremendous field experience very well, whereas researchers often lack good subject matter to analyse. By teaming up, NGOs and researchers could spark progress in developing replicable approaches to soil conservation.

Conclusions

This chapter has argued that soil erosion and conservation are as much socio-economic problems as technical ones. Socio-economic factors operating at multiple levels shape farmers' incentives to manage soil resources in ways that lead to erosion or conservation. At the national level, economic forces determine whether conservation investments will be profitable for the economy as a whole. If conservation is socially profitable, favourable economic policies will still be required to encourage private farmers to invest. At the household level, each farm family has its own socio-economic situation that determines both how much it will invest and what kind of practice it will adopt. At the watershed or village level, interactions between farmers and the transboundary nature of many erosion problems complicate conservation adoption because it requires collective action.

As a result of these complications, special efforts to develop new conservation methods or encourage their adoption will be fruitless if enabling economic conditions and policies are not in place, or if new techniques are not in harmony with local cultural and economic circumstances. Accordingly, the conservation practice that stands the best chance of success in reducing overall erosion levels may not be the one that is most effective from a technical standpoint.

Acknowledgements

The section on economic forces is adapted from material previously published in World Bank (1997). The author thanks the World Bank for permission to use it. He also thanks Kimberly Chung for commenting on an earlier draft.

References

Ashby, J.A. (1985) The social ecology of soil erosion in a Colombian farming system. *Rural Sociology* 50(3), 377–396.

Atwood, D.A. (1990) Land registration in Africa: the impact on agricultural production. *World Development* 18(5), 659–671.

Chambers, R., Saxena, N.C. and Shah, T. (1989) *To the Hands of the Poor: Water and Trees.* Oxford and IBH, New Delhi.

Cleaver, K. and Schreiber, G. (1994) *Reversing the Spiral: The Population, Agriculture and Environment Nexus in Sub-Saharan Africa.* World Bank, Washington, DC.

Feder, G., Onchan, T., Chalamwong, Y. and Hongladaron, C. (1988) *Land Policies and Farm Productivity in Thailand.* Johns Hopkins University Press, Baltimore.

Gill, G. (1991) Indigenous erosion control systems in the mid-hills of Nepal. In: Kerr, J.M., (ed.) *Farmers' Practices and Soil and Water Conservation Programs.* ICRISAT, Patancheru, India.

Hazell, P.B.R., Pomareda, C. and Valdes, A. (eds) (1986) *Crop Insurance for Agricultural Development: Issues and Experiences.* Johns Hopkins University Press, Baltimore.

Kerr, J.M. (ed.) (1991) *Farmers' Practices and Soil and Water Conservation Programs.* ICRISAT, Pantancheru, India.

Kerr, J.M., Sanghi, N.K. and Sriramappa, G. (1996) *Subsidies in Watershed Projects in India: Distortions and Opportunities. Gatekeeper Series No. 61.* Sustainable Agriculture Programme, International Institute for Environment and Development, London.

Kerr, J.M. and Pender, J.L. (1997) Farmers' perceptions of soil erosion and its consequences in India's semi-arid tropics. Unpublished draft. EPTD, IFPRI, Washington, DC.

Kerr, J.M. and Sanghi, N.K. (1992) *Indigenous Soil and Water Conservation in India's Semi-Arid Tropics. Gatekeeper Series no. 34.* Sustainable Agriculture Programme, International Institute of Environment and Development, London.

Mascarenhas, J., Jangal, J., Prem Kumar, P.D., Rathod, R., Naik, D. and Maidrappa (1991) Community organization and participatory learning methods in the watershed context. In: Kerr, J.M., (ed.) *Farmers' Practices and Soil and Water Conservation Programs.* ICRISAT, Patancheru, India.

Migot-Adholla, S., Hazell, P., Blarel, B. and Place, F. (1991) Indigenous land rights systems in sub-Saharan Africa: a constraint on productivity? *World Bank Economic Review* 5(1), 155–175.

Nelson, R. (1990) *Dryland Management: The Desertification Problem. World Bank Technical Paper 116.* Washington, DC.

Oldeman, L.R., Hakkeling, R.T.A. and Sombroek, W.G. (1990) *World Map of the Status of Human-Induced Soil Degradation: An Explanatory Note,* revised 2nd edn. International Soil Reference and Information Centre, Wageningen, the Netherlands.

Reij, C., Scoones, I. and Toulmin, C. (eds) (1996) *Sustaining the Soil: Indigenous Soil and Water Conservation in Africa.* Earthscan, London.

Sanghi, N.K. (1989) Changes in the organization of research on dryland agriculture. In: Chambers, R., Pacey, A. and Thrupp, L.A. (eds) *Farmer First: Farmer Innovation and Agricultural Research.* Intermediate Technology Publications, London.

Scherr, S. and Hazell, P. (1994) Sustainable agricultural development strategies in fragile Lands. EPTD Discussion Paper No. 1. IFPRI, Washington, DC.

Tiffen, M., Mortimore, M. and Gichuki, F. (1994) *More People, Less Erosion: Environmental Recovery in Kenya.* John Wiley, Chichester, UK.

Walker, T.S. (1989) Yield and household income variability in India's semi-arid tropics. In: Anderson, J. and Hazell, P. (eds) *Variability in Grain Yields.* Johns Hopkins, Baltimore.

World Bank. (1992) *World Development Report 1992: Development and the Environment.* Oxford University Press, New York.

World Bank (1997) *Promoting Sustainable Agriculture: Improving Policies for Natural Resource Management. Agriculture Technology Note 16.* Agricultural and Forestry Production Systems Division, The World Bank, Washington, DC. (Prepared by J. Kerr and W. Magrath.)

3

Bioeconomic Modelling for Analysing Soil Conservation Policy Issues

K. Menz and P. Grist

ACI Centre for Resource and Environmental Studies, ANU Canberra, 0200, Australia

The model SCUAF (Soil Changes Under Agro-Forestry, version 2), with an economic dimension added, was used to trace through the biological and economic consequences of soil based changes at plot and farm level. Soil erosion features in SCUAF are both a cause and a consequence of changes in farming system parameters. In this chapter, SCUAF was calibrated to conditions typical of smallholder shifting cultivation systems on moderate to steep slopes in the uplands of Indonesia. The model was then used to simulate changes in fallow length and also changes in land preparation method prior to cropping. The results of these changes in terms of soil erosion and other biological and associated economic parameters were assessed.

Introduction

SCUAF (Soil Changes Under Agro-Forestry, version 2) is a simple, deterministic biophysical model designed to predict the effect of various tree or crop combinations on soils and commodity outputs (Young and Muraya, 1990). SCUAF can model the broader features of farming systems adequately (Vermuelen *et al.*, 1993), and can do so in a manner comparable with more complex process models (Nelson *et al.*, 1996). An economic dimension was added to SCUAF, making it 'bio-economic'. SCUAF was calibrated and then used to model the effects of fallow length changes and burning as a method of land preparation in Indonesian shifting cultivation systems. Soil erosion and economic net returns are two key 'performance indicators' that can be gauged from the model, both on a per hectare and per farm basis. Nelson *et al.*, (1996) compared the results from SCUAF with a more detailed model, APSIM (Agricultural Production Systems Simulator). APSIM is a complex, dynamic process model that simulates the physical processes involved in a cropping system, including fluctuations in crop yields due to seasonal weather.

One of the most attractive features of SCUAF is the ease with which it can be applied. SCUAF avoids the demanding data requirements that can make complex process models, such as APSIM, difficult to apply. Nelson *et al.* (1996) found that SCUAF predicted similar trends in maize yields to APSIM in the medium term, but is unable to capture the seasonal variation due to climate.

The choice of SCUAF in this study followed an exploratory analysis with another model of shifting cultivation (Trenbath, 1984). Magcale-Macandog *et al.* (1996) detail an exploratory analysis using the Trenbath model. The Trenbath model requires two key parameters, the reduction in soil fertility per crop and the half recovery time of soil fertility during fallow. This critical information would not normally be readily available, but can be obtained from a model such as SCUAF. Also SCUAF explicitly models soil erosion and other soil characteristics not included in Trenbath. While SCUAF has its limitations, it was seen as being a more powerful model than Trenbath, while being similarly attractive in terms of user-friendliness. A new version of SCUAF has been produced, and will soon be publicly available (Young *et al.*, 1996); however, the results reported in this chapter are for version 2.

The general framework of the use of the model is shown in Fig. 3.1. For a detailed description of soil and climatic conditions for a typical Indonesian site with moderately steep slopes as used to calibrate the SCUAF model, see Menz and Grist (1996) and Grist and Menz (1997).

The Economic Model

An economic model was developed to evaluate the physical effects of the changes specified in SCUAF. This model determines the viability of each fallow system in terms of the economic returns from rice. It is a straightforward

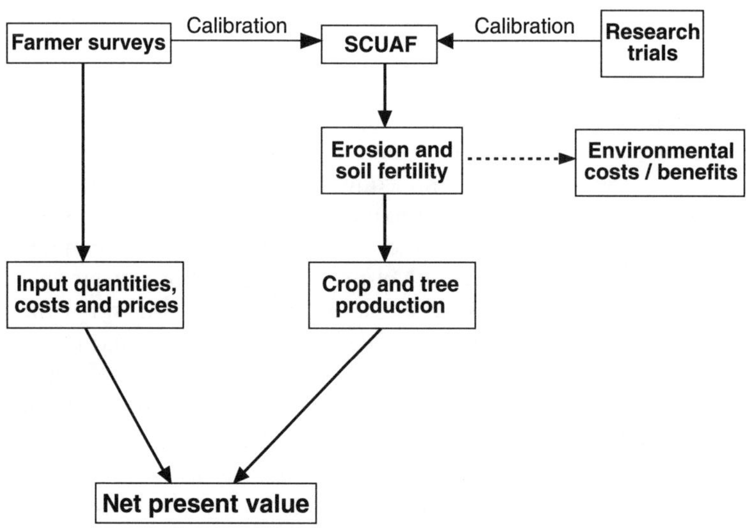

Fig. 3.1. General framework of the research.

cost–benefit analysis spreadsheet, providing the discounted net present value (NPV) corresponding to each fallow length.

The cumulative net present value of each fallow system over *n* years can be calculated as:

$$\mathrm{NPV} = \sum_{t=0}^{n} \frac{\left(B_t - C_t\right)}{\left(1+r\right)^t} \tag{3.1}$$

where B_t and C_t are the benefits and costs in year *t*, and *r* is a discount rate.

Labour availability on smallholder upland farms is assumed to be 450 man-equivalent days per year (Menz *et al.*, 1995), and the labour requirements for growing upland rice in *Imperata* areas under various conditions is presented in Table 3.1. Based upon this, the maximum area that can be cropped within each smallholding when clearing *Imperata* manually (as in Section 1) is 1.63 ha. In the first section of this analysis, labour is treated as a finite resource whose level cannot be exceeded. No charge is made per unit of labour used within that finite constraint. In the second section, labour is treated as an infinite resource, with the cost of labour included as a factor influencing the profitability of the system.

The discount rate used is 12%, which approximates real interest rates in Indonesia (Anon., 1996). All economic information is specified in Indonesian Rupiah (Rp). A price of Rp 400 per kg was used for rice in South Sumatra in 1995 (Grist *et al.*, 1995).

The focus of the analysis is on smallholder annual upland rice cropping/*Imperata* fallow systems on sloping lands – the predominant form of shifting cultivation in Indonesia. The study is presented in two sections:

- Section 1 – traces the consequences of population pressure (= shorter fallow periods); and
- Section 2 – contrasts burning with other methods of clearing in the context of relatively short fallows.

Model outputs are reported in both biophysical and economic terms.

Table 3.1. Labour requirements of three *Imperata* fallow/upland rice farming systems (per rice crop).

	Systems		
Activity	Clearing *Imperata* manually (days)	Clearing *Imperata* by burning (days)	Clearing *Imperata* with herbicide (days)
Land preparation	} 210	25	40
Weeding		65	5
Rice planting	25	25	25
Rice harvesting	40	40	40
Total	275	155	110

Section 1. Assessing the consequences of changes in fallow length from 20 years to zero

A series of *Imperata* fallow lengths between 20 and zero years (continuous cropping) are considered. The 'optimum length of fallow' is derived by reducing the amount of land deemed to be available per family, and then determining (by simulation) the fallow length that gives the maximum economic return for the amount of farm labour available. The annual area that can be cropped is a function of the land area and length of fallow. For example, if the total available land is 5 ha, then to achieve a 20-year fallow, the maximum area that can be cropped annually is 0.24 ha (5 ha/20+1 plots of land). In each fallow cycle, it is assumed that each plot has an identical cropping history. Thus the yield of each plot is the same within a given cycle. The total harvest produced, labour and input requirements are proportional to the area cropped (based on 1 ha).

Biological outputs

The 20-year *Imperata* fallow, or 'baseline' system, is modelled as being sustainable in terms of upland rice yield (Fig. 3.2), and in terms of the levels of labile soil carbon and nitrogen (Figs 3.3 and 3.4). Soil erosion is also relatively stable (Fig. 3.5). The negative impacts of the rice crop on soil are ameliorated during the 20-years of fallow. A long *Imperata* fallow is measured as being sustainable in terms of soil parameters by Shiddieq and Setyarso (1991), who found that soil parameters under a long *Imperata* fallow were similar to levels under forest. The sustainability of the 20-year fallow rotation system is also in accordance with the findings of Palm *et al.* (1994), who suggested that a stable/sustainable system could be achieved with a 16–17 year fallow.

In moving from a 20- to a 10-year fallow, a marked change is observed. Rice yield declines to under 50% of the original yield (Fig. 3.2). With this length of fallow, soil fertility is not able to recover sufficiently to provide a

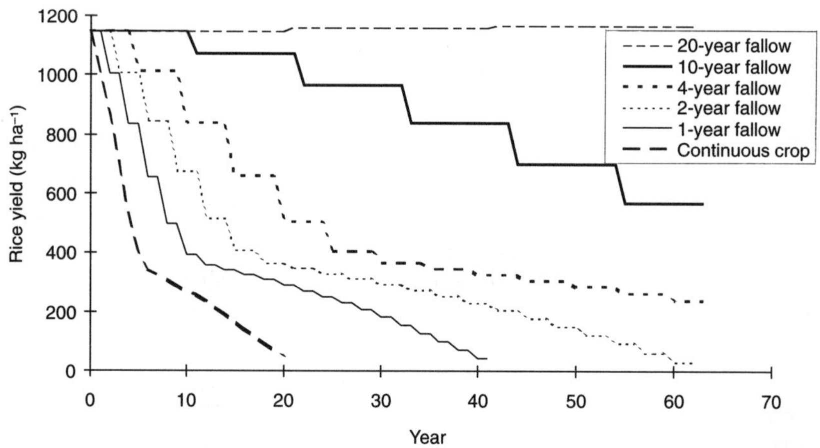

Fig. 3.2. Rice yield over time for various lengths of fallow.

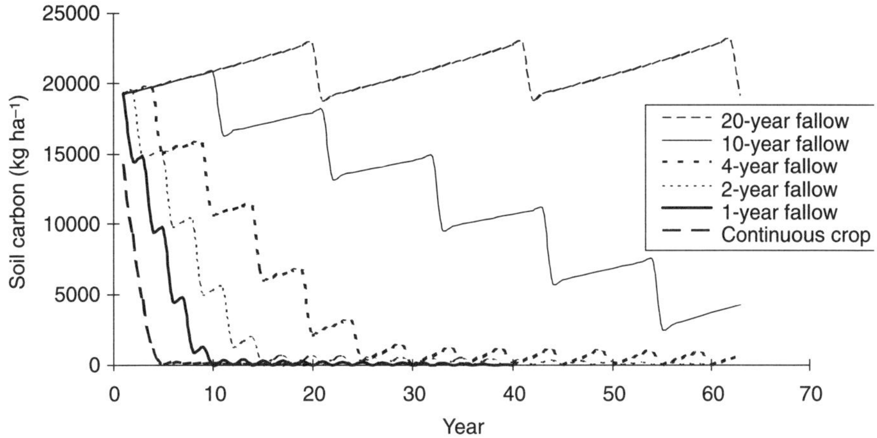

Fig. 3.3. Labile soil carbon over time for various lengths of fallow.

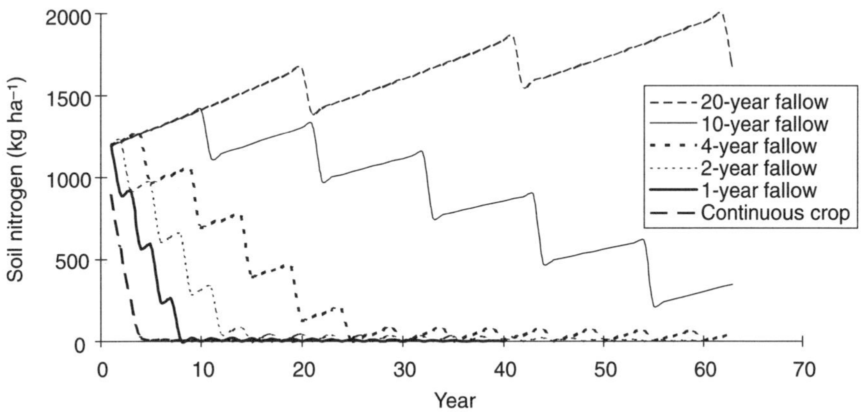

Fig. 3.4. Soil nitrogen over time for various lengths of fallow.

sustainable system. This is best observed in the labile soil carbon and nitrogen parameters, which fall to below 20% of their initial levels (Figs 3.3, 3.4). This deterioration in the soil factors causes a significant increase in soil erosion (Fig. 3.5).

With fallow periods shorter than 10 years (i.e. 4, 2 and 1 year), the decline in soil fertility (i.e. carbon and nitrogen levels) and the increase in soil erosion are even more pronounced. They are profoundly unsustainable systems by any obvious physical measurement, in the absence of input supplementation (see Figs 3.2–3.5). Yet a 4-year fallow (or less) represents the current experience of many upland farmers.

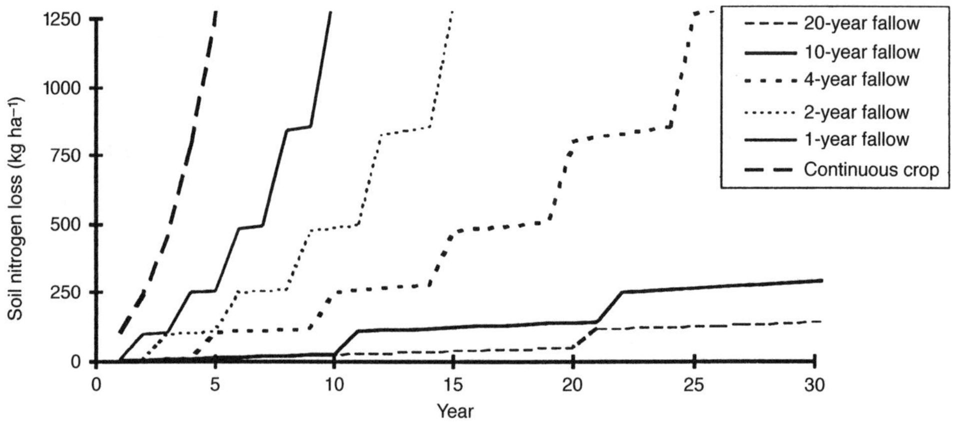

Fig. 3.5. Cumulative soil loss over time for various lengths of fallow.

Economic outcomes of shorter fallows

As fallow length falls below 20 years there is, at first, only a small decline in discounted economic returns (Fig. 3.6). When the fallow length is less than 10 years, further reductions in fallow length cause a significant decline in yield over time. Discounted economic returns show a concomitant faster rate of decline. For fallow lengths of 4 years or less, the effect on economic returns is dramatic, because the rate of crop yield decline over time is quite rapid.

So fallow lengths can be shortened to somewhat below 20 years, without undue sacrifice in terms of discounted revenue. The application of discounting ameliorates the negative effect of the shortened fallow in present-value economic terms. Despite the countervailing effect of discounting, when fallow lengths are reduced from 20 years to a level around 4 years or less, the impact on economic returns from the ensuing long-term rice yield decline is large.

In order to maintain a 20-year fallow, a smallholder is limited (by land availability) to cropping 0.095 ha year^{-1} when the farm size is 2 ha (Fig. 3.7). With a 1-year fallow, cropping area is limited to 1 ha. With smaller land areas, the biologically sustainable objective of maintaining a 20-year fallow length is contrary to the economic objective of maximum economic return. For example, with a 2-ha farm, a 20-year fallow results in a discounted revenue of Rp 412,000. However, the maximum discounted revenue that can be achieved is Rp 2.218 million from a 1-year fallow. This is so, despite the negative yield and other sustainability indicators (Figs 3.2–3.6) associated with the 2-year fallow. This fivefold difference in discounted revenue shows the economic sacrifice that a smallholder would have to incur in order to achieve biological sustainability. The fundamental reason why there is such a drop in NPV, with the longer fallow, is the restriction on cropping that is implied (Fig. 3.8).

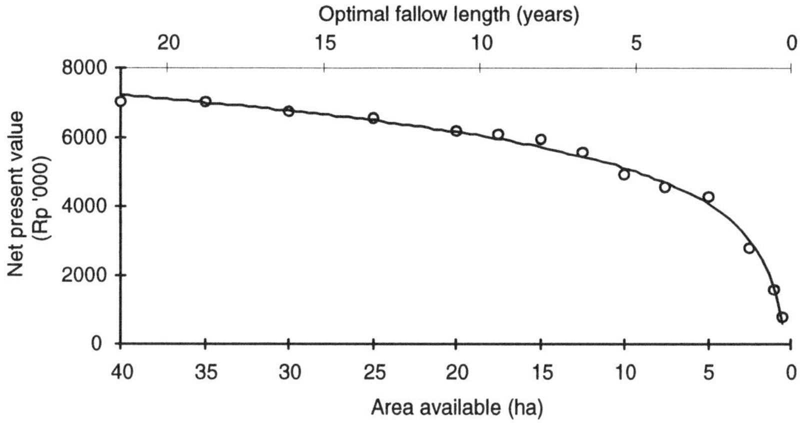

Fig. 3.6. Changes in discounted economic returns (NPV) with reducing available land area, 12% discount rate, 63-year time horizon.

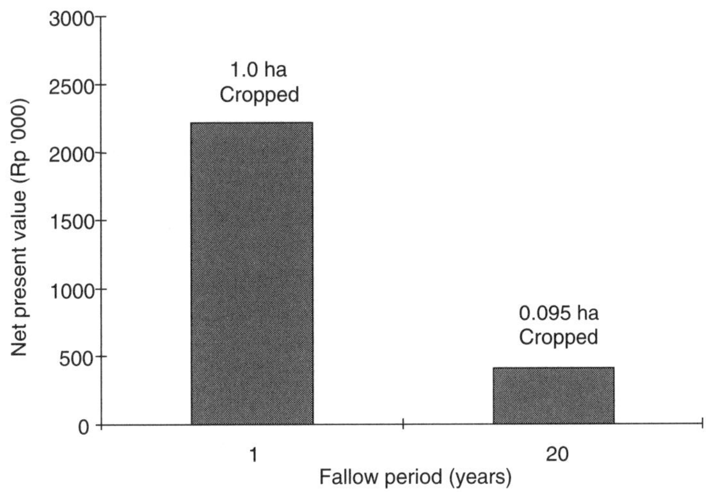

Fig. 3.7. Net present value and maximum area available for cropping that is consistent with fallow lengths of 1 year and 20 years for a 2-ha farm.

Although more intensive, unsustainable cropping levels have this associated future yield penalty, it is not sufficient to overcome the more immediate economic gains that can be made from intensive cropping (i.e. cropping a greater proportion of the land each year).

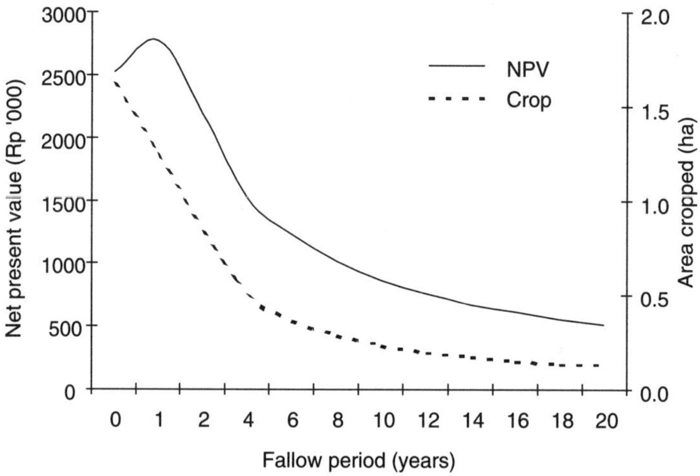

Fig. 3.8. Net present value and maximum area available for cropping that is consistent with various fallow periods for a 2-ha farm, 12% discount rate, 50-year time horizon.

Section 2. Fire versus alternative clearing methods in shifting cultivation systems with short fallows

Three different land clearing methods are analysed: clearing *Imperata* by burning, clearing *Imperata* manually, and clearing *Imperata* with herbicides. In this section of the analysis, a smaller range of fallow lengths is considered – 1 to 4 years, as representing the situation faced by many shifting cultivators today. Total available land area is taken to be 2 ha. Labour is regarded as being available at a cost. This change from Section 1 allows meaningful comparisons to be made of the clearing methods, which differ largely on the basis of labour usage. The assumption of a fixed amount of labour availability at a zero opportunity cost was thought to be the most relevant one for the historical context of the Section 1 analysis. However, the current situation is that most smallholders have some economic opportunities which can be pursued beyond the boundaries of their own farms, and the assumption of charging a daily rate to all labour used is more appropriate.

Soil and crop yield changes with different *Imperata* control methods

From a biophysical perspective, the main difference is expected between clearing *Imperata* with and without burning. The immediate impact of burning is the removal carbon and nitrogen contained in the burnt plant biomass from the plant/soil system. Consequently, there is an overall faster rate of depletion of soil carbon and nitrogen with burning. There are also some *indirect* effects of burning – less organic matter and nitrogen in the soil leading to less plant growth, so less ground cover and more soil erosion. Rice yield declines faster over time with burning. The amount of plant ground cover, either rice or *Imperata*, determines the susceptibility of the soil to erosion. When the levels of soil carbon and nitrogen are reduced by burning, the subsequent ground cover provided by crops is reduced. Therefore, soil erosion is exacerbated by burning.

However, the effect on soil nutrients, erosion and rice crop yield of reducing fallow length from 4 years to 1 year is more significant than the effect of burning. As an example, the graph of soil carbon, with and without burning, for a 4-year *Imperata* fallow is shown in Fig. 3.9. The full set of graphical output is available in Grist and Menz (1997). An examination of all combinations of fallow lengths and burning versus non-burning *Imperata* control treatments indicates that burning *Imperata* has approximately the same effect on soil quality as a 1-year reduction in fallow length.

Choice of *Imperata* control method

With daily wage rates at Rp 1600, burning is the most profitable method of clearing *Imperata* fallow in preparation for upland rice cropping (Fig. 3.10). The relatively higher profitability with burning occurs despite the negative long-term consequences for land degradation and crop yields. This stems from the low labour and cash requirements associated with burning compared with clearing manually or with herbicide. These results coincide with observations of actual farmer practice, where burning is the most common method of *Imperata* fallow removal, some smallholders use herbicide to control *Imperata*, and manual control of *Imperata* is rare.

Conclusions

The soil erosion impacts of shorter fallow lengths and burning on soil quality were clearly negative on moderately sloping land, the direct impacts of

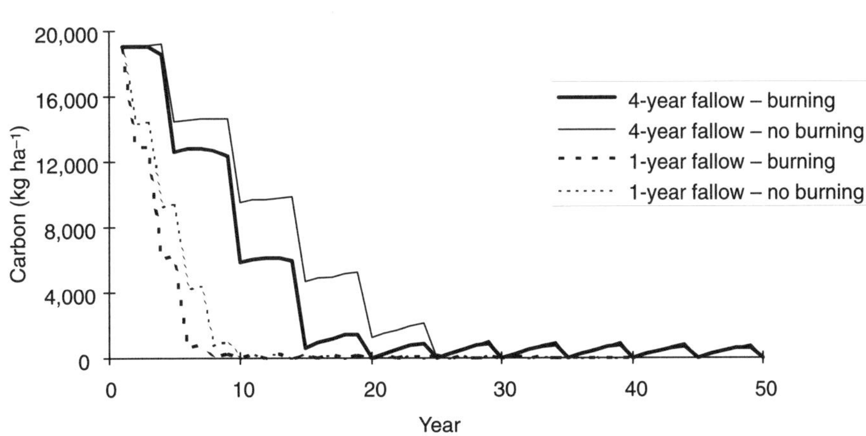

Fig. 3.9. Available soil carbon levels with and without burning.

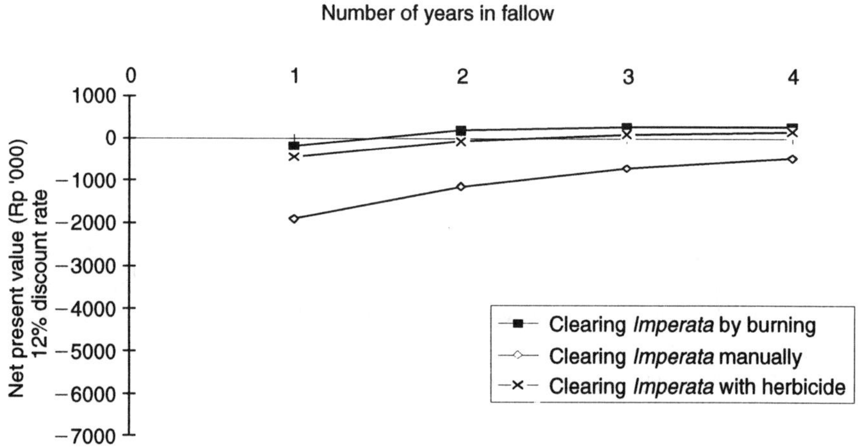

Fig. 3.10. NPV per 2-ha farm of three *Imperata* control methods, based upon a daily wage rate of Rp 1600.

shorter fallows on long-run soil fertility decline and soil erosion are much larger than the effects of burning.

It may well be that the more significant negative impacts of burning *Imperata* occur off-site (e.g. smoke haze, carbon dioxide release into the atmosphere, soil erosion leading to siltation of dams and rivers). Consideration of the off-site consequences of burning might provide an economic basis for the promotion of herbicide use via subsidies or other means, such as extension of knowledge to smallholders regarding herbicide spraying technology. Only the on-farm perspective was examined in this chapter. This perspective is important, however, since farmers will only change practices to the extent that their economic welfare is affected. Burning appears to be the most profitable method of clearing *Imperata* in a shifting cultivation system under prevailing biophysical and economic conditions.

The example of modelling shifting cultivation at the farm level allows policy implications to be drawn about:

- input or output pricing and the effect on soil erosion or other forms of environmental problems;
- research policy (e.g. running 'what if' scenarios); and
- broader social policy (e.g. stemming from the prediction of non-viability of shifting cultivation in the nearer term). Of course there are questions of extrapolation. How much is gained by disaggregating?

References

Anon. (1996) Zigzag: the great interest rate debate. *Economic Business Review Indonesia* 240, 6–11.

Grist, P. and Menz, K. (1997) *Burning in an* Imperata *fallow / upland rice farming system. Imperata* Project Paper 1996/7. CRES, ANU, Canberra.

Grist, P.G., Menz, K.M. and Thomas, R. (1995) *A modified version of the BEAM rubber agroforestry model: smallholders in Indonesia. Imperata* Project Paper 1995/3. CRES, ANU, Canberra.

Magcale-Macandog, D., Artiaga, C., Menz, K., Predo, C. and Albacea, E. (1996) *An application of the Trenbath model for studying shifting cultivation systems. Imperata* Project Paper 1996/5. CRES, ANU, Canberra.

Menz, K.M. and Grist, P.G. (1996) *The impact of changes in* Imperata *fallow length on upland cropping profitability. Imperata* Project Paper 1996/6. CRES, ANU, Canberra.

Menz, K.M., O'Brien, G.C. and Parilla, L.S. (1995) *A profile of upland farming systems on sloping land in the Philippines.* ACIAR Working Paper No. 46, Canberra, Australia.

Nelson, R.A., Grist, P.G., Menz, K.M., Paningbatan, E.P. and Mamicpic, M.A. (1996) A cost–benefit analysis of hedgerow intercropping in the Philippine uplands using SCUAF. *Imperata* Project Paper 1996/2. CRES, ANU, Canberra.

Palm, C.A, Swift, M.J. and Woolmer, P.L. (1994) Soil biological dynamics in slash and burn agriculture. *Transactions from the 15th World Congress of Soil Science (Supplement), Acapulco, Symposium ID-6.* pp. 78–92.

Shiddieq, D. and Setyarso, A. (1991) Reforestation of spear grassland: its impacts on soil characteristics and implications for forest management. In: Subarnurdin, M.S., Iswantoro, H. and Adjers, G. (eds) *Forestation of Alang Alang Grassland: Lessons from South Kalimantan.* Gadjah Mada University, Yogyakarta.

Trenbath, B.R. (1984) Decline of soil fertility and the collapse of shifting cultivation systems under intensification. In: Chadwick, A.C. and Sutton, S.L. (eds) *Tropical Rainforest: The Leeds Symposium.* Leeds Philosophical and Literary Society Ltd, Leeds, UK.

Vermuelen, S.J., Womer, P., Campbell, B.M., Kamukondiwa, W., Swift, M.J., Frost, P.G.H., Chivaura, C., Murwira, H.K., Mutambanengwe, F. and Nyathi, P. (1993) Use of the SCUAF model to simulate natural Miombo woodland and maize monoculture ecosystems in Zimbabwe. *Agroforestry Systems* 22, 259–271.

Young, A. and Muraya, P. 1990. *SCUAF: Soil Changes under Agroforestry.* ICRAF, Nairobi.

Young, A., Muraya, P. and Smith, C. (1996) *Soil Changes under Agroforestry.* Computer program, Version 4. ICRAF, Nairobi.

4

A Multiscale Approach for On-farm Erosion Research: Application to Northern Thailand Highlands

F. Turkelboom[1] and G. Trébuil[2]

[1]Department of Land Management, Catholic University of Leuven, K. Mercierlaan 92, 3001 Leuven, Belgium; [2]Agronomy, Plant Physiology and Agroecology Division, IRRI, Los Baños, Laguna, the Philippines

Over the last two decades, an impressive amount of erosion research work has been carried out in northern Thailand, but farmers' adoption of recommended erosion control measures is still very limited. The traditional biophysical approach relying on researcher-controlled runoff plots is no longer adapted to the actual agronomic, economic, social and policy circumstances faced by highland farmers. Market integration of agricultural production has been going on for years and led to a new farming context, characterized by an extensive diversity of small farmers' practices and strategies. At the same time, the spatial heterogeneity and temporal variability typical of this kind of mountainous environment must be better dealt with when designing alternative approaches.

A multiscale and interdisciplinary on-farm approach to erosion research was designed and tested in the context of an Akha village in Chiang Rai Province, upper northern Thailand. The selected methodology articulates complementary diagnostic analyses of erosion processes at field, farm and catchment levels.

At field level, causal relationships between physiographic, climatic and agronomic practices across diverse on-farm situations are presented and generate a series of key thresholds for decision-making purposes. Results of on-farm assessment of introduced and indigenous soil and water conservation practices are also shown. At catchment level, the emphasis is on understanding the spatial distribution of major erosion damage. The key role played by contour convexities and rural infrastructure is underlined. At farm level, different farmers' strategies and related erosion risks are first identified, and the way to use the farmer typology tool to target better and to improve the efficiency of soil and water conservation efforts is discussed. Finally, a critical evaluation of such an integrated erosion survey is made.

(eds F.W.T. Penning de Vries, F. Agus and J. Kerr)

Introduction: Defining the Problem

Limits of past research approach to erosion control

During the past twenty years, a very impressive amount of adaptive erosion research has been carried out on sloping land in northern Thailand. A common feature of the research is the implementation of researcher-controlled runoff plots on stations. These small plots are planted to one of the main local annual crops, in the presence or absence of introduced erosion control measures (mainly buffer strips with hedgerows or grass strips, terraces and contour ditching). Such an approach leads to an incomplete understanding of actual erosion processes, and consequently a low spontaneous adoption of the station-proven erosion control techniques is commonly observed. This lack of impact can be explained by the very different conditions between those long-term standard trials on 'homogeneous' small runoff plots and the very heterogeneous farmers' fields as well as the diversity of farming circumstances at household level. Small farmers' strategies are dominated by short-term economic objectives, which result in an increasing diversity of cropping systems as production units are integrated into the market economy. In such a heterogeneous and diverse farming context, any standard technical recommendation is condemned to a limited impact in farmers' fields (Turkelboom *et al.*, 1996). A more effective approach should be based on a preliminary in-depth understanding of actual erosion processes under the diversity of existing cropping and farming systems.

The need for an integrated on-farm R&D approach

To be more successful than past efforts, new research methods must deal with the whole complexity of the soil erosion problem and present the following key properties.

- Interdisciplinary: especially by articulating preliminary agronomic and socio-economic diagnoses prior to any technical recommendation.
- Multiscale: complementary analyses at field, farm and catchment levels are to be carried out.
- On-farm: diagnostic analyses to be based on phenomena observed in actual conditions of production.
- Dynamic: adaptable to rapid changes in land use practices.

To satisfy these requirements, two challenges to be dealt with are the modalities for data collection and their integration. An approach combining the three following tools was assumed to be adapted to such needs.

1. An erosion survey in which the on-farm diversity of situations is not considered as an obstacle but as an opportunity to assess the influence of key landscape features and main existing cropping systems on runoff generation and soil erosion. When needed, the survey approach can be combined with complementary on-farm or on-station experiments to improve the understanding of some important processes.

2. A catchment survey to analyse and rank the causes of larger erosion symptoms.
3. A farming household typology built on farmers differentiated socio-economic strategies. Based on their available resources, constraints and farm history, farmers pursue differing strategies to achieve their socio-economic objectives. Such a diversity of strategies must also be taken into account if one is aiming at limiting land degradation at village catchment level as the importance and perceptions of soil erosion can differ significantly among the different household groups (Thong-Ngam *et al.*, 1997; Trébuil *et al.*, 1997).

Based on these three complementary research tools, it was assumed that such an on-farm diagnostic analysis of susceptible situations for soil erosion at different scales (Fig. 4.1) would allow the exploration of adapted ways for the better management of soil erosion at field, farm and catchment levels.

Key characteristics of northern Thailand highlands

The biophysical conditions of upper northern Thailand are dominated by spatial heterogeneity along slopes (slope angles, slope lengths, slope shapes; soil types; mosaics of small, cultivated and fallowed fields on composite toposequences) and temporal variability between days (changing characteristics of successive rainy events) and seasons. Any attempt at understanding erosion processes has to be carried out according to a methodology specifically designed to manage such key conditions.

Pressed by high population growth, immigrations from neighbouring areas and national policies for watershed conservation, the farming systems are forced to transform rapidly from self-subsistence shifting cultivation to cash-crop-dominated semipermanent cropping systems. As symptoms of land degradation are obvious in the landscapes and public as well as government awareness of the problem is on the rise, sustainable sloping land management is still in high demand.

Approach and Methodology

Field level erosion survey

In the recent past, erosion surveys have mainly been used in a descriptive way and not so much as an analytical tool. A few remarkable exceptions are Boiffin *et al.* (1988) and Vandaele *et al.* (1996) who used such an approach in the case of the gentle loess slopes of north-western Europe, and Kooiman (1987) who studied the impact of cover on erosion features in Sumatra. Because a carefully designed erosion survey can deliver a rapid assessment of local erosion problems under actual on-farm conditions, there exists a renewed interest for this alternative erosion research approach (Bergsma, 1992; Herweg, 1996). However, there are still very few examples of such surveys which integrate data collected at multiple scales, especially under tropical conditions.

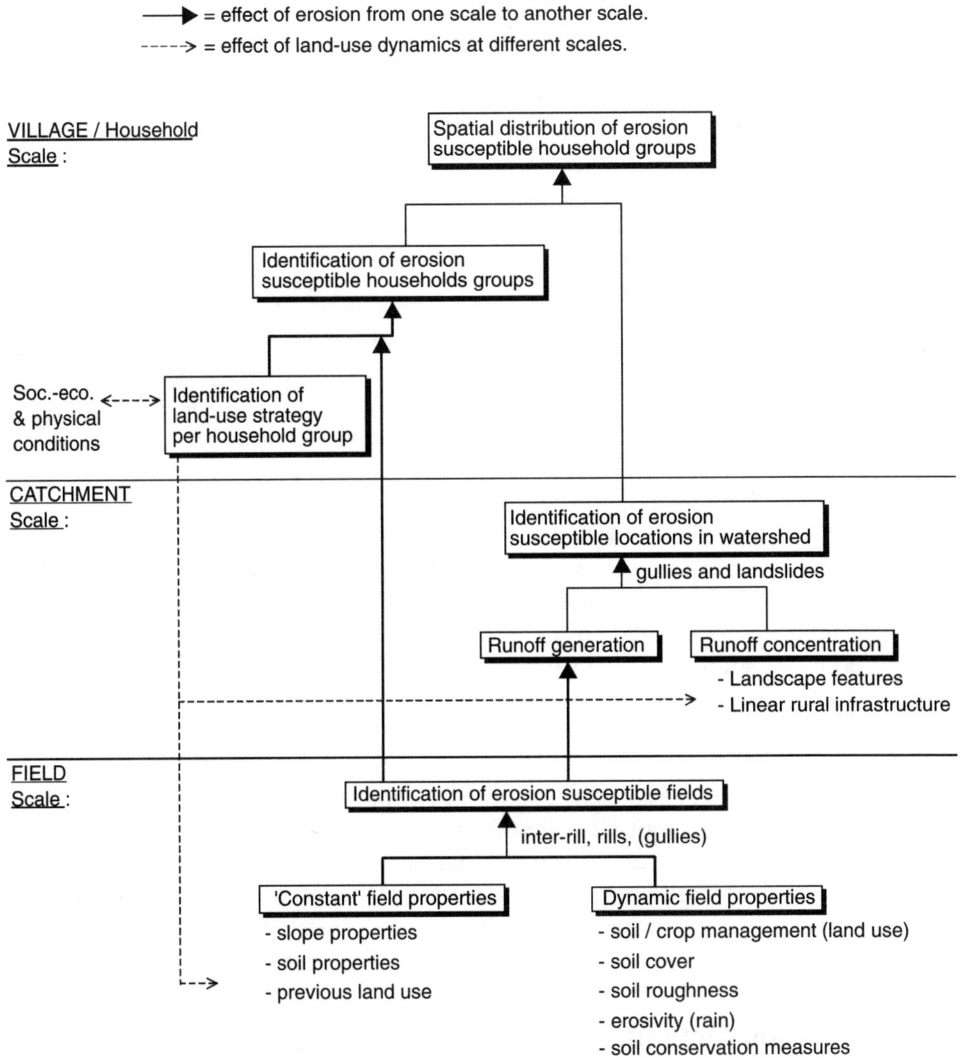

Fig. 4.1. A multiscale approach for on-farm erosion research.

The basic principle of the erosion survey is to link the appearance of erosion features to rainfall characteristics, slope conditions, erodibility and cropping systems. The last are evaluated via their effects on soil cover and soil surface states (roughness, shear strength and compaction).

Hydrologically isolated, homogeneously managed and well-delimited monitoring zones in farmers' fields (10 m wide × 10–67 m long) are visited repeatedly after each significant rainy event or cultivation practice (tillage, seeding, weeding) which modify the cover and soil surface state. Due to time limitations, all the pertinent and variable parameters are assessed using rapid and non-destructive measurements or by typologies (such as for erosion symptoms). A subsequent original characteristic of this approach is that, thanks to the large number of rapid and repeated observations made, comparisons between an extensive range of actual field situations and quantification of the erosion risk can be carried out. The period of observation is limited to the 'critical' periods for soil erosion, which are defined as the phases of the crop cycles during which low soil cover coincides with erosive rains. Observations are terminated when crop and weed cover make rapid field assessments difficult and reduce erosion to negligible amounts.

It is assumed that the upper parts of these monitoring zones generate excess runoff water and overland flow; while in the lower parts, runoff is conveyed into a network of preferential channels. As a consequence, observations on soil cover and soil surface states are performed in the upper half of the homogeneous zone, while erosion features are monitored in the lower part. Characterization of erosion features was derived from Bergsma (1992). Govers (1991) and Evans (1993) found that results from well-managed erosion surveys compare well with each other as well as with runoff plot-derived data. However, the accuracy of this technique decreases as the vegetative cover, roughness or the number of rills increases (Herweg, 1996). Even if a significant margin of error needs to be kept in mind, one has to consider that these survey results are used mainly to make comparisons between very differing field situations to spatialize the erosion risk in the watershed agroecosystem.

Apart from rain-induced erosion processes, tillage erosion also has been addressed. Land preparation by hoe is an increasingly important technique as farmers have to till their land more often to cope with increasing weed pressure related to rapidly decreasing lengths of fallow periods.

Catchment survey

At catchment scale, larger erosion symptoms were surveyed, namely gullies and landslides. Each of the three catchments had a size of about 100 ha and was made of a patchwork of small fields, fallows of varying age and remnants of secondary forest. To understand the occurrence of gullies, the area of runoff generation was located and the causes of runoff concentration were detected. Integration of these two factors enables the identification of critical situations generating major erosion damage at catchment scale.

Household analysis and farmer typology

At village level, participatory rural appraisals (PRA) and a farm land-use survey as well as many informal farmer interviews were conducted to identify and to analyse the functioning of various kinds of family-based farming units managing different amounts of land, labour and capital resources. This led to the identification of a limited number of different management strategies, resulting in specific land-use combinations for each of them. Based on the results obtained from the field-level erosion survey, the varying level of susceptibility to erosion risk among these main farm types was assessed. Thanks to land-use maps prepared at watershed level, fields managed by the most erosion susceptible types of farms can then be spatially located and integrated into the catchment survey results (Trébuil *et al.*, 1997).

Site selection criteria and key site characteristics

Pakha village is an Akha hill tribe village of Chiang Rai Province, upper northern Thailand. Pakha was selected because it is located in an area where, for the past twenty years, highland agriculture has been integrated gradually in the market economy while population pressure has increased, leading to a preoccupying level of land resource degradation. With a population density of 65 inhabitants km^{-2}, and most of the land no longer accessible for agricultural production, fields are cultivated on a semipermanent basis. Traditional subsistence crops such as upland rice and maize are still grown by farmers, but increased land pressure and improved market access have led to the introduction of paddy rice on irrigated terraces, as well as a series of annual cash crops such as soybean, and high-input, high-risk horticultural crops like common cabbage, ginger and fruit trees. A local NGO (Hill Area Development Foundation, HADF) has been promoting alley cropping since 1987 with only limited success. Most of the farmers' fields are very steep (30–70%) and the dominant parent material is phyllite. The elevation ranges from 800 to 1100 m; annual rainfall varies between 1600 and 2200 mm. Based on rainfall data of the nearby Doi Tung station, the 1994 rainy season is seen as a rather wet year (return period of 4.9 years), particularly during its first phase (return period of 32 years for June), while the 1995 wet season corresponds to a more average rainy year (return period of 2.8 years).

The field-level erosion survey was carried out in several catchments surrounding Pakha village and a total of 48 fields were monitored during the 'critical' periods of these 1994 and 1995 rainy seasons. The variation in soil properties between the monitored fields was minimized by selecting only those located on the predominant phyllite-derived soils (which are preferred by farmers). The selected fields were planted to the main four types of cropping systems (upland rice, maize, soybean or cabbage) and were distributed across a range of slope conditions, both in terms of angle (31–71%) and length (11–67 m). When possible, both indigenous (contour mulch strips and declining ditches) or introduced erosion control measures (alley cropping) were included in the survey.

Selected Results at Different Scales

Field plot scale

Importance of tillage erosion

In the late 1980s, weed pressure increased as a consequence of declining fallow periods. Farmers reacted by tilling the land more intensively. The effect of hoe tillage on soil displacement was quantified by combining an on-farm tillage experiment with a tillage survey (Turkelboom *et al.*, 1997). 'Tillage erosion' was measured on five slopes (32–82%) by monitoring tracers and by measuring the dimensions of a tillage step, which remains as a residue of tillage at the top of a field. Little difference in soil flux was observed among slopes up to 60%, but on steeper slopes, soil movements due to tillage increased significantly as the angle of repose of clods was exceeded and many macro-aggregates started to roll down the slope. On a typical upland rice field (slope of 30–50%, slope length of 30–50 m), where tillage is usually intensively performed to achieve a fine seedbed, rates of soil loss from 8 to 18 t ha^{-1} hoe tillage pass^{-1} were calculated. So tillage erosion is a significant contributor to total soil losses and will become even more important as land pressure continues to increase. Figure 4.2 shows that tillage erosion is the dominating process in fields shorter than 20 m, whereas in longer fields water-induced erosion is responsible for the bulk of soil loss. Tillage erosion in fields with contour hedgerows and grass strips contributes to the gradual formation of terraces.

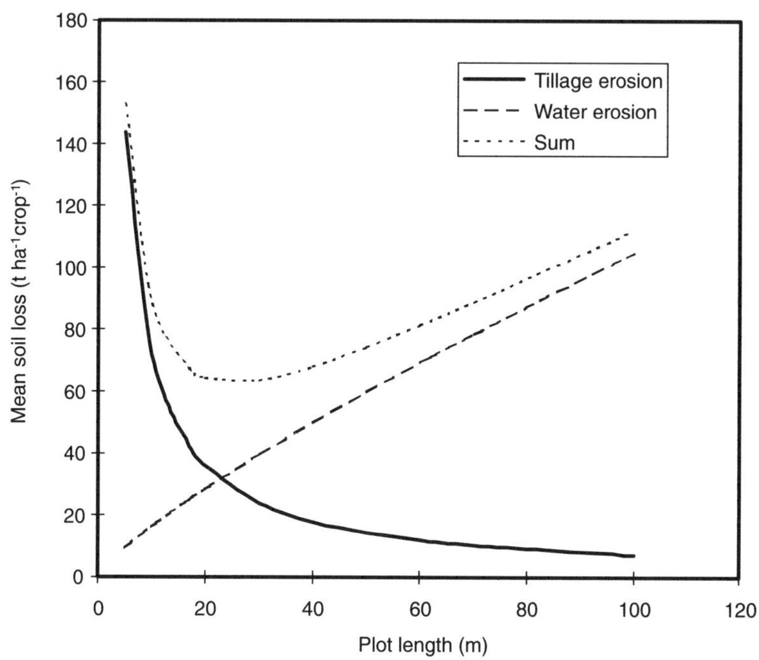

Fig. 4.2. Comparison of tillage erosion and water erosion as a function of slope length for an upland rice field with a 40% slope. (For rain-induced erosion, a mean annual soil loss of 50 t ha^{-1} year^{-1} was assumed for a plot on a 40% slope with a length of 40 m and cultivated with annual crops. This rate was extrapolated for shorter and longer plots using the L-factor for a high rill to inter-rill ratio according to RUSLE (McCool *et al.*, 1997). For a 40% slope, this corresponds with the equation: L-factor = (plot length (m) / 22.13 m) $^{0.81}$).

Typology of erosion symptoms

Because the appearance of erosion features can vary for different locations, a typology of erosion features was constructed early in the erosion survey (Table 4.1). As erosion processes at field level were dominated by concentrated rill erosion, observations were focused on linear erosion features such as pre-rills, rills, 'wide rills' (due to widening or coalescing rills), 'red rills' (rills cutting into the reddish untilled soil) and 'plough layer erosion'. The last one is caused by coalescing wide rills or by a lateral slide of the topsoil. This was mainly observed in the case of fine seedbeds on very steep slopes and represents the most severe type of erosion at this scale as the complete ploughed layer is lost. The evolution from initial erosion features (pre-rills) to very severe ones (plough layer erosion) represents a development from a transport-limited erosion process to a detachment-limited one. By analysing the time series of the erosion features, moments of accelerated erosion (AE) can be detected. AE scores (1, 2 or 3) correspond to increasingly severe types of erosion features since the preceding field visit.

At the end of the cropping season, the maximum development of erosion features was used for estimating the total soil loss. Absolute amounts of soil loss per field were obtained by calculating the rill volume and assessing the inter-rill contribution. Based on literature results (Zacher, 1982; Govers and Poesen, 1988), a contribution from inter-rill erosion of 20% was added for heavily rilled fields (i.e. more than one rill every 2 m) and 30% for fields with few rills (i.e. less than one rill every 2 m). As rill features were destroyed several times during the crop cycle, the total soil losses per plot were obtained by summing up the successive amounts of soil lost at each stage of maximum development of erosion features along the crop cycle. Stages of severity of observed erosion features relate well to these amounts of calculated soil losses (Table 4.1). In the absence of any linear erosion features the total soil loss is estimated at less than 2 t ha^{-1} $crop^{-1}$, while erosion is assessed to be higher than 150 t ha^{-1} $crop^{-1}$ when plough layer erosion took place. The correspondence found between the diagnosed five stages of erosion severity and calculated soil losses is a useful tool for roughly, but rapidly, assessing soil losses at field level in this study area.

Least erosive rainy event

A storm of at least 11 mm, a minimum erosivity value of 53 MJ mm ha^{-1} h^{-1}, a minimum I_{30} (30 min intensity) intensity of 15 mm h^{-1} and a duration of at least 37 min is necessary to generate visually detectable erosion symptoms at Pakha. Erosivity on a per storm basis was calculated by using the Rosewell formula (1986) for storm energy and EI_{30} (storm kinetic energy $\times$ I_{30}) formula for storm erosivity. For daily amounts of 20 mm and more, new erosion symptoms were always observed. Based on these results, a minimum daily amount of 10 mm can be adopted as a rule of thumb threshold for the least erosive rainy event in this area.

Thresholds for slope angle and length

Rills could be observed in nearly every field, even on the least steep and shortest slopes. This means that the surveyed slope conditions are above the

Table 4.1. Relationship between calculated soil losses and maximum development of observed erosion features in Pakha (Chiang Rai Province, upper northern Thailand, 1994 and 1995 wet seasons).

Observation year	Crop	Calculated soil loss (t ha^{-1} crop^{-1})	Plough layer erosion	Red rills	Rill network	Pre-rill network	Pre-rills or occasional rills	
1995	UR	350	x	x	x			**Stage 5: Very severe erosion (150–350 t ha^{-1} crop^{-1})**
1994	UR	270	x	x	x	x		Plough layer erosion at the bottom of the field by widening rills or planar
1994	Mz	229	x	x	x	x		slides
1994	UR	187	x	x	x			
1994	UR	145	(x)	x	x	x		**Stage 4: Severe erosion (100–150 t ha^{-1} crop^{-1})**
1994	UR	108	(x)	x	x	x		Several active red rills and plough layer erosion at some local spots
1994	Bn	64			x	x		**Stage 3: Moderate erosion (20–100 t ha^{-1} crop^{-1})**
1995	UR	61			x			Presence of rill network (> 5 per 10 m contour) or presence of red
1994	UR	60		x	x	x		rills
1994	Mz	36		x	x	x		
1994	Bn	20				x		**Stage 2: Mild erosion (5–20 t ha^{-1} crop^{-1})**
1995	Bn	18				x		Presence of a pre-rill network (> 5 pre-rill per 10 m contour) or several
1994	UR	17				x		active rills (> 3 per 10 m contour)
1995	Mz	13				x		
1995	UR	10				x		
1995	Mz	10				x		
1994	Bn	10				x		
1995	UR	7				x		
1994	Mz	6				x		
1995	UR	6					x	
1995	Bn	5				x		
1995	Mz	4				x		**Stage 1: Very mild erosion (0–5 t ha^{-1} crop^{-1})**
1995	UR	2					x	There are occasional rills (1–2 per 10 m contour) and there is a low
1995	Mz	2					x	density of pre-rills (< 5 per 10 m contour)
1995	UR	2					x	

UR, upland rice; Mz, maize; Bn, bean.

threshold for rill erosion. Therefore, it was investigated whether a threshold could be found for more severe erosion features (Table 4.2). For each slope combination in this table, the maximum observed erosion feature and soil loss is shown. Plough layer erosion started to be observed in fields with slope angles and lengths superior to 47% and 25 m, respectively, but occurred more frequently when the angle was over 57%.

Importance of field history: 'the fallow effect'

In most of these slope situations, it was observed that erosion symptoms differed significantly between fields created after clearing a fallow and older fields. Complementary laboratory and rainfall simulator experiments showed that such differences can be explained by a high content of fallow vegetation roots in new fields causing a very high macro-aggregate stability (Ryan, 1995). Soil erodibility therefore depends in the first place on the field history and much less on the inherent soil properties. On very steep slopes (65% and more), this effect does not make any difference as the angle of repose of these aggregates is being exceeded and many of them start rolling down the slope. Survey data show that this 'fallow effect' also disappears as soon as the second year of cultivation. Under certain conditions, maintaining a short fallow period may thus have a beneficial effect on erosion control.

Delimitation of 'critical' periods for the main cropping systems

In order to identify the 'critical' periods for erosion per cropping system, an erosion efficiency score (EE score) was calculated. For each storm, the sum of the accelerated erosion scores (AE) obtained was divided by the sum of the maximum AE scores that could have taken place. An EE score of 100% means that the maximum increase of erosion features (AE score of 3) was observed in all the monitored fields, while an EE score equal to zero means that no visual changes in erosion symptoms could be detected following that specific rainy event.

When applied to the main cropping systems grown in Pakha, this approach shows that upland rice can be clearly distinguished from the other shorter-duration crops (Fig. 4.3). Upland rice displays a long period of susceptibility to erosion of some 4 months (from early May to early September),

Table 4.2. Relationship between calculated soil losses and observed erosion features in Pakha.

Slope length	Slope angle > 57%	Slope angle 47 – 57%	Slope angle < 47%
> 25 m	Plough layer erosion: high risk (Max obs: 350 t ha^{-1} $crop^{-1}$)	Plough layer erosion: medium risk (Max obs: 270 t ha^{-1} $crop^{-1}$)	Rill network + 'Red' rill (Max obs: 61 t ha^{-1} $crop^{-1}$)
< 25 m	Rill network (Max obs: 64 t ha^{-1} $crop^{-1}$)	Pre-rill network (Max obs: 20 t ha^{-1} $crop^{-1}$)	Pre-rill network (Max obs: 10 t ha^{-1} $crop^{-1}$)

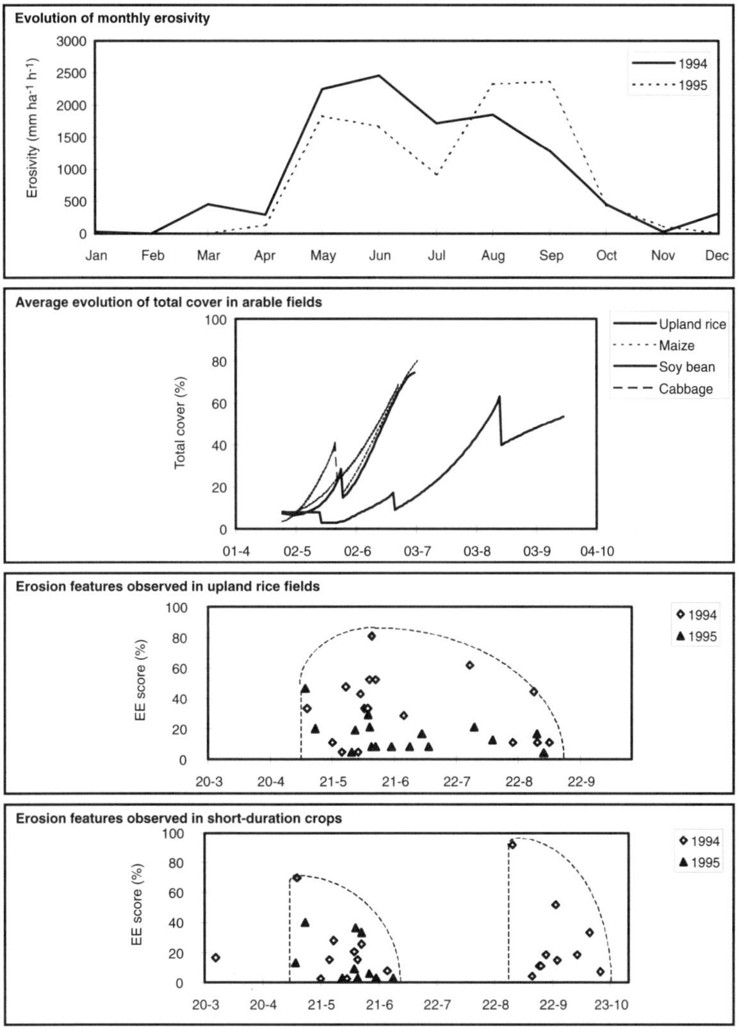

Fig. 4.3. Identification of critical periods for soil erosion in Pakha for 1994 and 1995 wet seasons.

while for maize and beans, a maximum duration of the 'critical' period of 1.5 months is observed. The peak during the first month, followed by a steady decrease of the erosion efficiency, is due to the evolution of soil cover.

Thresholds for soil cover

All the different types of soil cover were monitored separately during the erosion survey. In order to facilitate a comparison between soil cover and the observed erosion features, the different cover components were aggregated into 'total cover' and 'contact cover'. Total cover is defined as the percentage of the soil surface which is protected against rain drop detachment. This includes the canopy cover and a fraction of the residue and/or stone cover which is not covered by the canopy. Contact cover is defined as the percentage of the soil surface which is protected against overland flow. Total cover evolution for upland rice or maize fields is very different (Fig. 4.4). This is not only due to a slower growth rate of the rice canopy itself, but it is also due to more intensive weed management practices in these fields.

The analysis of the relationship between soil cover and the generation of erosion symptoms allowed the identification of two new thresholds. Above 50% total cover and 30% contact cover, erosion is negligible. The duration of the 'critical period' for each field crop can then be calculated by applying these cover thresholds to the cover evolution diagrams. For fields planted to short-duration crops, the critical period for erosion varies around 40 days with a variation of plus/minus 10 days, while upland rice fields are prone to erosion for about 120 days. During these critical periods, weeds, residues and stones provide the major part of the total/contact cover, especially for maize and beans. This indicates that field management practices

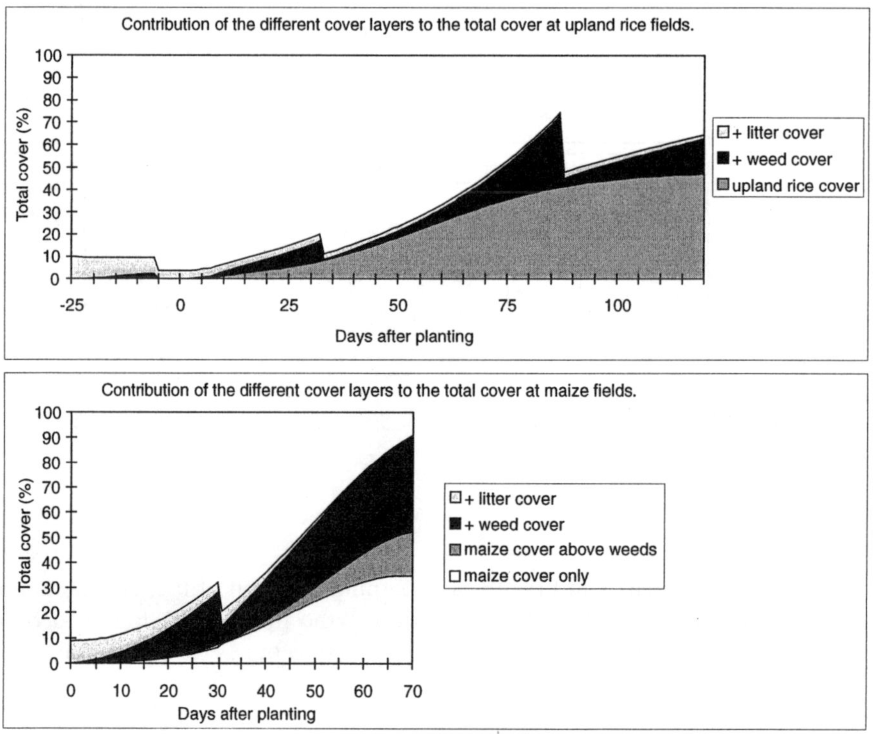

Fig. 4.4. Average evolution of total cover in an upland rice and maize field in Pakha during 1995 wet season.

can have a major impact on erosion, via their effects on weed and residue cover.

Integration of the thresholds and delimitation of the high-erosion-risk domain

The integration of the above-mentioned thresholds enables the identification of the field combinations corresponding to high erosion risk (Table 4.3). For example, a severe storm on an old field with a steep and long slope, while soil cover is still under the critical thresholds, will result in accelerated erosion in 50% of the cases. This kind of matrix is useful in planning appropriate improvements in land-use with farmers and to provide local stakeholders and decision-makers with a spatialization of the erosion risk measured at the field plot level. It can also be used as a local reference in mediation between conflicting interests. In the case of Pakha, these are represented by the Royal Forestry Department promoting reforestation and the farming community, supported in the negotiation by a local NGO (HADF), wishing to maintain local agricultural production.

Evaluation of conservation measures

The on-farm erosion survey allowed an *in situ* identification of indigenous and introduced soil conservation practices. Table 4.4 summarizes farmers' perceptions and field observations for ten different techniques encountered in Pakha. The declining ditches and (cultivated) mulch strips are considered by farmers as the most suitable ones because of the short-term benefits they bring while being easy to handle without necessitating major disturbances in the current functioning of their production sytems. They can be easily implemented, modified and removed from the field, and result in little competition with the crops. Nevertheless, the off-site effects of the declining ditch technique were assessed in a complementary catchment survey.

Table 4.3. Effects of slope characteristics and field history on the risk of erosion when soil cover is under the critical threshold in Pakha ($n = 258$, 1994 and 1995 wet seasons).

		Slope characteristics			
		< 57%		> 57%	
Soil cover	Field history	< 25 m	> 25 m	< 25 m	> 25 m
< Critical cover	Fallow clearing	12%	7%	25%	<u>50%</u>
	Old field	29%	29%	<u>33%</u>	<u>51%</u>

Underlined % represents situations with very high erosion risk.
For each combination of slope properties and field history the following equation is used:

$$\% = \frac{\text{No. of visits when accelerated erosion was observed (AE)}}{\text{Total no. of visits after storms}} \times 100$$

Table 4.4. Assessment of farmers' perceptions concerning soil conservation measures at Pakha (source: farmer interviews and field survey).

Conservation measure	Short-term benefit	(0–10)	Efficiency of erosion control (0–10)	Free of competition effects (area, light, nutrients, H_2O) (0–10)	Implementation (labour, feasibility, flexibility) (0–10)	Average score (0–10)
Indigenous methods						
Mulch lines	Weed residue management	6	6	8	9	7
Cultivated mulch lines	Yield + weed residue management	10	7	6	7	7
Weed strips	Border at bottom of field	3	8	6	10	7
Gully plugs	Weed residue management	5	4	10	9	7
Stone walls	Physical soil improvement	4	5	9	3	5
Declining ditches	Avoid seedling erosion	10	7	8	9	9
Irrigation channels	Irrigation	10	7	9	2	7
Furrows and beds	Drainage	10	2	6	7	6
Introduced methods						
Alley cropping (+ Nft / grass)	Land claim + incentives	1	7	3	5	4
Bench terraces	Land claim + incentives	1	10	4	1	4

Score of farmer perceptions: 10 = most appropriate, 0 = least appropriate; Nft, nitrogen-fixing trees.

Catchment scale

When we broaden the observation scale up to the catchment level, other factors come into the picture. Tillage erosion at catchment level displays its major effects on ridges and plan convexities, as soil export by tillage exceeds the soil influx at these sites (Govers *et al.*, 1994). Runoff in a catchment is mainly generated in fields planted to annual crops, as areas occupied by forest and fallows produce only modest amounts of runoff (5–18% versus 2–3% of the annual rainfall, based on a literature survey for northern Thailand, SFC, 1993). How this runoff is concentrated into gullies depends on the land-use pattern and the shape of the slope (Fig. 4.5). Rills can develop into a gully when fields are long or when runoff can flow from one field into another. This does not take place very often (10% of the observed gullies) because of the mosaic of land use in small fields and the occurrence of a rural infrastructure interrupting runoff flows. Runoff concentration takes place mainly as a result of contour concavity of the landscape (36%) and the occurrence of rural infrastructure (49%) such as paths, field border ditches, irrigation channels, paddy terraces, declining ditches, and furrows. In some cases, the flow of declining ditches was controlled by the filtering effect of large enough fallow fields. In other cases, they caused severe gullies down the slope. Paddies can also be a source of gullies when uncontrolled water inflow starts flowing over the bunds. Landslides are observed only during the peak of the rainy season (August–September). Two types of landslides were identified: the first type being related to saturation of the soil profile as a result of high infiltration (e.g. paddy fields and bamboo fallows), the second type being the result of slope instability caused by incising streams or roads.

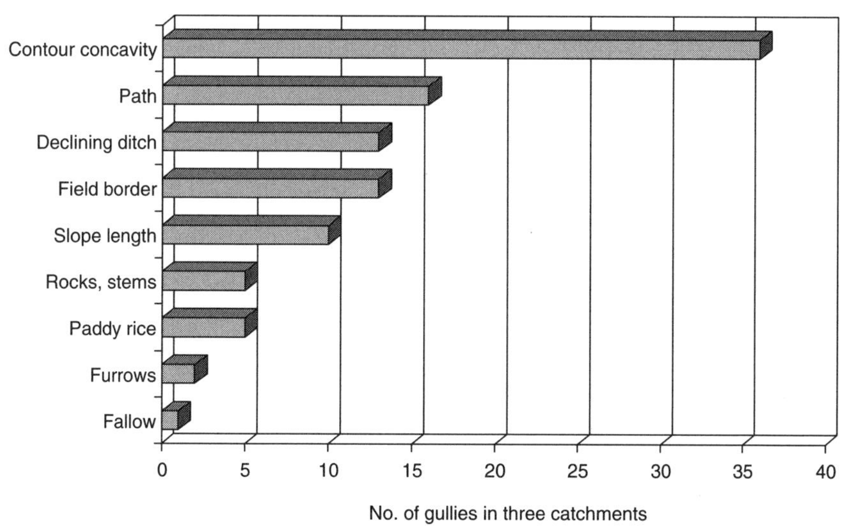

Fig. 4.5. Identification of factors causing gullies at catchment scale in Pakha during 1995 wet season.

The catchment survey is complementary to the field level one and leads to the spatialization of major erosion hazards in the catchment. As large erosion symptoms are a result of landscape, field mosaics and rural infrastructure, it is important to take the diversity of farmers' strategies and practices into account in order to propose acceptable land-use strategies at this scale.

Household livelihood system level

Before designing an adapted strategy for soil conservation at catchment level, information obtained at field level must be matched to insights on the functioning and evolution of the diversity of whole farming household systems (Papy and Boiffin, 1988, 1989).

The strategies of farming households in Pakha

As assumed at site selection stage, an extensive socio-economic differentiation among farming households was found in Pakha. A typology made of four different household groups was built based on their strategies and available resources. Their specific characteristics and land-use choices are shown in Table 4.5. The combination of these land-use patterns with results obtained through the erosion survey at field and catchment level leads to a better targeting of soil and water conservation efforts (Table 4.6).

Table 4.5. Differentiation of strategies among farming households in Pakha during 1995 cropping season (based on land-use survey and participatory rural appraisal (PRA) methods).

Farm type		Group 1: Secure investors	Group 2: Profit maximizers	Group 3: Diversifiers	Group 4: Survivors
Average land use in 1995					
Area of cultivated land	(ha)	2.2	1.0	1.5	0.9
Total land holding area	(ha)	5.7	3.8	5.0	2.0
Cropped land / total owned land	(%)	40	27	31	44
Upland rice	(ha)	0.5	0.3	0.5	0.3
Paddy rice	(ha)	0.8	0.1	0.2	0.0
Maize	(ha)	1.1	0.7	0.8	0.6
Beans + mix	(ha)	0.2	0.1	0.2	0.1
Cabbage	(ha)	0.5	0.7	0.3	0.0
Fruit tree cultivation	(ranking PRA)	High	Medium	Low	Medium
Welfare indicators					
Welfare ranking	(ranking PRA)	High	Medium	Low	Low
Rice sufficiency	(months year^{-1})	10–12	0–9	0–9	0–3
Dependency on off-farm income	(% of group)	0	0	60	92
Percentage of households	(% of village)	14	31	20	35

Table 4.6. Identification of potential erosion risk, average steepness of fields and application of conservation strategies per household group in Pakha during 1995 cropping season.

		Secure investors	Profit maximizers	Diversifiers	Survivors
Steepness of fields[a]		Medium	Steep	Steep	Very steep
Calculation of potential annual erosion					
Erosion / cropped land[b]	(t ha^{-1} year^{-1})	23	28	28	31
Cropped land / total owned land	(%)	40	27	31	44
Erosion / total owned land	(t ha^{-1} year^{-1})	9	8	9	14
Erosion / household	(t ha^{-1} year^{-1})	52	29	42	28
Application of conservation measures[a]					
Declining ditches	(% of fields)	7	20	17	27
Alley cropping	(% of fields)	22	19	15	13
Total	(% of fields)	29	39	32	40

[a]Measured during the field survey.
[b]Weighting factors for: upland rice = 60, paddy = 5, maize = 18, beans = 10, cabbage = 30 and ginger = 60 (based on field survey).

Relationships between household strategies and potential erosion risk

The level of erosion risk associated with each of these main types of farming system can be summarized as follows.

SECURE INVESTORS. They represent a minority (14%) of relatively well-off farmers, managing the largest holdings in Pakha. Because most of them were early settlers, they had access to the most suitable farmland, less steep and well located to build rice terraces. As a consequence, they are almost self-sufficient in rice and can invest securely in cash farming. The cultivation of paddy rice also explains why the annual soil loss per unit of cropped land was the lowest for this type of farm. Now that fruit trees are becoming popular in the area, these secure investors have enough land and capital to convert part of their non-terraced farmland from annual cropping to perennial plantations. Some of them are also investing their available capital into off-farm activities, such as transport and trading. The present soil loss per household is the highest, but in the future it is expected to decline due to these new types of activities. Consequently, they should not be the target of soil conservation efforts, except for the risks involved with paddy rice. The relatively high adoption of alley cropping among these farmers was probably a way to improve their claim on the land.

SURVIVORS. This is an important group (35%) of very poor farmers with tiny holdings where they maintain a low production of self-subsistence crops (mainly upland rice and maize). As these farmers cultivate the steepest slopes and the possibility of fallowing their land is very limited, they produce the highest annual soil loss per unit cultivated/owned land. This situation explains the relatively high frequency of soil conservation practices recorded in their fields (40%). Off-farm activities play a very important role for the economic survival of these households. Off-farm incomes allow a limited number of these farmers to switch to a 'profit maximizer' strategy by starting the production of high-input annual cash crops (such as common cabbage). However, in the medium term, it is expected that a large majority of these poor farmers will become almost full-time wage earners. Their tiny plots will probably be used for crops with a low labour requirement, such as maize and fruit trees. For these reasons, conservation efforts targeting this group should focus on labour-extensive methods requiring minimal maintenance costs, such as declining ditches.

PROFIT MAXIMIZERS. These represent almost one-third of the local farming community. They manage relatively small and fully market-oriented farms on steep land, and are intensively involved in the production of high-input annual cash crops, such as common cabbage and ginger. Their key economic management criterion is the maximization of the net margin per cultivated land unit. Here, the economic risk is quite high as repeated crop failures or important drops in market price can occur. In such a case, these farmers are forced to adopt alternative 'diversifier' or, in the worst cases, 'survivor' strategies. In contrast, a series of successful crops in a period of high market prices (such as for ginger production in 1996) allowed some of them to invest in fruit tree plantations or transport (usually pick-up truck or motorbike), like new 'secure investors'. The cash crops, which are commonly grown in areas equipped with a well-developed network of rural infrastructure for easy marketing of cash crops, make this group very vulnerable to soil erosion. On the other hand, the remaining part of their land holding is less intensively used or even left fallow. Although these fallows reduce the erosion risk on a per household basis, farmers do not fully benefit from the above-mentioned 'fallow effect' as they always cultivate fields located close to paths and roads.

DIVERSIFIERS. Farmers belonging to this type manage agronomic and economic risks through diverse on-farm as well as off-farm activities. The local self-subsistence crops (upland rice and maize) remain important ones and are grown on steep land. As their strategy is dominated by near term objectives, they do not seem to be able to introduce perennial crops in their production systems. Soil erosion is assessed to be relatively high, both on a per cropped area unit or per household basis. As part of the family labour is involved in off-farm activities, the implementation of soil conservation measures is more limited than in the case of 'profit maximizers'. A relatively high erosion risk characterizes these last two household groups representing more than half of the local farming community, and this situation is expected to continue in the absence of interventions to improve it. Therefore, they are seen as the most appropriate target groups for participatory soil and water conservation efforts in this study area.

Conclusion: Evaluation of the Approach and Methodology

The erosion survey approach is a fast and cheap way to generate new knowledge about erosion in the context of very diverse farming situations. This was possible without having to rely on costly and difficult to maintain equipment. As all observations are made in actual conditions of agricultural production and aim at understanding the key processes at work, the generated outputs are very relevant to address local issues. This approach also provides insights in the actual on-farm performances of recommended and indigenous conservation measures at both field (technical efficiency) and farm (acceptability by the different household groups) levels. The determination of thresholds related to major causes of soil erosion, combined with an understanding of the diversity of household strategies, makes such an approach very useful for participatory land-use planning as well as in identifying the domain of extrapolation of a given advice to farmers. The role of the on-farm researcher is (i) to integrate data gathered at complementary scales and from several relevant disciplines and (ii) to assist changes proposed by farmers. She/he is also in a position to facilitate the dialogue between different stakeholders expressing conflicting interests by providing them with relevant information on locally valid agronomic references on soil conservation and land-use issues.

Progress has still to be made to improve the methodology. While setting up the survey, the management of field heterogeneities is very important and can be based on field zoning techniques to delimit relatively homogeneous monitoring zones. The precision of measurements made in these erosion surveys can be questioned, as they are based on semiquantitative observations and are usually conducted by different observers. A preliminary group training in the field about scoring procedures is therefore essential. On the other hand, the high number of observations and the lack of artificial border conditions improve the accuracy of data generated through such on-farm surveys (Hudson, 1993; Stocking, 1996). A well-conducted survey usually produces outputs which are satisfactory to make relative comparisons between different field situations. The analysis of the survey data is a time-consuming activity, as there is no standard procedure for the complete analysis of semiquantitative datasets. However, one has also to consider the wealth of information that can be generated in a limited period of time. This is essential in areas where upland and highland agriculture are undergoing very fast and profound changes.

Overall, the proposed approach based on semiquantitative diagnostic surveys allows the erosion issue to be put in a more holistic context by integrating physiographic properties, cultivation effects and socio-economic conditions at different scales. The authors think that significant methodological improvements could be made if more case studies under tropical conditions were developed using similar approaches. Training of national researchers and development workers will then be needed to allow this kind of on-farm erosion research approach, methodology and tools to be accessible to a larger audience.

Acknowledgements

The authors would like first to thank the farmers of Pakha for their friendly hospitality and patience in making us understand the challenges of steep highland farming, as well as the Hill Area Development Foundation (HADF) who facilitated the research work in this village.

They also appreciated the significant contributions to this research made by Prof. Somchai Ongprasert (Department of Soils and Fertilizers at Mae Jo University), Prof. K. Vlassak, Prof. J. Poesen, Chirawat Vejpas, Sean Healy, Ilse Ohler, Inneke Peersman, Cynthia Ryan and Madeleine Mason-Spanoghe.

The support provided by the Belgian Agency for Development Cooperation (BADC), the National Research Council of Thailand (NRCT), the Flemisch Inter-University Council (VLIR), Mae Jo University (MJU), the Catholic University of Leuven, IRRI, CIRAD and the Vlaamse Leergangen is also gratefully acknowledged.

References

Bergsma, E. (1992) Features of soil surface microtopography for erosion hazard evaluation. In: Hurni, H. and Kebedo, T. (eds) *Erosion, conservation and small-scale farming.* Geographica Bernensia, Walsworth Publishing Company, Marceline, Missouri.

Boiffin, J., Papy, F. and Eimberck, M. (1988) Influence des systèmes de culture sur les risques d'érosion par ruissellement concentré. I: Analyse des conditions de déclenchement de l'érosion. *Agronomie* 8(8), 663–673.

Evans, R. (1993) On assessing accelerated erosion of arable land by water. *Soils and Fertilizers* 56(11), 1285–1293.

Govers, G. (1991) Rill erosion on arable land in central Belgium: rates, controls and predictability. *Catena* 18, 133–155.

Govers, G. and Poesen, J. (1988) Assessment of the interill and rill contributions to total soil loss from an upland field plot. *Geomorphology* 1, 343–354.

Govers, G., Vandaele, K., Desmet, P., Poesen, J. and Bunte, K. (1994) The role of tillage in soil redistribution on hillslopes. *European Journal of Soil Science* 45, 469–478.

Herweg, K. (1996) *Field Manual for Assessment of Current Erosion Damage.* Soil conservation programme, Ethiopia and Centre for Development and Environment, University of Berne, Switzerland.

Hudson, N.W. (1993) Field measurement of soil erosion and runoff. *FAO Soils Bulletin No. 68.* FAO, Rome, Italy.

Kooiman, A. (1987) Relations between landcover and landuse, and aspects of soil erosion, specifically for the upper Komering catchment, South Sumatra, Indonesia. MSc thesis, ITC, The Netherlands.

McCool, D.K., Foster, G.R. and Weesies, G.A. (1997) Slope length and steepness factors (LS). In: Renard, K.G., Foster, G.R., Weesies, G.A., McCool, D.K. and Yoder, D.C. (eds) *Prediction of Soil Erosion by Water: A Guide to Conservation Planning with the Revised Universal Soil Loss Equation (RUSLE).* Agriculture Handbook No. 703, USDA-ARS, pp. 101–141.

Papy, F. and Boiffin, J. (1988) Influence des systèmes de culture sur les risques d'érosion par ruissellement concentré. II: Evaluation des possibilités de maîtrise du phénomène dans les exploitations agricoles. *Agronomie* 8(9), 745–756.

Papy, F. and Boiffin, J. (1989) The use of farming systems for the control of runoff and erosion. In: Schwertmann, U., Rickson, R.J. and Auerswald, K. (eds) *Soil Erosion Protection Measures in Europe.* Soil Technology Series 1. Catena Verlag, Cremlingen-Destedt, Germany, pp. 29–38.

Rosewell, C.J. (1986) Rainfall kinetic energy in Eastern Australia. *Journal of Climate and Applied Meteorology* 25(11), 1695–1701.

Ryan, C. (1995) An assessment of the factors contributing to the aggregate stability and the erodibility of tropical highland soils. BSc thesis, Carleton University, Ottawa.

SFC (1993) *Research Report 1989–1992.* Soil Fertility Conservation project, Mae Jo University, Thailand.

Stocking, M. (1996) The rapid appraisal of land degradation – from biophysical to socio-economic impacts. In: Rerkasem, B. (ed.) *Montane Mainland Southeast Asia in Transition.* Chiang Mai University, Chiang Mai, Thailand.

Thong-Ngam, C., Shinawatra, B., Healy, S. and Trébuil, G. (1997) Resource management and decision-making in the Thai highlands. *Journal of Contemporary Asia* 27(2), 179–197.

Trébuil, G., Kam, S.P., Turkelboom, F. and Shinawatra, B. (1997) Systems diagnoses at field, farm and watershed levels in diversifying upland agroecosystems: towards comprehensive solutions to farmers' problems. In: Teng, P.S. (ed.) *Applications of Systems Approaches at the Farm and Regional Levels.* Kluwer Academic Publishers, UK, pp. 99–114.

Turkelboom, F., Trébuil, G. and Vejpas, C. (1996) Starting from the farmers' fields: on-farm analysis and development of conservation strategies on steeplands. In: Sombatpanit, Samran, Zöbisch, M.A., Sanders, D.W. and Cook, M.G. (eds) *Soil Conservation Extension: From Concepts to Adoption.* Soil and Water Conservation Society of Thailand, Bangkok, pp. 142–152.

Turkelboom, F., Poesen, J., Ohler, I., van Keer, K., Ongprasert, S. and Vlassak, K. (1997) Assessment of tillage erosion rates on steep slopes in northern Thailand. *Catena* 29, 29–44.

Vandaele, K., Poesen, J., Govers, G. and van Wesemael, B. (1996) Geomorphic threshold conditions for ephemeral gully incision. *Geomorphology* 16, 161–173.

Zacher, D. (1982) *Soil Erosion. Developments in Soil Sciences No. 10.* Elsevier, Amsterdam, The Netherlands.

5

Building the Smallholder into Successful Natural Resource Management at the Watershed Scale

D.P. Garrity, A. Mercado Jr and M. Stark
International Centre for Research in Agroforestry, Southeast Asian Regional Research Programme, PO Box 161, Bogor 16001, Indonesia

Spontaneous adoption (without persuasion, without coercion, without subsidy) is the real test of a technical innovation in conservation farming. This chapter explores the issues surrounding successful conservation farming as a process defined by spontaneous adoption. It assesses the components of a low-labour hedgerow system as an approach to a technical solution, and addresses the issue of how to disseminate information about a conservation technology cost-effectively at a sustainable rate. We observed rapid adoption of a low-labour, zero-cash-cost conservation practice based on natural vegetative strips in Claveria, northern Mindanao, the Philippines that led to examination of each component of the process of establishing and maintaining low-labour hedgerow practices. We responded to farmer interest by combining our technical expertise with the extension skills of a technician from the Department of Agriculture and the practical knowledge of a motivated farmer to provide meaningful extension services. This Contour Hedgerow Extension Team (CHET), composed of these three individuals, initially worked with individual farmers who requested their assistance. This evolved into supporting a people's conservation organization (Land Care Association) that took on the main responsibility for technology dissemination. Local government became involved in supporting the effort financially, with active involvement of the village leaders (a process led by local government). We are currently planning to scale-up the effort to the watershed and regional levels, and evaluate whether it is practical in the context of agroecological domains.

Introduction

During the deliberations of the Symposium, it was stated that the technical problems of soil conservation were effectively solved and excellent methods to control erosion had been identified. The problem was how to get farmers to adopt them. Unfortunately, this is not the case. Identifying practices that control erosion is *not* solving the technical problem. The technical problem is solved only when a soil erosion practice is sufficiently cheap and requires so little labour and management, in addition to being effective, that it is likely to increase short-term farm profitability. This is proved only by witnessing farmers spontaneously adopting the practice. Spontaneous adoption (without persuasion, without coercion, without subsidy) is the real test of a technical innovation in conservation farming. But it is rare. We've been fortunate to have had the chance to observe this in Claveria. The lessons learned have been enlightening.

Two other companion chapters in this volume discuss aspects of the natural vegetative strip (NVS) technology in Claveria. Pandey and Lapar (Chapter 6) discuss the adoption issues, and Cramb and Nelson (Chapter 7) examine modelling outputs to assess the long-term profitability and conservation issues. This chapter carries on from these important analyses by delving further into the following issues.

- What are the components of a very low-labour hedgerow system as an approach to a technical solution (as defined above)?
- How do you disseminate information and back-up about a conservation technology cost-effectively at a sustainable rate?

Contour hedgerows of pruned leguminous trees have been promoted for almost two decades in several countries in South-East Asia as a solution to the problems of sustainable crop production in the uplands. This farming system aimed to provide effective soil erosion control, organic fertilizer to the companion annual food crops, fodder for the ruminant animals, fuel woods for the farm families, restore water quality and quantity in the watershed, and other benefits. Despite these benefits, farmer adoption has not been widespread. After years of on-farm research and working closely with farmers, we identified some key constraints and their solutions. The constraints include:

- high labour requirements to establish and maintain the hedgerows;
- limited value added to the farm income;
- unanticipated problems in soil fertility due to hedgerow competition;
- irregular width of the alley;
- too dense hedgerows in moderately to steeply sloping farms;
- poor species adaptation and lack of planting materials; and
- insecure land tenure.

We were probably very fortunate when we started working in Claveria in 1985 to have had no experimental station upon which we might have conducted our trials on tree legume hedgerows. If we had, we might still be a couple of cycles behind where we are now in our learning experience. When we worked with farmers on experiments superimposed on contour hedgerows that they installed themselves, it was clear that pruned tree hedgerows were too labour intensive and productive forage grass hedgerows were too competitive with the associated crops. Adoption of these technologies was not taking off. However, we saw that the concept of contour hedgerows was a popular idea. Some farmers experimented with the concept by placing their crop residues in lines on the contour to form 'trash bunds'. These rapidly revegetated with native grasses and weeds and soon formed stable hedgerows with natural front-facing terraces. Other farmers tried laying out contour lines but didn't plant anything in them. These lines evolved into natural vegetative strips, which appeared superb in soil erosion control and reduced maintenance labour to a minimum (Fujisaka, 1989; Agus, 1993; Fujisaka *et al.*, 1995; Garrity, 1996).

These latter innovations caught the imagination of many more farmers. By about 1994, it was estimated that over 250 farmers had adopted contour hedgerow systems, while the number of pruned tree hedgerow fields decreased after 1990. The new wave of hedgerow systems was predominantly natural vegetative strips. We also observed a broad-based change in tillage systems. Claveria is a rolling area of acidic (pH 4.5–5.0) soils, predominantly oxisols and ultisols, with cropping systems dominated by two crops of maize per year (annual rainfall about 2000 mm) (Garrity and Agustin, 1995). When research had first begun in Claveria in 1985, virtually all farmers ploughed up and down the slopes. Contour ploughing was unheard of. By 1995, it was evident that nearly all farmers had converted to the idea of contour ploughing, or were at least attempting to do so.

Evolving the Components of a Successful Conservation Farming System

Interest in NVS continued to increase. Since it is quite uncommon for an effective soil conservation structure to be adopted by large numbers of farmers spontaneously, and without public subsidies, we took note that perhaps we were witnessing the kind of low-labour, zero-cash-cost alternative that might have widespread applicability. We began to examine each component of the process of establishing and maintaining low-labour hedgerow practices. The establishment of NVS requires only a fraction of the labour needed for conventional contour hedgerow of tree legumes. The only labour required is the laying out of contour lines (about 2 person-days per hectare). NVS are narrow contour strips of field area left unploughed and allowed to vegetate naturally. The total amount of time required to plough is reduced according to the proportion of the unploughed strips, thus offsetting the labour expended on laying out these contour strips. The amount of labour required to prune or maintain the NVS is proportional to the density of hedgerows per hectare. Mercado *et al.* (1997) found that NVS spaced 6 m apart and dominated by *Chromolaena odorata* required 15 person-days per cropping per hectare or 30 person-days per year. This was less than a quarter of the time required for conventional tree-legume based contour hedgerow systems (ICRAF, 1997). For low-statured NVS like *Paspalum* spp. or *Digitaria* spp., it requires even less (3 to 10 days per cropping season) (Mercado *et al.*, 1997).

Our surveys of those farmers who had not yet installed contour hedgerow systems, but desired to do so, indicated that their overriding reason for not contouring was that they lacked the technical know-how to do it right. We had recently uncovered an extremely simple and practical means of laying out contours without equipment even as sophisticated as an A-frame: the cow's back method (ICRAF, 1997). The cow's back method involves ploughing across the slope and maintaining the angle of the cow's back on the level. When the animal is heading upslope, its head is higher than its back; when it is off-course downslope, the rear part of the animal is elevated compared with the front. Stark *et al.* (1996, unpublished data) found that this cow's back method was on the average less than 2% off the real contour compared with either the A-frame method or the hose level method. This is perfectly good for practical purposes; particularly in light of the fact that most farmers don't bother with A-frames at all, but simply judge their contours by eye (which is much less accurate).

Feedback from farmers isolated another factor that causes many smallholders to hesitate in installing contour hedgerow systems: conventional recommendations that hedgerows be separated by only a 1 to 1.5 m drop in elevation. On steep slopes the crop area lost may be 15 to 20% or more. Crop yields cannot be expected to increase enough to counterbalance this amount of lost area. This is compounded by the increased labour in establishing and maintaining many hedgerows in each field. We therefore conducted trials to determine how reducing the density of hedgerows affects the expected control of soil loss. We found that hedgerows spaced at a 4 m vertical drop are still effective in reducing soil loss (Mercado *et al.* 1997). Even a single NVS strip placed on the contour halfway down a 60 m long slope reduces soil loss to 40% of that on the open slope. We concluded that farmers could space their hedgerows at much wider intervals than the conventional rule-of-thumb recommendation suggests, even up to 8 to 12 m apart on such slopes. Erosion control will not be quite as good, but the adoptability of the practice is greatly increased. More hedgerows can be added between the original ones after the farmer has gained more confidence. This wider spacing is also particularly appropriate when the farmer intends to convert his or her NVS strips into fruit or timber trees. There is now great interest in Claveria in establishing fruit and timber trees on NVS.

A wider spacing of NVS is very useful for farms where it is desired to continue growing food crops as the fruit and timber trees mature. However, farmers with larger farm sizes tend to opt for somewhat closer hedgerow spacing, and move food crop cultivation to other parcels once the tree canopies shade the annual crops. These fast-growing timber tree systems have a 6 to 8 year cycle. Farmers that establish cash perennial hedgerows (like pineapple) tend to want to have closer-spaced hedgerows in order to have more rows of these cash crops, as they often earn more from the hedgerow component than from the maize or other annuals planted in the alleys.

NVS can evolve into many forms of agroforestry systems. Farmers in Claveria are planting fodder grasses and legumes, timber trees and fruit trees, and other cash perennials on their NVS fields. The fodder grasses used include *Setaria* spp., *Pennisetum purpureum* and *Panicum maximum*. The forage legumes include *Flemingia congesta* and *Desmodium rensonii*. Timber species cultivated include *Gmelina arborea*, *Eucalyptus* spp., *Sweitenia* spp.,

Pterocarpus indicus and others. The fruit species include mangoes, rambutan and durian. Other perennial crops grown in or just above the strips include pineapples, bananas and coffee.

The groundswell of enthusiasm among hundreds of Claveria farmers and the rich store of farmer experience with a wide range of prospective contour hedgerow management options stimulated consideration of how the public sector research and extension institutions might evolve more effective techniques to diffuse NVS technology rapidly to much larger numbers of interested farmers within the municipality and elsewhere. The adoption and technology modification process has been documented (Fujisaka, 1989; Fujisaka *et al.*, 1995), but aside from documentation and wider sharing of experiences very little extension follow-up was undertaken thereafter. Extension methods can be basically classified as either the individual/household approach or the group approach. The individual approach is most effective for activities to be undertaken within the full control of the individual farmer or household (e.g. establishing contour hedges), while working with groups or the community at large is more suitable for matters related to the whole community (e.g. postharvest public grazing) or if activities will be undertaken (more cheaply) by a group (e.g. group nursery). The group approach is particularly suitable where group work is common, like the Philippine Bayanihan, the farmer work groups based on voluntary work contribution for a common benefit.

The Conservation Hedgerow Extension Team – the Individual Approach

Late in 1995, ICRAF was approached by farmers for assistance in installing contour hedges to prevent soil erosion. We responded by combining our technical expertise with the extension skills of a technician from the Department of Agriculture and the practical knowledge of a motivated farmer. This Contour Hedgerow Extension Team (CHET) is composed of three individuals. They initially worked with individual farmers who requested their assistance (Fig. 5.1). Subsequently, group training was conducted to reach more farmers and involved five to seven participants from each of the seven villages in which the team was working. Before the end of the training the participants decided to organize themselves into a people's self-help organization on conservation farming. Officers were elected and the organization came to be known as the Claveria Land Care Association (CLCA).

People's Organization Model – The Group Approach

After the CLCA was formed, the participants grouped themselves according to the villages they represented. Groups of farmers from each individual village formed separate chapters, each chapter being a Land Care Centre (LCC), a subgroup of the CLCA. Chapters elected their officers and expanded their membership within their respective villages. Experienced chapter members spread the NVS technology to other neighbouring farmers in their village.

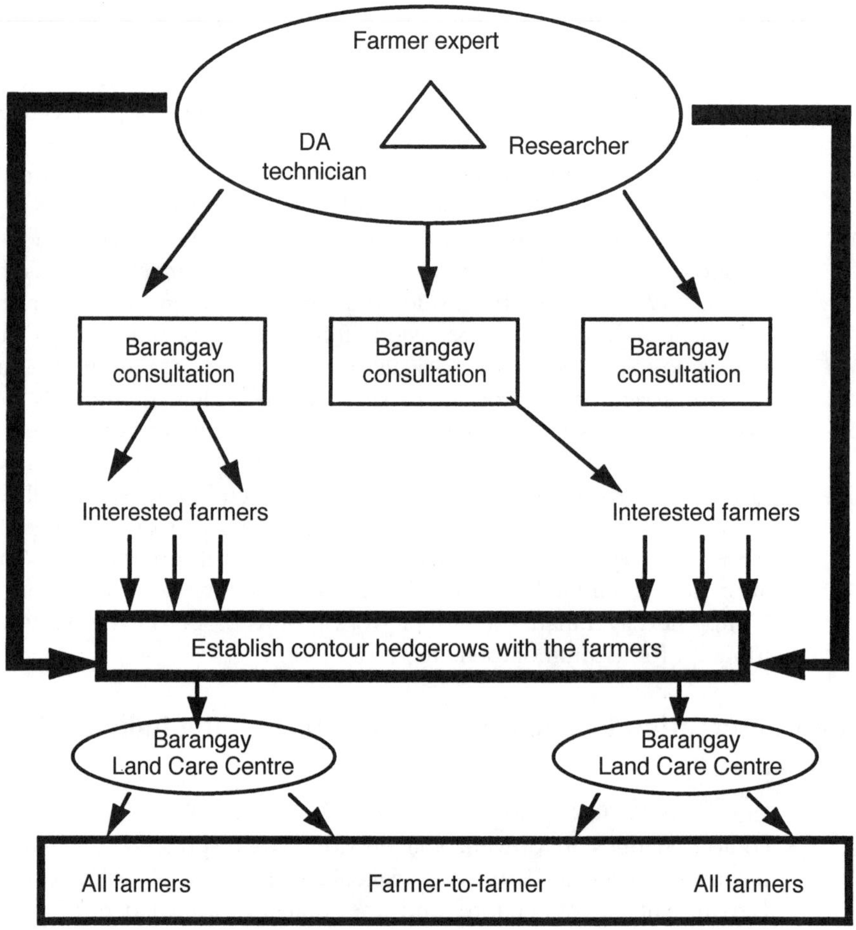

Fig. 5.1. Conservation team approach.

Subsequent group training was organized upon request from the chapters; this was conducted in the village where the requesting chapter was located. The newly trained farmers joined their respective village chapters, thus increasing the LCC membership. The conservation team's role shifted from working with individual farmers to supporting the chapters as they pursued their objectives in disseminating the NVS technology.

Twenty-four chapters of the CLCA have now been activated in villages across the municipality. The CLCA has a monthly meeting attended by the chairmen of each of the different chapters. Chapter chairmen discuss issues and problems in their respective chapters, thus giving regular feedback to the CLCA and the conservation team. The chapters each have regular meetings.

One of the key issues that emerged in various meetings was the establishment of cash perennials on the NVS. Although farmers appreciate the role of NVS in controlling soil erosion, most want to optimize the hedgerow space. They are keen to establish timber and fruit trees on their NVS. *Gmelina arborea* has been widely planted and farmers were looking for other species. We scheduled visits to a wood processors and tree plantations. After the visit farmers were interested in a new species, *Eucalyptus deglupta*, because of its better market potential for poles and lumber. The CLCA put up a central nursery. It was agreed that the chapters would contribute the labour required and costs of the establishment and maintenance. ICRAF provided the improved seed. Nursery establishment and management training was conducted with the chapter chairmen, select members and *barangay* (village) councillors. The training included lectures and hands-on experience with the very different nursery practices required for *deglupta*.

Twenty volunteer village nurseries have now been set up and are producing timber and fruit tree seedlings for the NVS. The seedlings raised are *Eucalyptus* spp. such as *deglupta*, *robusta*, *camaldulensis* and *torelliana*. Chapter members provided the nursery sheds, fencing, cellophane bags, and potting material and implemented all activities in the nurseries. Maintenance of the nurseries, as watering and cleaning, was carried out by members in a rotation. The nursery activities did not compete with hedgerow establishment. NVS are established during the land preparation period, which is therefore a seasonal activity only. The demands for NVS establishment assistance are high during the months of February, March, April, May, September and October. The NVS are proving to be a foundation for the evolution of more productive timber- or fruit tree-based agroforestry systems.

The Local Government Model – Towards Self-reliance

As the *barangay* officials became increasingly aware of the ongoing activities, interest grew in participating in the programme. With their involvement, a local government unit (LGU)-led, technology dissemination model was conceived (Fig. 5.2). The decentralization programmes of the national government gave increased power to the local government units to manage their natural resources. Many national government programmes have been devolved to the municipal level, such as agriculture, health and nutrition, natural resources management and police. The *barangays* are given funds (called *barangay* internal revenue allotments, IRA) to maintain administrative and infrastructure maintenance costs. One of the components of the IRA is human and ecological sustainability (HES). HES programmes are skewed toward environment-related projects such as soil and water conservation, tree planting and waste management.

In the model for LGU-led technology transfer, the conservation team at the municipal level trains or works with the *barangay* captains and *barangay* councilmen designated as chairman of the committee on agriculture and with the other members of the council (a municipality is composed of 15 to 30 *barangays*). The conservation team ensures that these core people understand

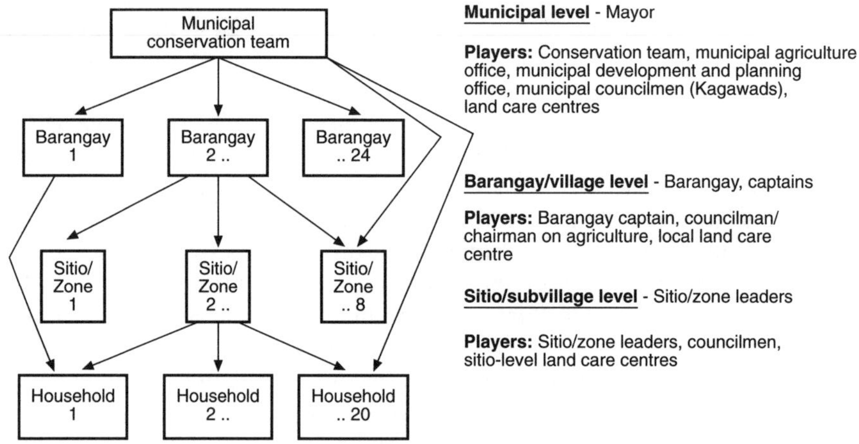

Fig. 5.2. Local government unit (LGU)-led conservation team.

the technology through village meetings, slide shows and subsequent small-group training on NVS establishment in farmers' fields. These core *barangay* leaders then work with the *sitio* or zone leaders (a *barangay* is composed of 5 to 10 *sitios*, a *sitio* is a subvillage), ensuring that these *sitio* leaders understand and can be effective in advising their farmer neighbours on implementation. The *sitio* leaders disseminate new knowledge to the farm families within the *sitio*. A *sitio* is usually composed of 10 to 20 households.

This structure ensures that the prospective spread of the technology occurs on an ever radiating basis. The breakthrough in this new paradigm occurs when the community leaders assume their role in the diffusion of the technology transfer, as is now happening in Claveria. This new paradigm may be viewed as a structured radiation of the farmer-to-farmer extension method. In a farming community the leaders are farmers themselves. The conservation team is linked with core people who have resources and influence in the community. In the Claveria experience we found that it is easier and more effective to work in a community where there are core people rather than trying to conduct the extension tasks independently. We have been deeply encouraged by seeing how proud the farmer leaders are in turning in lists of names of the people they have assisted in establishing NVS. Soil and water conservation that includes tree planting is now a common topic during the *barangay* or *sitio* assemblies or meetings.

Building on Past Experiences: a Two-pronged Approach

The LGU-led and people's organization (Land Care Association) models have both positive and negative aspects. The LGU model has financial and human resources, and can draw upon them as necessary for effective technology dissemination. But the LGU-led model is affected by political uncertainties (such as a change in political administration). The people's organization is more stable but has limited resources. It is more slow but sure.

The current dissemination efforts on NVS and establishment of village nurseries are done by the *barangays* and the CLCA chapters in collaboration. They complement each other. The CLCA is receiving support from the *barangay*. The conservation team works through both the people's organization and the LGU. Two kinds of nurseries are now evolving: one managed by the CLCA chapters (at *sitio* level) and the one managed by the *barangay* (*barangay* level). The CLCA is obtaining support from *barangays* particularly on the nursery establishment.

Two people's organizations recently joined the CLCA: the Angela – a group of ethnic people living on the forest margins; and the Claveria Tree Growers' Association (CTGA) – a local association of tree farmers. Each group is considered as another chapter. The interest of Angela in joining the association is for their members to gain more exposure to natural resource management through soil and water conservation and tree planting. The interest of the CTGA in joining the CLCA is for members to learn more about tree planting and silvicultural practices, and to share their experiences in tree growing. The participation of the CTGA in the association is beneficial to the CLCA because CTGA members are mostly professionals who have a strong interest in tree planting and can share their managerial and leadership skills with the CLCA members who are taking a leadership role at the *sitio* level.

The Need for Monitoring and Evaluation

For monitoring purposes, ICRAF has been keeping records of all those who have attended a training course or have been assisted with establishing NVS on their farms, as well as of farmers who requested assistance. Details on farming and conservation practices, training and follow-up needs are recorded on a diagnostic card, which is updated on regular follow-up visits by the ICRAF staff. The leaders of the *sitios* or local land care centres (CLCA chapters) have been supporting this activity by facilitating the distribution and collection of the diagnostic cards to and from the *sitios* and new CLCA members. As a preliminary evaluation, a survey on the adoption and dissemination progress is now being conducted, with an emphasis on farmers' technology modification and the reasons behind their decision-making. This will occur approximately 1.5 years after the start of the extension programme.

Participatory Technology Development and Dissemination – What's Next?

We have learned that making the conservation farming technology simpler, easier and more convenient to the farmers, and disseminating it in a participatory way, optimizes the involvement of the members of the community at different levels to enhance rapid technology adoption. NVS has proved to be a foundation for farmers of sloping upland to evolve into a wide range of sustainable agroforestry systems. The institutions providing technical back-up to them need to provide a full range of options that farmers can select from and adapt to their exceedingly diverse circumstances and objectives.

The participation of the CTGA to the CLCA provides a window for the members to learn managerial and leadership skills that are lacking in the village chapters. Further strengthening to the CLCA chapters is necessary so that they will become a coherent and sustainable group in the village.

We are planning to scale-up our efforts to the villages in the upper watershed of Claveria and in adjoining municipalities that are requesting involvement. We see the need to continue to adapt the methods, and evaluate whether they are practical in the context of other areas. The conservation team role does not end when farmers establish the NVS, rather it is the beginning of a long-term relationship.

References

Agus, F. (1993) Soil processes and crop production under contour hedgerow systems on sloping oxisols. PhD thesis, North Carolina State University, Raleigh, NC, USA.

Fujisaka, S. (1989) The need to build upon farmer practice and knowledge – reminders from selected upland conservation projects and policies. *Agroforestry Systems* 9, 141–153.

Fujisaka, S., Mercado, A. and Garrity, D. (1995) *Farmer Adaptation and Adoption of Contour Hedgerows for Soil Conservation.* International Rice Research Institute, Manila, the Philippines.

Garrity, D.P. and Agustin, P.A. (1995) Historical land use evolution in a tropical upland agroecosystem. *Agriculture, Ecosystems and Environment* 53, 83–95.

Garrity, D.P. (1996) Conservation tillage: Southeast Asian Perspective. Paper presented at the *Conservation Tillage Workshop,* Los Baños, the Philippines, 11–12 November.

ICRAF (1997) *Annual Report for 1996.* International Centre for Research in Agroforestry, Nairobi, Kenya.

Mercado, A.R., Garrity, D.P., Sanchez, N. and Laput, L. (1997) Effect of natural vegetative filter strips density on crop production and soil loss. Paper presented at the *13th Annual Scientific Conference of the Federation of Crop Science Societies of the Philippines, Hotel Supreme, Baguio City.*

6

A Microeconomic Analysis of Adoption of Contour Hedgerows in the Philippine Uplands

S. Pandey and M.A.L. Lapar
Social Sciences Division, International Rice Research Institute, PO Box 933, 1099 Manila, the Philippines

Soil degradation in the sloping uplands of Asia is a serious problem that threatens the sustainability of agriculture. Although several soil conservation technologies have been developed and promoted, their adoption has not been widespread. A microeconomic analysis of adoption of contour hedgerows by upland farmers in the Philippines was conducted to identify factors that determine adoption. A number of farmer characteristics were found to condition adoption. The relative importance of these factors, however, differed across sites. Farmers did not perceive yield increase to be the major benefit from adoption. The high cost of establishment and maintenance of hedgerows and the loss of land to hedgerows were considered to be the major disadvantages by non-adopters. The economics of the contour hedgerow system was found to improve substantially if crop intensification or cash cropping is possible. In addition to the need to develop a range of more cost-effective technologies, the study points out the need to develop a better understanding of the effects of production system characteristics on incentives to adoption. The results also highlight the need for a targeted approach to technology development and dissemination.

Introduction

One of the major factors leading to unsustainability of agriculture in the sloping uplands is soil erosion (Blaikie and Brookfield, 1987). Although the extent of anthropomorphic soil erosion is debatable (Oldeman *et al.*, 1991), activities such as deforestation and intensified land use in the uplands undoubtedly have led to increased soil erosion. In the humid tropics of Asia, farmers grow a range of subsistence crops in sloping and marginal uplands using practices which are often highly erosive (Garrity *et al.*, 1993). In addition to reducing the *in situ* productivity and sustainability, these practices also reduce the

 Soil Erosion at Multiple Scales
(eds F.W.T. Penning de Vries, F. Agus and J. Kerr)

sustainability of lowland agriculture through siltation and damage to irrigation infrastructures (Francisco, 1994).

What are the conditions under which farmers will adopt practices which conserve soil and enhance the long-term sustainability of agricultural production systems in these sloping uplands? Experiences from a large number of projects indicate that the problem is often not the lack of technology *per se* but an ill-fit of the technology promoted to the socio-economic conditions under which farming is carried out (Fujisaka, 1989; Anderson and Thampapillai, 1990; Baum *et al.*, 1993; Lutz *et al.*, 1994). Engineering solutions such as rock walls, check dams, and terraces have had very little success in generating wider impact because of their high costs. Other less costly technologies such as contour hedgerows, contour ploughing and cover management have also been adopted, but sporadically. Overall, the technological research and promotional work done so far seem to have had a limited impact in increasing the adoption of conservation technologies.

The limited success of these efforts highlights the need to develop a better understanding of conditions which encourage the adoption of conservation practices. Such factors could be farmer specific, farm specific or technology specific. Farmer-specific factors include the goals of the farmer and the resource base. At a broader level, the socio-economic milieu under which production takes place determines the resource base by allowing resource augmentation through market participation. Farm-specific factors are related to the biophysical characteristics of the production systems such as soil characteristics and climate, as well as the broader characteristics of the production system. Technology-specific factors are features of the technology available to farmers and which are designed to alter the farmers' production systems. In this chapter, a conceptual model describing how the broader characteristics of the production system determine farmer incentives for adoption of conservation practices is first presented. This is followed by a microeconomic analysis of adoption of contour hedgerows by upland farmers in the Philippines.

The Nature of Production Systems and Incentives for Soil Conservation

The characteristics of the production system at any point in time are the result of the interplay between farmers' attempts to change them in a particular way to meet their goals and the environmental factors which determine what is feasible. As such, the study of technology choice cannot be divorced from the study of the production system and its evolution.

The three major determinants of the nature of the production system are population density, market access and the agroclimatic conditions (Boserup, 1965, 1981; Pandey, 1996). A simplified construct based on population density and market access is presented in Fig. 6.1. When the population density is low and no outside market demand exists, returns to labour are maximized by adopting an extensive land-use strategy, such as shifting cultivation based on long fallow periods (Type I). The farmers' primary objective in this land

		MARKET ACCESS	
		Low	High
POPULATION DENSITY	Low	* Traditional * Subsistence * Sustainable Type I	* Perennial or food crop * Sustainable? Type III
	High	* Intensified * Subsistence * Sustainable?? Type II	* Diversified * Food and cash crops * Sustainable?? Type IV

Source: Pandey (1996)

Fig. 6.1. Effect of market access and population pressure on land use and sustainability.

surplus situation is to allocate labour to expand the area, not necessarily to improve the yield. As a result of lack of investment, land quality suffers over time until the field is finally abandoned and the farmer moves to a new area. During the long fallow period, the natural regeneration may restore soil productivity sufficiently to allow recultivation of the field. Incentives for soil conservation under this situation will be minimal.

When the land frontier is closed, further increases in population provide incentives for an intensification of agricultural production, and sedentary farming systems evolve (Type II). In the absence of markets, farmers attempt to be self-sufficient by producing a range of agricultural output under a subsistence mode of production. As there is no scope for increasing the area, the only option available is to improve the yield per unit area by a greater use of labour and other inputs for current production. Labour-intensive methods of soil conservation are likely to be more appropriate for Type II systems. Although investment in soil quality is also a way of improving the productivity, whether or not farmers will actually do so will depend on several factors which determine economic incentives. Where such incentives are lacking, soil degradation will continue until migration reduces the population pressure so that a steady-state equilibrium can be achieved or the area itself is completely abandoned.

When population density is low but market access is well developed, plantation-type commercialized production systems are likely to evolve (Type III). Depending on marketing conditions, annual or perennial food/fibre or industrial crops predominate. A commercialized production system based on food and cash crops is likely to evolve when both population density and market access are high (Type IV). Such production systems may be diversified or specialized depending on a range of conditions. The opportunity to interact with the market in these systems can influence soil conservation decisions in two ways. First, to the extent that cash crops are more profitable per unit area than food crops, an increase in cash crop production would increase the marginal value product of soil and encourage the adoption of conservation practices. Income from cash crops will also help relax liquidity constraints that may restrict such investments. Second, market access will also increase the opportunity cost of labour used for conservation investment. This will make labour-intensive conservation practices more expensive, thus encouraging a preference for labour-saving methods. If such techniques are not available or limited security of tenure restricts the planning horizon, farmers may allow land to degrade.

As soil conservation is an investment to enhance future productive capacity of soil, a microeconomic model of investment analysis can be utilized to study adoption behaviour. Several specifications of such models are available in the literature (McConnell, 1983; Clarke, 1992; La France, 1992). These models indicate that the major factors that determine profitability of investment on soil conservation are the length of the planning horizon, the discount rate, the land area available for production, the price of output, the marginal productivity of soil quality, the cost of investment in soil quality improvement, the natural rate of soil regeneration and the rate of soil depletion associated with agricultural production. Some of these factors define the initial conditions and are given for a farm at a point in time, while others can be manipulated through technology and policy interventions. Economic gains are larger if the planning horizon is longer, the discount rate is lower and the cost of investment for improvement in soil quality is lower.

Some Microeconomic Evidence from the Philippines

Study area and sampling design

Claveria and Cebu are the two major sites in the Philippines where the contour hedgerow technology was promoted actively in the early 1980s. In Cebu, the World Neighbors in collaboration with the Department of Environment and Natural Resources promoted a range of soil conservation practices including contour hedgerows in the early 1980s. Overall, the contour hedgerow technology has spread around the initial target area and the Cebu case is often cited as a successful example of the adoption of soil conservation technology (Garcia and Gerrits, 1995).

In 1985, the International Rice Research Institute initiated farming systems research in the acid uplands of Claveria, Misamis Oriental in collaboration with the Department of Agriculture. A contour hedgerow-based farming system was promoted by farmer-to-farmer extension based on the approach of the World Neighbors in Cebu (Fujisaka and Garrity, 1988). Sixty-four farmers, out of 182 trained, established contour hedgerows by the end of 1990 (Fujisaka, 1993). The status of the contour hedgerow systems of these farmers in 1995 has been documented by Cenas and Pandey (1996). While contour hedgerows were considered not useful and abandoned by about 25% of the initial adopters, others modified and/or maintained these conservation devices.

The specific conservation practice examined in detail in the present study is based on the contour hedgerows. A sample of 130 farmers consisting of 74 adopters and 56 non-adopters was selected from Cebu, Visayas and Claveria, Misamis Oriental, Mindanao. Cebu City provides a good market outlet for commercial production from the peri-urban sloping land. Compared with Cebu, the market for the products from Claveria is somewhat limited to the smaller city of Cagayan de Oro. The population densities in the Cebu province and Misamis Oriental are 574 and 285 persons km^{-2},respectively (NCSO, 1997). The study area in Cebu comes under Cebu City which has a population density of 2358 persons km^{-2}. Thus, in terms of Fig. 6.1, Cebu represents an area with good market access and high population pressure, whereas Claveria represents an area of low population pressure and somewhat limited market access.

The sampling design used to select farmers was stratified purposively. As the objective of the study was to carry out an in-depth analysis of adoption behaviour, farmers were divided into two strata, namely adopters and non-adopters. Adopters were defined as farmers who have established contour hedgerows in at least one parcel of their farms, were cultivating and maintaining their contoured parcels, and had crop production during the year prior to the survey year. The sample consisted of 39 adopters and 21 non-adopters from nine *barangays* (villages) of Claveria. In Cebu, 35 adopters and 35 non-adopters from six *barangays* were selected. In order to reduce the effect of environmental variations, non-adopters were selected from the same *barangays* and, where possible, were farmers with fields adjacent to contour hedgerow fields of the adopters. Given the purposive nature of the sampling design, caution should be exercised in extrapolating the farm and farmer characteristics of the sampled farmers to the overall population.

A structured questionnaire was used for the field interviews. Detailed information on production systems, input use, cost and returns, nature and the extent of contour hedgerow adoption, adoption of soil conservation practices other than contour hedgerows, and farmers' perceptions regarding advantages/disadvantages of contour hedgerows were collected. Interviews with key informants including government officials, non-government organizations and other agencies engaged in soil conservation were also conducted.

The farm-level data on production systems collected during the survey pertain to the 1995 cropping season. The basic characteristics of the production systems in these two locations are summarized in Table 6.1. Corn is the dominant crop in both locations with cash cropping (mainly vegetables and flowers) being more important in Cebu. The average farm size in Claveria (3.0 ha) is much larger than in Cebu (1.7 ha). Due to the importance of cash crops, the gross farm income per household is higher in Cebu than in Claveria.

Characteristics of adoption and adopters

The extent of adoption, as measured by the percentage area under contour hedgerows of households who have adopted the technology and the percentage of farmers establishing contour hedgerows in the whole farm, is higher in Cebu than in Claveria (Table 6.2). Adoption started earlier in Cebu with almost a quarter of the sampled farmers having adopted contour hedgerows for more than 10 years.

Adopters of contour hedgerows in both Cebu and Claveria have a greater proportion of area under cash crops than under food crops (Table 6.3). The adoption of contour hedgerows may have encouraged a shift towards cash cropping.[1] The heads of the households adopting contour hedgerows are younger and have more years of schooling. More of the adopting households are members of local labour-sharing group (*alayon*) and have received training on soil conservation for a slightly longer period than the non-adopting households. More of the adopting households own at least a parcel of the land they are operating. The houses of adopters are closer to the main road than those of non-adopters.

Characteristics of contour hedgerows

Based on the combined data for Cebu and Claveria, the average height of the riser was estimated to be 0.8 m. It was found to be positively related to the number of years of adoption. The average annual increase in the height of the riser was 5 cm year^{-1}. The average width and average length of hedgerows per hectare contoured were 1.3 m and 1300 m, respectively. Hedgerows in nearly 50% of the fields were based on natural grasses. Napier grass was the second most important, with 30% of the fields having hedgerows based on this forage grass. Other major hedgerow species were *Gliricida sepium, Setaria* sp. and *Leucaena leucocephala.*

Earlier work in Claveria indicated that farmers gradually moved towards

[1]If the causality runs in this direction, the opportunity cost of not adopting contour hedgerows is the income from cash cropping forgone. Benefit–cost analysis of conservation practice based solely on the difference in yield of food crops with and without contour hedgerow will underestimate the economic value of adoption in such cases.

Table 6.1. General characteristics of the production systems in the study area.

Features	Cebu	Claveria
Average area per household (ha)	1.7	3.0
Average parcel size (ha)	0.7	1.24
Cropping intensity (%)	150	190
Average slope (%)	32	24
Area under corn (%)	54	84
Area under cash crops (%)	14	5
Average yield of corn (kg ha^{-1})	902	1284
Average gross income ($ $household^{-1}$)	1620	1470

Table 6.2. Extent of adoption of contour hedgerows.

	Cebu	Claveria	All
Percentage share of contour hedgerow area to total area operated by adopters[a]	86.8 (23.2)	69.4 (33.0)	77.6 (30.0)
Percentage of full adopters[b]	62.8	46.2	54.0
Partial adopters:			
≥ 50% but < 100%[c]	28.6	25.6	27.0
< 50%[d]	8.6	28.2	19.0

[a] Average share for all adopters. Figures in parentheses are standard deviations.
[b] Adopters who have established contour hedgerows on 100% of their farm.
[c] Adopters who have established contour hedgerows on at least half but less than the whole area of their farm.
[d] Adopters who have established contour hedgerows on less than half of their farm.

Table 6.3. Characteristics of adopters and non-adopters of contour hedgerows.

Characteristics	Adopters	Non-adopters
Mean age of the household head (years)	45	47
Mean years of schooling (years)	6	4
Farmers with membership in *alayon*[a] (%)	13	2
Mean duration of training on soil conservation obtained (weeks)	1.4	0.6
Average distance of house from road (km)	0.47	0.55
Farmers owning at least a parcel of land (%)	54	37
Mean area under cash crop (%)	12	5

[a] *Alayon* is a local organization of farmers in which labour exchange among members is practised.

using natural grass-based hedgerows (Cenas and Pandey, 1996). Only about 8% of the fields started out with natural grass on hedgerows and remained that way. However, in 35% of the fields, the change has been in favour of natural grasses. Natural domination of grasses and high mortality of the planted species were the major reasons cited by farmers for a shift towards a natural grass-based system.

Farmer perceptions regarding contour hedgerows

Since farmers have had exposure to the contour hedgerow technology in the study area, we attempted to elicit their perceptions regarding benefits as well as problems with the technology. An open-ended interview format was used to elicit the responses.

More than 50% of the respondents indicated that the major benefit from contour hedgerows was the reduction of soil erosion. About 16% of the respondents replied that hedgerows improve soil fertility. Obviously, one would expect the reduction in soil erosion and improvement in soil fertility to help ultimately in improving the yield. However, yield increase ranked only third among the responses (11%). Although there may have been visible improvements in soil attributes, farmers were probably not able to observe a very clear yield gain. This indicates that yield gain is probably not very high and is confounded by other factors. If farmers do not in fact consider yield gain to be the major benefit, the extension message for promoting soil conservation is likely to be better received if couched in terms of other benefits that farmers easily perceive. Some of these other benefits were the ease of land preparation as alleys become flatter (5%), hedgerow species serving as fodder (8%) and the avoidance of physical damage to crops due to gully formation and landslide (4%). These other benefits, however, tend to be somewhat location specific.

Adopters of contour hedgerows were asked about what they perceived as the single major problem with the technology. While 50% of the respondents mentioned that there was no problem with the technology, 35% considered the high cost of labour for establishment and maintenance to be the major problem. Other problems elicited were increased labour time for land preparation (3%), destruction of hedgerows by stray animals (4%), shading of crops by hedgerows (2%) and the persistence of soil erosion despite hedgerows (2%).

We also elicited responses from non-adopters about the reasons for not adopting contour hedgerows. High labour intensity needed for establishment and maintenance was again the major reason (38%). Lack of technical knowledge and loss of cultivable area were considered to be the reason for non-adoption by more than 10% of the respondents. Other reasons cited were difficulties in land preparation in the presence of contour hedgerows (5%), non-ownership of land (5%), lack of capital (2%) and the perception that soil erosion is not a problem (5%).

The non-adopters were asked what would encourage them to adopt contour hedgerows. Almost one-third of the farmers considered the need for technical assistance and training for establishing contour hedgerows. The need to have sufficient labour and capital was the second most important factor (20%). A wider choice of hedgerow species (11%), the ownership of land (5%) and the clear evidence of loss of soil on their farms (5%) were other major factors that non-adopters felt would lead to increased adoption. About 15% of the respondents felt that there was no need for them to adopt contour hedgerows as they did not perceive soil erosion to be the major problem.

The non-adopters were also asked about what they consider to be the major erosion-related problems on their farms. Destruction of crops and loss of seedlings and inputs were mentioned as the major consequence of erosion by more than 50% of the farmers. About 16% of the farmers considered gully formation as the major problem. Judging from these responses, it appears that visible damage to fields and physical losses of seedling and crops loom large in farmers' minds. No farmer mentioned the reduction of yield due to soil erosion as a problem.

When asked if they have taken some measures to reduce soil erosion from the fields, 55% of the non-adopters of contour hedgerows responded in the affirmative. Of these farmers, 23% constructed diversion/drainage canals, another 23% practised contour ploughing and about 16% responded simply by using more fertilizers. Fallowing the land and placing other physical barriers was practised by 12% of the farmers. These responses indicate that farmers use other practices that may be cheaper or more suitable to specific conditions of their fields than contour hedgerows. The implication is that promoters of soil conservation in the uplands must not view contour hedgerows as the universal solution for all farmers but as one of the options. If a wider range of methods is promoted, farmers will have the flexibility to choose practices that they consider most appropriate to their fields.

Factors determining adoption

The adoption data were analysed using a probit model to identify the factors determining the adoption of contour hedgerows for soil conservation. The details of variable selection and expected effects on adoption are discussed in Lapar and Pandey (1997).

Adoption in Claveria is influenced significantly by the age of the household head, tenure status and slope (Table 6.4). The results suggest that a farmer who is younger, owns the farm and operates a field with steeper slope is more likely to adopt the contour hedgerow technology. The likelihood of adoption in Claveria for an average farmer is estimated at 46%. For Cebu, adoption is influenced significantly by education status, membership in *alayon*, slope and market access. The effect of all these variables on adoption is positive. The likelihood of adoption in Cebu is estimated at 51% on average. If we consider only the significant factors in these two sites, they are farmer age, education status, land ownership, access to markets and slope.[2]

[2]Given the nature of the data collected, it is not possible to explain the reasons for different sets of factors being important in these two locations. A more in-depth study of the differences in the production systems in these two sites is needed to provide a satisfactory answer to this question.

Table 6.4. Estimated coefficients of the probit model and the marginal probabilities of the independent variables.

	Cebu		Claveria	
	Probit coefficient	Marginal probabilities	Probit coefficient	Marginal probabilities
Constant	−1.78		0.52	
Age of household head	0.01	0.003	−0.03**	−0.012
Education (years)	0.17***	0.067	−0.03	−0.012
Tenure (dummy)[a]	−0.31	−0.124	1.02***	0.402
Member (dummy)[b]	0.84**	0.335	0.34	0.134
Land/labour ratio	0.10	0.040	−0.94	−0.371
Slope (%)	0.02**	0.008	0.05***	0.020
Non-farm income	0.01	0.002	−0.01	−0.002
Distance from the road	−0.94**	−0.373	−0.10	−0.039
−2 Log likelihood ratio	192.54***		165.51***	
Probability of adoption	0.51		0.46	

*** Significant at 1%, ** Significant at 5%, * Significant at 10%.
[a] Dummy variable = 1 if owner; 0 otherwise.
[b] Dummy variable = 1 if *alayon* member; 0 otherwise.

Household heads with longer schooling may be more aware of the causes and consequences of soil erosion and may be more able to access technologies for reducing soil erosion than their counterparts with shorter schooling (Hoover and Wiitala, 1980; Ervin and Ervin, 1982; Rahm and Huffman, 1984). As mentioned earlier, access to market increases the possibility of cash cropping, which can generate greater returns than subsistence crops. Farmers may be more willing to adopt hedgerows in steeper fields as soil erosion could be more readily perceived in such fields than in flatter fields. The positive effect of membership in *alayon* may be due to its role in relaxing the labour constraint (Gabunada and Barker, 1995).

Returns to adoption

Rather than conducting a full-fledged cost–benefit analysis, given the difficulties in quantifying the erosion–productivity relationship, here we have taken a somewhat simpler approach of relying on sensitivity analysis to highlight the major factors that determine the profitability of investment in soil erosion control. We have ignored the temporal patterns of changes in yield both with and without contour hedgerows and assumed yield to remain constant at their respective levels. The average cost of establishing contour hedgerows (49 man-days), the average annual cost of maintenance (14 man-days) and the proportionate area lost to hedgerows (20%) were obtained from the survey data. The percentage increases in yield of corn required to recoup these costs at two different internal rates of return were calculated assuming the life of contour hedgerows to be 10 years. As adoption of contour hedgerows may encourage more intensification of production, two levels of cropping intensity were also considered. The value of unity for cropping intensity means single cropping while the value of 2 indicates double cropping.

The results presented in Fig. 6.2 indicate that the break-even increase in percentage yield declines rapidly with the increase in base yield without conservation. In very poor environments where the corn yield is only 200 kg ha^{-1}, the break-even increase in yield required is as high as 160% under the assumed internal rate of return of 40% and the cropping intensity of unity. Under the most favourable scenario of corn yield of 1600 kg ha^{-1}, the internal rate of return of 20% and with the cropping intensity of 2, the required yield increase is 26%. The results highlight a paradox. A much higher increase in productivity is needed to persuade farmers to adopt conservation practices in relatively low-yielding (or more degraded) environments where investment in soil conservation may be most needed to prevent further erosion. On the contrary, yield gains required are much lower in relatively high-yielding (or less degraded) environments where the need to conserve soils may not be so apparent. The results suggest that agricultural researchers face a much tougher challenge in making investment in soil conservation economically attractive to farmers in areas with low initial productivity. Of course, the scenario presented above captures only the on-site costs and benefits which are the normal concerns of the farmer. From the society's point of view, consideration of off-site effects and other externalities may change the balance in favour of low productivity areas.

The break-even increment in yield is lower if adoption of contour

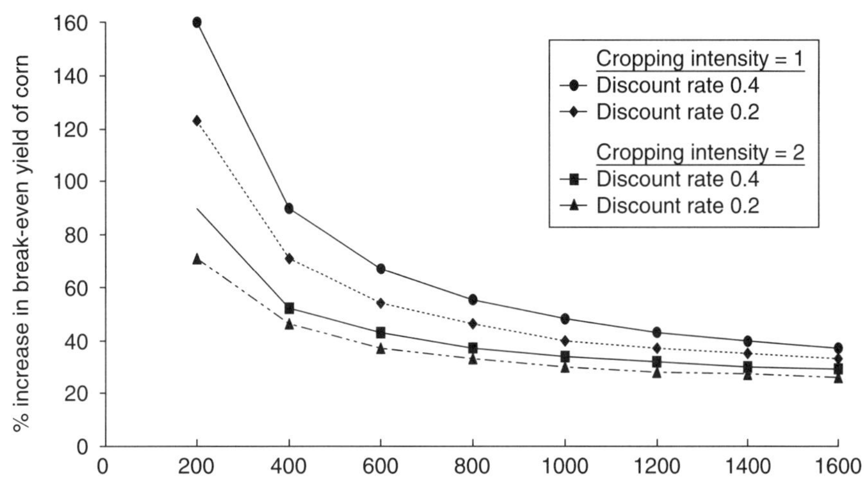

Fig. 6.2. Percentage increase in yield of corn needed for investment in contour hedgerows to break-even at different discount rates and cropping intensity.

hedgerows does encourage intensification of cropping (or generates more income through switching to high value crops). In areas with more favourable climatic conditions, an improvement in market access may thus encourage adoption by facilitating intensification. Yield-increasing technologies will also similarly encourage adoption.

To identify the component of the cost that affects the profitability the most, the analysis presented above was repeated for different scenarios of cost reduction. The base values assumed were corn yield of 400 kg ha^{-1}, an internal rate of return of 20% and the cropping intensity of unity. The analysis was done by changing one of the cost components at a time while keeping the other two fixed at their base values. The results presented in Fig. 6.3 indicate that the break-even increment in yield declines with a reduction in all costs. What seems to matter most is the reduction in annual cost of maintaining the hedgerows. The reduction in establishment cost ranks the second and the reduction in area loss the last. The farmer perception that maintenance of hedgerows is the most problematic corroborates the economic analysis presented here. The implication is that research aimed at reducing the annual maintenance cost of hedgerow ranks higher in priority. Careful selection of hedgerow species that require minimal maintenance would improve the economic benefits from adoption.

Natural vegetative strips are believed to be the least costly in terms of establishment as well as maintenance. Although our estimate of the maintenance cost of natural vegetative strips is similar to that of planted hedgerows, the establishment cost of the former is much lower (Table 6.5). It is not sur-

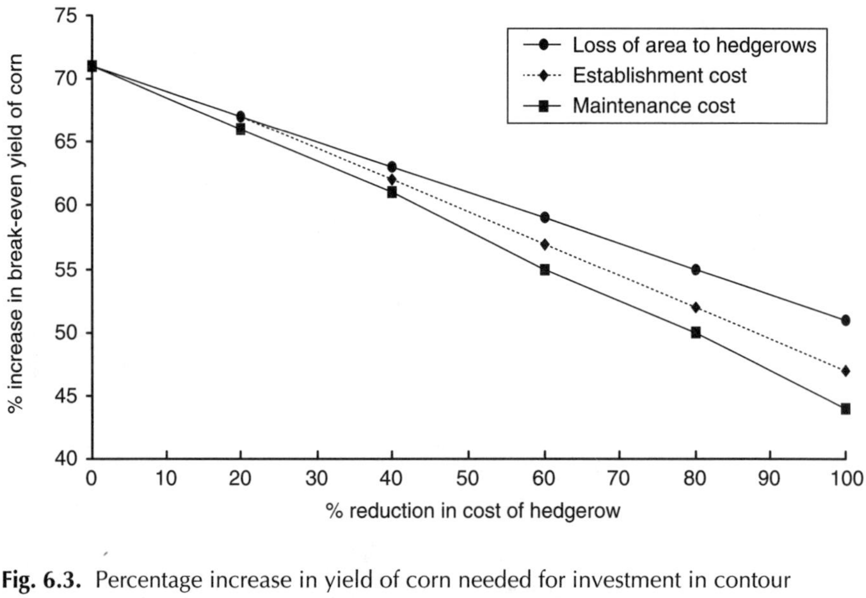

Fig. 6.3. Percentage increase in yield of corn needed for investment in contour hedgerows to break-even under different percentage reduction in cost.

Table 6.5. Average labour (man-days ha^{-1}) for construction and maintenance of contour hedgerows, by type of hedgerows.

	Natural vegetative strip	Planted species[a]	All
Construction	4.3	55.1	48.9
A-frame construction	0.0	0.5	
Layouting	1.6	9.5	
Ploughing	2.0	8.1	
Shovelling	0.7	22.5	
Hauling of planting materials	–	5.3	
Seeding/planting	–	9.2	
Maintenance	13.2	13.5	13.5
Fertilizer application	–	0.1	
Pruning and weeding	13.1	10.7	
Replanting	–	0.9	
Gully repair	0.1	1.8	

[a] Includes leguminous species, forage grasses and perennial species.

prising that natural grass-based systems are becoming more popular. The major cost of establishment of planted species is the cost of the planting material, the cost of transportation and the cost of shovelling to prepare the planting area. Farmers who use natural vegetation do not incur these costs – hence the overall cost of establishing hedgerows is much lower. If the cost of maintenance of natural vegetation and the loss of land could be reduced to 3 man-days and 7%, respectively, the break-even increment in yield would be only 15%. This would certainly make contour barriers based on natural grasses very attractive compared with other species. Grasses which are not too tall and whose growth can be checked effectively by small quantities of herbicides could reduce the maintenance cost substantially. Similarly, loss of land could be reduced by widening alleys. Research to find ways to reduce these costs under a range of environmental conditions is hence warranted.

Concluding Remarks

An important implication of the analysis presented in this paper is that policy support to increase farm production as well as to encourage a switch from low-value subsistence crops to high-value cash crops would improve the returns to investment in soil conservation. Without access to improved technologies and better marketing infrastructures, farmers are unlikely to view investment in soil conservation as being economically worthwhile. Investment in rural infrastructures and policies to facilitate the development of an efficient marketing system will, therefore, encourage adoption. Promoters of soil conservation technology have to consider it not in isolation but as an integral component of interventions designed to increase the profitability of the overall production system. A paradigm shift from that of finding a technical fix to the problem of soil erosion towards that of improving the farmers' incomes by facilitating a transition to more remunerative land use systems is needed. Otherwise, our efforts are unlikely to make a significant dent in the problem of continuing erosion in the sloping uplands of Asia.

The results also highlight the need to reduce the cost of adopting contour hedgerows, the need to provide a range of options and the need for better targeting. Contour hedgerows, especially of planted species, are too expensive for most farmers. Reducing the cost of establishment and maintenance will certainly improve the economics of contour hedgerows. The recent popularity of hedgerows of natural vegetation is due mainly to the cost advantage.

Contour hedgerows are but one of the ways of reducing soil erosion. Farmers in the study area use many other complementary and substitute methods ranging from contour plowing, mulching, other physical barriers, diversion canals and fallowing. One or a combination of these methods may be more suitable in a specific situation depending on field conditions and farmers' resource constraints. Promotion of these other methods and improvement in their effectiveness will provide more choices for farmers so that they can select methods that they consider to be most appropriate to their situations. Often promotional activities narrowly target specific technology and ignore other options that may be equally or more effective than the one being promoted. An output-oriented (such as area covered by different soil conservation methods, the extent of soil erosion reduced) rather than a physical target-oriented (metres of contour hedgerows established) incentive structure may encourage extension agents and technology promoters to take on a broader perspective.

A range of field-, farm- and farmer-specific factors condition the adoption of soil conservation technology. Factors such as low population density, low slope, insecurity of tenure, low initial productivity and limited access to market reduce incentives to conserve soil. Incentives are higher in the contrary situation. Careful targeting of technology in areas with favourable conditions, provision of information about the consequences of continued soil erosion, demonstration and adaptation of soil conservation technologies under farmer conditions, and provision of technical assistance and training could encourage wider adoption. In areas where available technologies are less likely to be adopted, efforts are needed to widen the domain through further improvements in technology.

The recent changes in the production system in the Asian uplands indicate that factors that lead to the adoption of erosive land-use practices often originate outside the upland systems. Macro- and sectoral policies that encourage a subsistence mode of agricultural production and reduce labour absorption outside agriculture are major contributing factors to land degradation in the uplands. In addition, poorly defined and unenforceable property rights to land can aggravate the degradation problem. Hence, institutional and policy reforms as well as technologies that not only reduce soil erosion but also directly increase farmers' incomes are needed. A greater understanding of the dynamics of land use changes in the uplands is likewise essential in designing technological, institutional and policy interventions for reducing soil erosion and enhancing sustainability.

References

Anderson, J. and Thampapillai, J. (1990) *Soil Conservation in Developing Countries: Project and Policy Intervention. Policy and Research Series No. 8.* Agriculture and Rural Development Department, The World Bank.

Baum, E., Wolff, P. and Zobisch, M.A. (1993) *Acceptance of Soil and Water Conservation: Strategies and Technologies.* DITSL, Federal Republic of Germany.

Blaikie, P. and Brookfield, H. (1987) *Land Degradation and Society.* Methuen, London.

Boserup, E. (1965) *The Conditions of Agricultural Growth.* Allen & Unwin, London.

Boserup, E. (1981) *Population and Technological Change: A Study of Long-Term Trends.* University of Chicago Press, Chicago.

Cenas, P. and Pandey, S. (1996) Contour hedgerow technology in Claveria, Misamis Oriental. *Philippine Journal of Crop Science* 21, 1–6.

Clarke, H. (1992) The supply of non-degraded agricultural land. *Australian Journal of Agricultural Economics* 36, 31–56.

Ervin, C. and Ervin, D. (1982) Factors affecting the use of soil conservation practices: hypotheses, evidence, and policy implications. *Land Economics* 58, 277–292.

Francisco, H. (1994) *Upland Soil Resources of the Philippines: Resource Assessment and Accounting for Soil Depreciation.* INNOTECH, Manila, the Philippines.

Fujisaka, S. (1989) A method for farmer-participatory research and technology transfer: upland soil conservation in the Philippines. *Experimental Agriculture* 25, 423–433.

Fujisaka, S. (1993) A case of farmer adaptation and adoption of contour hedgerows for soil conservation. *Experimental Agriculture* 29, 97–105.

Fujisaka, S. and Garrity, D.P. (1988) Developing sustainable food crop farming systems for the sloping acid uplands: a farmer-participatory approach. Paper presented at the *4th Southeast Asian Universities Agroecosystem Network (SUAN) Research Symposium, Khon Kaen, Thailand, Agricultural Economics Department Paper No. 88–16.*

Gabunada, F. and Barker, R. (1995) *Adoption of contour hedgerow technology in Matalom, Leyte, Philippines.* Visayas State College of Agriculture, Leyte, Mimeo.

Garcia, J.N.M and Gerrits, R.V. (1995) *Soil Conservation in an Upland Farming System in Cebu: A Socio-economic Survey.* SEARCA-UQ Uplands Research Project, Survey Report No.1. Los Baños, the Philippines.

Garrity, D.P., Kummer, D.M., and Guiang, E.S. (1993) The Philippines. In: *Sustainable Agriculture and the Environment in the Humid Tropics.* Committee on Sustainable Agriculture and the Environment in the Humid Tropics, Board of Agriculture and Board of Science and Technology for International Development and National Research Council. National Academy Press, Washington, DC.

Hoover, H. and Wiitala, M. (1980) *Operator and Landlord Participation in Soil Erosion in the Maple Creek Watershed in Northeast Nebraska.* US Department of Agriculture, Washington, DC.

La France, J.T. (1992) Do increased commodity prices lead to more or less soil degradation. *Australian Journal of Agricultural Economics* 36(1), 57–82.

Lapar, M.L. and Pandey, S. (1997) A microeconometric analysis of adoption of soil conservation in the Philippine uplands. Paper presented at the *23rd International Conference of Agricultural Economists, Sacramento, California, 10–16 August.*

Lutz, E., Pagiola, S. and Reiche, C. (1994) The costs and benefits of soil conservation: the farmers' viewpoint. *The World Bank Research Observer* 9, 273–295.

McConnell, K.E. (1983) An economic model of soil conservation. *American Journal of Agricultural Economics* 65, 83–89.

NCSO (1997) *Philippine Statistical Yearbook 1996.* National Census and Statistics Office, Manila, the Philippines.

Oldeman, L.R., Hakkeling, R.T.A. and Sombroek, W.G. (1991) *World Map of the Status of Human-Induced Soil Degradation: An Explanatory Note.* International Soil Reference and Information Center, UNEP, Nairobi.

Pandey, S. (1996) Socio-economic context and priorities for strategic research in Asian upland rice ecosystems. In: Piggin, C., Courtois, B. and Schmit, V. (eds) *Upland Rice Research in Partnership.* International Rice Research Institute, Manila, the Philippines.

Rahm, M. and Huffman, W. (1984) The adoption of reduced tillage: the role of human capital and other variables. *American Journal of Agricultural Economics* 66, 405–413.

7

Investigating Constraints to the Adoption of Recommended Soil Conservation Technology in the Philippines

R.A. Cramb[1] and R.A. Nelson[2]

[1]Department of Agriculture, University of Queensland, Brisbane, Australia; [2]Forestry Technical Services (FORTECH), Canberra, Australia

This chapter discusses some of the conceptual and practical issues involved in studying the farm-level constraints to adoption of recommended soil conservation technology. The problem is approached in two stages: the benefits and costs of adopting soil conservation technology, and the technical and socio-economic factors that constrain farmers from adopting otherwise profitable technologies. Applications of cost–benefit analysis and erosion–productivity simulation models to assess the profitability of soil conservation technology are reviewed. Applications of rapid rural appraisal (RRA), participatory rural appraisal (PRA) and formal questionnaire surveys to investigate the technical and socio-economic factors constraining adoption are also reviewed. The applications are drawn from research conducted through various projects which focused on the adoption of hedgerow intercropping by farmers in the Philippine uplands. An alternative perspective is advocated for viewing adoption which gives less emphasis to a specific, recommended technology and more to the diverse and complex process of farming systems development.

Introduction

In this chapter we discuss some of the conceptual and practical issues involved in studying the farm-level constraints to adoption of recommended soil conservation technologies.[1] We approach this problem in two stages. First, we consider the benefits and costs of adopting soil conservation technology

[1]The chapter arises out of a research project commissioned by the Australian Centre for International Agricultural Research (ACIAR Project 9211) to evaluate soil conservation technologies for upland farming systems in the Philippines (Cramb and Saguiguit, 1994; Cramb, 1998). We are grateful to Frits Penning de Vries, Jim Sumberg, Stephen Biggs and Frank Ellis for comments on earlier drafts.

in a relatively unconstrained setting (i.e. investigating whether the technology is profitable). This is applied to the evaluation of hedgerow intercropping in the Philippines. Second, recognizing that economic profitability is necessary but not sufficient for adoption to occur, we consider the variety of technical and socio-economic factors constraining farmers from adopting what may be an otherwise profitable technology (i.e. investigating whether the technology is adoptable). This is illustrated with case study material on the adoption and adaptation of soil conservation technology (principally hedgerow intercropping) in the Philippines. In examining both these aspects (i.e. profitability and adoptability), we question whether investigating adoption/non-adoption of a given technology is in fact the best way to view the problem of soil conservation. We conclude by advocating an alternative perspective, which gives less emphasis to a specific, recommended technology and more to the diverse and complex process of farming systems development.

Farm-level Benefits and Costs: Is the Technology Profitable?

Cost–benefit analysis

Cost–benefit analysis can be used to assess whether the costs of implementing conservation farming are offset by the returns from sustained crop yields or reduced off-site damage (Nelson, 1996). Over a defined time period, farming practices with a net present value greater than zero are profitable, and farming methods with high net present value dominate those with low net present value. The ranking of alternative farming methods provided by cost–benefit analysis needs to be interpreted in terms of other decision criteria used by farmers. For example, it is widely recognized that farmers are cautious and risk averse, whether they live in developing or industrialized nations (Anderson and Thampapillai, 1991). In low-input farming systems, adoption decisions of farmers operating close to the margin of economic survival may be heavily influenced by the possibility of negative returns in any year, even if expected net present value is positive in the long term. The relative risk of negative returns from alternative farming methods can be assessed by analysing the probability distribution of net returns resulting from uncertainty in prices, costs and yields.

The field-level costs and returns of farming can be derived from production budgets (Pagiola, 1994). A production budget sums input quantities and costs for each farming operation, for comparison with the value of production over a specified time period such as a crop season or year. Net present value is the sum of net returns from production budgets for each year of analysis, discounted to present value terms. The cost side of crop production budgets is relatively easy to obtain through surveys with experienced farmers. The costs of farming are relatively standard for specific farming methods under similar agronomic conditions, and can be obtained from relatively few farmers using a key-informant survey approach rather than more extensive surveys.

The potential benefits of conserving soil over long time periods are more difficult to predict, because the influence of erosion on the capacity of soil to support plant growth is difficult to quantify (Stocking, 1984; Littleboy *et al.*, 1992a). Seasonal variation in climate and spatial variation in soil conditions make it difficult to identify relationships between changing soil quality and crop yields (Gardner *et al.*, 1992). Changing management adds to the complexity of agricultural production systems. Farmers constantly make decisions to vary inputs and cultural practices in response to changes in the physical environment and their socio-economic circumstances (Hammer and White, 1992).

There are two related methods for quantifying a relationship between erosion and crop production: field experimentation and simulation modelling (Stocking 1984; Littleboy *et al.*, 1992b). Field experimentation involves comparison of crop yields from fields subject to different rates of soil erosion, or monitoring soil loss and crop yields from individual fields over time. Field experiments are costly and difficult to conduct across a representative sample of environments and management strategies for sufficiently long periods of time (Littleboy *et al.*, 1992a). It is impractical to establish long-term field trials at many locations (Gardner *et al.*, 1992). Soil types, topography, climate and cropping methods can vary greatly, making it impractical to measure erosion under all possible conditions (Freebairn *et al.*, 1989).

Simulation models can be used to summarize and mathematically represent erosion and crop growth processes (Freebairn *et al.*, 1989; Silburn and Loch, 1992). They can be used as a means of integrating knowledge from field trials and extrapolating it across a range of environments and management strategies (Gardner *et al.*, 1992). Once parametrized and validated, models provide a fast and inexpensive way of exploring the long-term consequences of different farming methods in a wide range of environments (McCown, 1989). In environments with seasonally variable climate, modelling may be the only practical way of exploring the relationship between soil erosion and crop production (Hammer and White, 1992). Historical climate records can be used in dynamic process models to explore changes in cropping systems under a realistic range of seasonal climate variability (McCown, 1991).

Figure 7.1 illustrates the difficulty of establishing a relationship between erosion and crop yields using limited time-series data. The figure shows the variability of maize yields from a comparative field trial of open-field farming and hedgerow intercropping established by ACIAR Projects 8551 and 9201 at Tranca, Laguna, in the Philippines. Between 1990 and 1994, soil loss averaged 107 t ha^{-1} $year^{-1}$ from open-field farming compared with 6.4 t ha^{-1} $year^{-1}$ from hedgerow intercropping. Despite much higher rates of soil loss from open-field farming, there was no significance difference in maize yields.

Crop yields from the comparative field trial at Tranca were influenced by seasonal variability in climate, changes in management and environmental perturbations as well as erosion-induced changes in soil quality. For example, a change in the variety of maize from a landrace, Lagkitan, to a hybrid, IPB193, was responsible for a dramatic increase in yields from 1993 onwards. Lower yields in 1991 were attributed to restricted flowering because of ash

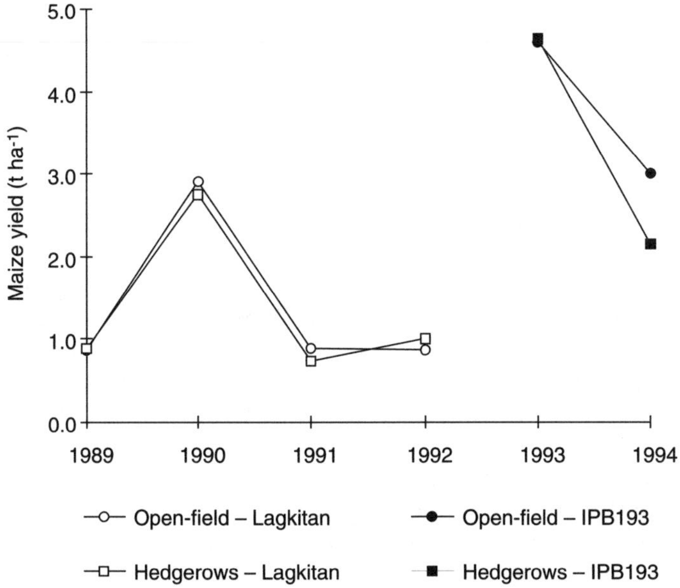

Fig. 7.1. Maize yields from a comparative field trial of open-field farming and hedgerow intercropping at Tranca, Laguna, the Philippines (Paningbatan, 1995).

falls during the eruption of Mt Pinatubo. Delays in funding interrupted the timing and intensity of management operations in 1992. The Tranca trial demonstrates that short measurement periods may not reveal much about the long-term consequences of erosion on crop yields.

Erosion–productivity modelling

To estimate the benefits of soil conservation at the field or farm level, a model is required that can be parametrized with minimal data to simulate soil loss and crop yields accurately in the farming system being studied. To provide realistic results from limited time-series data, the model needs to make an accurate prediction of variability in crop yields resulting from seasonal variability in weather conditions and other factors. However, the choice of model involves a trade-off between the precision and reliability of simulated outcomes, and the amount of data required to derive model parameters. This trade-off requires the complexity and data requirements of each model to be matched with the data available to derive parameters for the model. There are four types of modelling that can be used to establish relationships between erosion and crop production: statistical or empirical models, productivity index models, component process models, and cropping systems models.

The APSIM and SCUAF models were used in this research to estimate the benefits of conservation farming and represent different intensities of erosion–productivity modelling. APSIM (McCown *et al.*, 1996) is a relatively precise cropping systems model that requires large quantities of data to derive and validate input parameters. Processing this data and parametrizing the model is time-consuming, and for this application took over a year and required collaboration with a team of researchers. However, APSIM simulates the physical processes within a cropping system in some detail, accurately predicting the interaction between changing weather, management and soil conditions on crop yields (Fig. 7.2). In contrast, SCUAF (Young and Muraya, 1990) is a relatively imprecise model, though appealing because relatively few input parameters need to be derived. However, the level of aggregation in SCUAF limits the extent to which input parameters can be derived from physical measurements of cropping systems. SCUAF models long-term average soil loss and yields, and validating the model requires long time-series data that are rarely available. SCUAF was used in this research because it has been extensively 'prevalidated' to agroforestry systems and soil processes in the tropics.

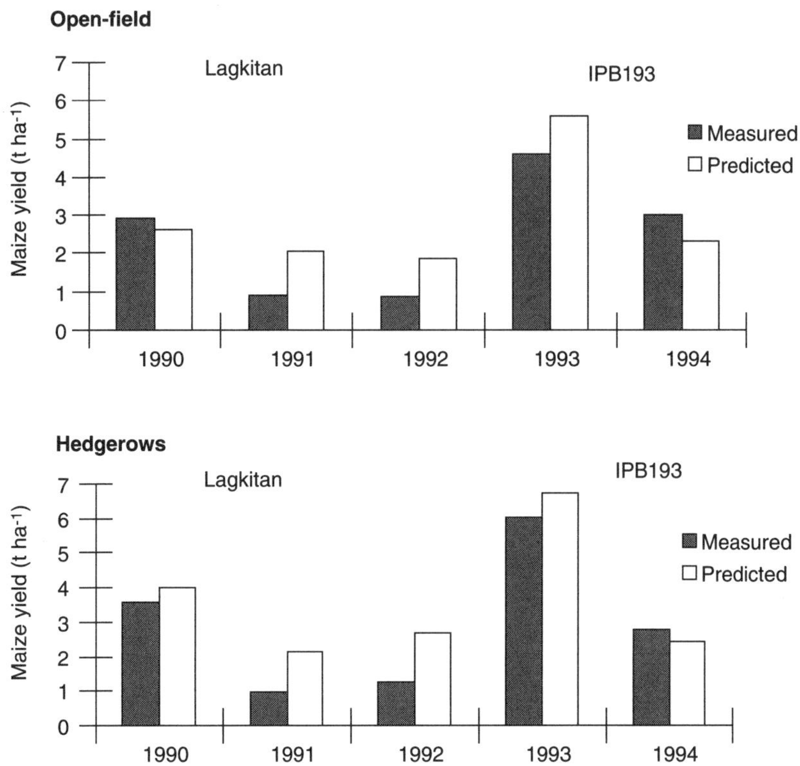

Fig. 7.2. Maize yields predicted using APSIM, for open-field farming and hedgerow intercropping at Tranca, Laguna, the Philippines.

Profitability of hedgerow intercropping

Cooperative research in ACIAR Projects 9211 and 9201 with the International Centre for Research into Agroforestry (ICRAF) enabled cost–benefit analysis to be used to evaluate the economic incentives for farmers to adopt SALT, a form of hedgerow intercropping widely promoted in the Philippine uplands. The cost–benefit analyses combined economic data from production budgets for maize farming in two communities, Timugan and Claveria, with maize yields simulated using two erosion–productivity models, APSIM and SCUAF. The APSIM model was chosen as the preferred model because of its capacity to simulate accurately the response of maize yields to weather and other factors that vary seasonally. The SCUAF model was chosen because of its capacity to simulate erosion and maize yields from hedgerow intercropping in the humid tropics, and the ease with which it can be applied. Details of the modelling work have been presented in Nelson *et al.* (1996a–f), and Nelson (1996).

Economic surveys with experienced maize farmers were used to obtain estimates of the quantity and cost of labour and material inputs for hedgerow intercropping and open-field farming of maize. The economic surveys were based on the representative farming practices used in the comparative field trials of hedgerow intercropping adjacent to the two communities, which were used to parametrize APSIM and SCUAF. This ensured that simulated maize yields were consistent with the amount of labour and inputs invested in maize farming under the same agronomic conditions. The results of these surveys have been presented by Nelson *et al.* (1996g) and Nelson (1996).

The APSIM model was used to investigate the economic viability of intercropping maize between shrub legume hedgerows on erodible soils of moderate fertility under the economic conditions prevailing in relatively inaccessible upland areas. Marketing opportunities for maize in relatively inaccessible communities are limited to grain for livestock feed, which is a relatively low-value crop. APSIM was parametrized to simulate intercropping of maize between hedgerows of shrub legumes on the moderately fertile, but erodible, soils of a research trial at Tranca, Laguna.

In this setting, hedgerow intercropping has potential to sustain maize production relative to traditional methods of open-field farming, by controlling erosion and contributing nitrogen to the cropping alleys (Fig. 7.3). However, the economic viability of hedgerow intercropping is influenced by farmers' planning horizons and discount rates. In the short term, both continuous and fallow open-field farming provide higher economic returns to farmers despite high rates of soil loss (Fig. 7.4). The value of sustained maize yields from hedgerow intercropping are not realized rapidly enough to compensate farmers with limited planning horizons for establishment and maintenance costs. In the long term, the ability of hedgerow intercropping to conserve soil and contribute nitrogen provides an economically attractive method of sustaining cropping relative to continuous open-field farming.

The SCUAF model was used to investigate the economic viability of hedgerow intercropping with shrub legumes for farmers in peri-urban

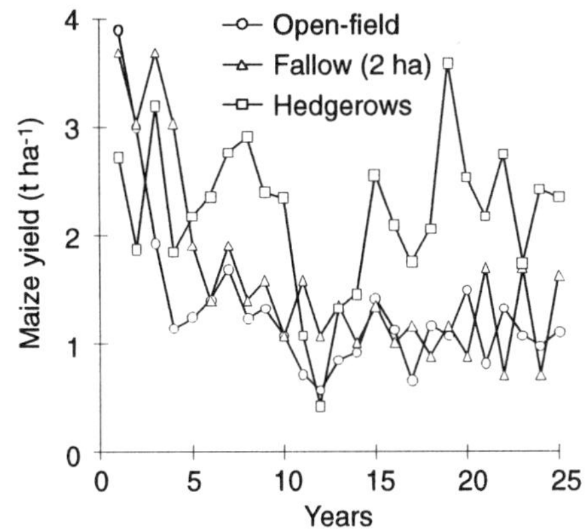

Fig. 7.3. Annual maize yields predicted using APSIM on a relatively erodible soil.

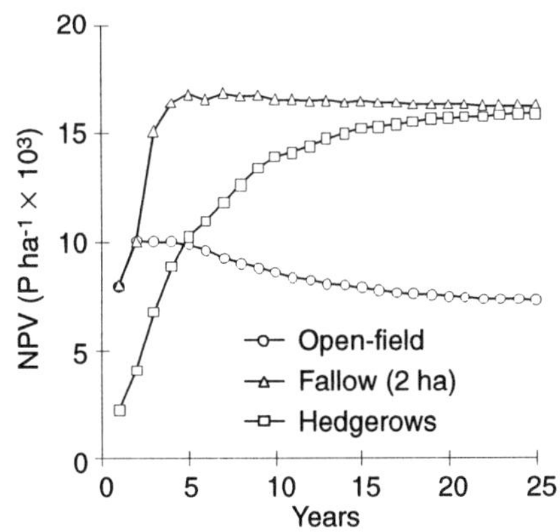

Fig. 7.4. Expected net present value predicted using APSIM on a relatively erodible soil, with a discount rate of 25%.

communities, again for moderately fertile soils of relatively high erodibility. High returns were predicted from hedgerow intercropping in the long term because of the high value of maize when marketed as a fresh vegetable, and the potential productivity gains on highly erodible soils (Figs 7.5 and 7.6). However, hedgerow establishment costs are also high in peri-urban upland communities because of high labour costs, reducing the economic viability of hedgerow intercropping in the short term. As a consequence, traditional methods of open-field farming are economically attractive to farmers with limited planning horizons.

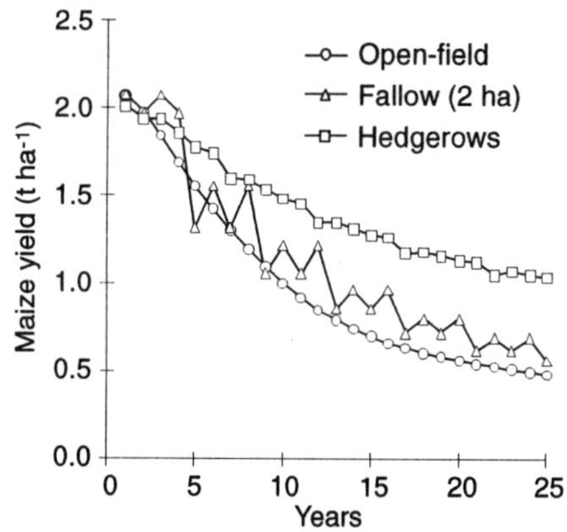

Fig. 7.5. Annual maize yields predicted using SCUAF on a relatively erodible soil.

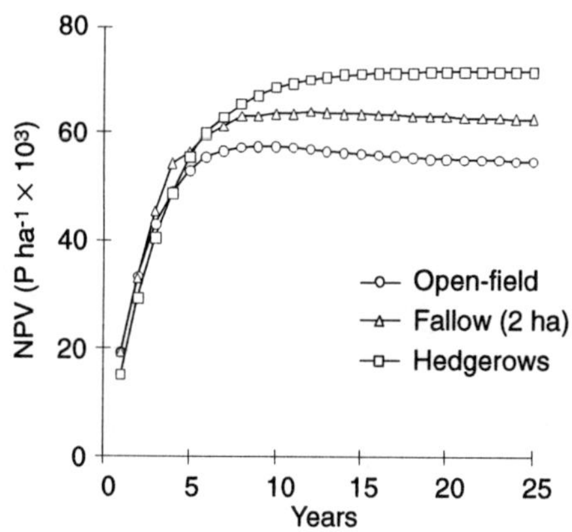

Fig. 7.6. Expected net present value predicted using SCUAF on a relatively erodible soil, with a discount rate of 25%.

In addition to changes in relative economic viability over time, other decision criteria used by farmers were used to interpret the results of the cost–benefit analyses. In particular, the predicted distributions of annual net returns over time were used to assess the relative risk of negative annual returns from the alternative farming methods. Using APSIM to predict seasonal variability in maize yields, it was found that hedgerow intercropping with shrub legume hedgerows reduced the risk of negative returns in the long term. However, a high risk of negative returns was predicted in years when hedgerow establishment coincided with low maize yields due to unfavourable seasonal conditions. The analysis using SCUAF also indicated that hedgerow intercropping reduced the risk of negative returns in the long term. Stochastic dominance analysis indicated that allowing for risk aversion did not alter the ranking of the alternative farming methods based on expected net present value.

Farm-level Constraints: Is the Technology Adoptable?

Rethinking the adoption framework

Economic viability is necessary but not sufficient to ensure that new technology is permanently incorporated into existing farming systems. A variety of technical and socio-economic factors may constrain farmers from adopting technologies that otherwise appear profitable (Lutz *et al.*, 1994; Current *et al.*, 1995). There is a long tradition of socio-economic research into the factors affecting the diffusion and adoption of agricultural technology (Jones, 1967; Feder *et al.*, 1985; Lovejoy and Napier, 1986; Lindner, 1987; Smit and Smithers, 1992; Rogers, 1995). In Lindner's (1987) view, the essence of the technology adoption process is that it involves the progressive acquisition of information (learning) about the technology in question, and a sequence of risky decisions (in that uncertainty exists about whether the decision-maker will be made better or worse off by adopting the technology). As the decision-maker learns more about the technology, either by experience or observation, his/her decision to adopt or not is periodically reviewed. From this perspective, the factors affecting adoption of a given technology are those which influence the decision-maker's state of knowledge about the available options and his/her evaluation of that knowledge in terms of utility maximization. As Parton argues, however, there remains a need to identify and analyse the many observable socio-economic factors which affect the decision-maker's subjective utility as 'it is these underlying influences which are the targets of policy intervention and extension efforts' (1992: 113).

The view of adoption as a process of learning and decision-making leads naturally to the delineation of stages of adoption (Lindner *et al.*, 1982; Sinden and King, 1990; Rogers, 1995). Anderson (1993) emphasizes the need to examine the steps in adoption decision-making in order to determine appropriate points of intervention, so as to achieve higher pay-offs to research and better outcomes for farmers. He distinguishes three stages: establishing a mindset for change (influenced, for example, by input suppliers, extension services, orchestrated field tours and farm visits); learning the particularities of the technology (e.g. by observing neighbours, pilot testing, seeking other people's experience and views); and implementing the new technology (including, for example, planning for adjustments, arranging credit, and adopting some elements in a package).

Implementation itself, however, can involve several substages which may need to be distinguished. In the case of agroforestry technology such as hedgerow intercropping, three levels of adoption can be identified: willingness to test the technology on-farm; willingness to maintain and manage the technology; and extension of the technology to other pieces of land, or re-establishment at the end of its life cycle (Scherr and Muller, 1991). Frequently, adoption in the first sense, usually within the context of a development project, is taken as evidence of adoption in the second or third senses. Strictly speaking, however, adoption means 'the integration of a new concept, activity or input as part of the pattern of normal practice .... It implies repeated usage over time, and adaptation to farm specific requirements' (Scherr and Muller, 1991: 244).

This raises the issue of the relationship between technology adaptation and adoption. In the conventional view, technology emanates from 'upstream' activities in the formal research system and is adapted by 'downstream' research until it is ready for dissemination to farmers (Biggs, 1990; Biggs and Clay, 1981). Anderson and Hardaker (1979) use an analogy from home economics rather than hydrology, speaking of quarter-baked (notional), half-baked (preliminary) and fully-baked (developed) technology, while Scherr and Muller (1991) refer to the same process of technology development in terms of experimental, prototype and off-the-shelf technology.

However, as Anderson (1993) notes, adoption and adaptation are intertwined in that adaptation of the technology frequently occurs in the process of implementing it on-farm, a phenomenon which Rogers (1995) terms 'reinvention'. Sumberg and Okali (1997) go further, arguing that such adaptation is the norm, resulting from an on-going process of 'farmer experimentation'. This experimentation is not confined to a few 'research-oriented' farmers, but is the process by which almost all farmers incorporate technology into their farming systems. Technology supplied by the formal research and extension system thus becomes 'raw material' for farmer experimentation (Sumberg and Okali, 1997). In terms of the above analogies, rather than acquiring a 'fully-baked' technology 'off-the-shelf', farmers can be viewed as shopping around for 'ingredients' or technological components which they incorporate into their own recipes. In other words, technology is only fully developed or adapted as part of a specific, operational farming system.

The conventional adoption framework explains the adoption decision and its timing (early or late) primarily in terms of the decision-maker's perceptions and inherent characteristics, with 'innovators' at one extreme and 'laggards' at the other (Rogers 1995). However, farmer decision-making is generally more complex than this implies. As Scherr (1995) emphasizes, farmers have multiple objectives (including food security, adequate cash income, a secure asset or resource base, social security) and select 'livelihood strategies' to pursue these objectives with the resources available to them (see also Ellis, 1997). Both the objectives and the available resources vary between farmers and change over the life cycle of the farm household (e.g. some farmers at some time may rely on off-farm work as a major source of livelihood, restricting their capacity to invest in labour-intensive conservation measures). Thus farmers in the same environment may have different objectives and livelihood strategies, and so respond differently to a given conservation technology. Hence Biot *et al.* (1995: 24) suggest that 'different behaviour [with respect to soil conservation] may be as much a function of different opportunities and constraints as of different perceptions'.

The adoption framework further simplifies the analysis of the adoption decision by its implicit assumption of an individual 'decision-maker'. Within the farm household, the ability to make decisions regarding resource use and technology varies according to age, gender and other categories, and actual decisions can depend on a complex bargaining process among household members (Ellis, 1993: ch. 9; Biot *et al.*, 1995; Jackson, 1995). Beyond the household, group processes and the ability to harness them can play a crucial role in adoption decisions, particularly with regard to conservation practices (Chamala and Mortiss, 1990; Frank and Chamala, 1992; Pretty and Shah, 1994; Chamala and Keith, 1995). Moreover, decisions about new technology are frequently prompted by an intervention of some sort, typically in the form of a project. As Long and Van der Ploeg (1989) argue, such interventions involve a variety of social actors, with diverse histories and agendas, from both within and beyond rural communities. Hence a project intervention needs to be recognized as part of 'an ongoing, socially-constructed and negotiated process, not simply the execution of an already-specified plan of action with expected outcomes' (Long and Van der Ploeg, 1989: 228).

These considerations lead us to expect a range of responses to the promotion of a soil conservation technology, not merely a clear-cut decision to adopt or not. Differences between the environment in which the technology was developed and the environment of the 'target' community will prompt farmers to adapt the technology in the process of adopting it. Differences within a given community in farmers' goals and circumstances, hence livelihood strategies, and the complexity of intrahousehold, group and project interactions and decision-making, will result in a variety of adoption–adaptation behaviours, which should be investigated on their own terms and not prejudged by labelling them as 'poor adoption' or 'non-adoption'.

Investigating adoption and adaptation

The ACIAR project investigated farm-level factors associated with the adoption of recommended soil conservation technologies in eight case-study sites in the Philippines, seven of which have been analysed sufficiently to report on here (Table 7.1). The 'adoption process' in each study site began with the commencement of a development project which sought, among other things, to promote conservation technologies. Though many farmers had already invented or adopted various conservation practices (including ploughing across the slope and piling rocks or crop debris in lines across the slope), none had adopted contour hedgerows or dryland bench terraces (with or without rock walls), the two principal technologies promoted. Hence the role of the project in initiating the adoption process was crucial and needs to be taken into account in analysing farmers' adoption decisions. In some cases the prior decision was to participate in the project in order to receive whatever benefits were on offer (e.g. livestock dispersal, stewardship contracts [a conditional lease of public land], farm tools and inputs, subsidies) and a concomitant of that decision was the implementation of recommended conservation technologies.

The meaning of 'adopters' obviously varies somewhat from site to site,

Table 7.1. Case study villages and projects.

Village	Province	Project/organization	Main technologies
Tabayag	Cebu	Mag-uugmad Foundation Inc. (MFI)	Bench terraces, rock walls, and contour hedgerows
Guba	Cebu	Mag-uugmad Foundation Inc. (MFI)	Contour hedgerows, bunds and canals
Pananag	Davao del Sur	Mindanao Baptist Rural Life Centre (MBRLC)	Contour hedgerows (Sloping Agricultural Land Technology; SALT)
Managok	Bukidnon	MUSUAN Project, Central Mindanao University (CMU)	Contour hedgerows
Salogon	Palawan	Upland Stabilisation Project (USP), Department of Environment and Natural Resources (DENR)	Contour hedgerows bench terraces
Magdungao	Iloilo	Magdungao Agroforestry Project, DENR	Contour hedgerows and contour bunds
Domang	Nueva Vizcaya	Integrated Social Forestry Project, DENR	Contour hedgerows

but in general the term is used to refer to farmers who at some point implemented contour hedgerows and/or bench terraces (sometimes with rock walls) on all or a substantial part of their farms. With contour hedgerows in particular, there was considerable on-farm adaptation by project staff and individual farmers, affecting such aspects as hedgerow species, hedgerow spacing, pruning frequency and cropping pattern. The logic behind this adaptation is considered below. It is likely that the adoption process was largely complete in most sites, the studies being conducted 10–14 years after the commencement of the conservation project. However, in many cases (even in Guba, the site of the most widespread and long-term implementation), confirmation of adoption through extension and re-establishment had not always occurred and a process of discontinuance may have been underway.

The research methods involved a combination of less formal 'rapid rural appraisal' (RRA) or 'participatory rural appraisal' (PRA) methods (Chambers *et al.*, 1989; Mikkelsen, 1995; Chambers, 1997) and a formal questionnaire survey of a sample of farm households from within and around the project area. The less formal methods included: a review of existing municipal and village documents, farm walks and direct observation of farms, semistructured interviews with community groups and key informants, resource mapping, time lines, seasonal diagrams, life histories, and community histories. The formal survey was administered to a random sample of farm households in each of three categories: 'adopters' of the recommended soil conservation measures;

'non-adopters' within the same setting, that is, other farmers in the project village(s) who were thus exposed to the conservation project but chose not to implement the technologies; and (where feasible) a 'control group', that is, farmers in a similar setting but from outside the project area, most of whom were not aware of the conservation technologies being promoted and hence had not yet commenced the adoption decision process. The survey typically involved a single, hour-long interview in the respondent's home, with husband and wife both present, some questions being directed specifically to the wife. In some cases farm inspections were also made. Details of research methods and analytical procedures for each site can be found in the original survey reports (Garcia *et al.*, 1995–1997).

RRA and PRA methods are frequently portrayed as being clearly superior to formal surveys, not just because they are supposedly more effective in obtaining data but because they can in some sense empower the participants by giving them greater control over the generation and application of knowledge. Our experience has been that both kinds of method have their strengths and weaknesses and need to be viewed as complementary rather than competing approaches. The problems of the survey approach have been well summarized by Chambers (1983). The problems of participatory approaches have been less well documented, partly perhaps because of the ideological nature of 'participation' (Biggs and Smith, 1998). Yet participatory appraisal, to be done well, can be very demanding, both in terms of the skill and effort required of the facilitator and the investment of time on the part of the participants. Moreover, group or community methods in particular can produce a 'narrative' which glosses over contentious issues of intracommunity disagreement and conflict, or which privileges the viewpoint of a politically dominant subgroup. Insight into such issues is more likely to emerge in private household interviews than in participatory appraisal sessions, which are often predicated on an idealistic notion of 'community'. Case histories can be particularly useful in this respect but are also time-consuming and often difficult to write up and analyse. In brief, though the two kinds of methods overlapped and thus reinforced each other, RRA/PRA methods were particularly useful for gaining an overview of the characteristics, current status and trends in the local environment and farming system, and a factual account of project interventions, while the formal survey was most helpful in documenting, quantifying and correlating the range of farmers' circumstances and responses, and pursuing specific questions in more detail.

In analysing the factors associated with adoption and adaptation of the recommended technologies, comparisons were made between 'adopters' and 'non-adopters' within the project site, though individual and group perceptions of changes over time were used to supplement these cross-sectional comparisons. The adoption factors were analysed sequentially, the main form of quantitative analysis being pairwise comparison of means and proportions for the two groups, though interactions between factors were also examined where this seemed relevant. Full multivariate analyses of adoption were conducted for particular sites (Shively, 1996; Garcia, 1997) and these tended to confirm the conclusions drawn from the simpler approach summarized here.

The advantage of this less formal approach is that it enables the qualitative data to be woven into the analysis of the quantitative data, producing a more coherent overall explanation of farmers' responses. In the interests of space, only a brief synopsis of the main findings is presented, without supporting tables or other data (see Cramb, 1998).

Attributes of adopters

Two broad sets of attributes affecting adoption are considered: (i) the personal attributes of the principal decision-maker and his/her perceptions, both of the problem of erosion and of the recommended technologies; and (ii) the attributes of the farm, including physical and economic attributes.

Personal attributes and perceptions

In general, readily measurable personal attributes of farmers (age, education, etc.) were not important in explaining adoption. However, age and gender were factors where strenuous work was required to implement the technology (i.e. terracing and rock walls), older farmers and female farmers being less inclined to adopt such measures. Less easily measured personality traits related to 'innovativeness' and 'managerial ability' were clearly important, particularly among early adopters. Such traits are usually readily identified by community members and experienced extension workers.

Awareness of soil erosion was relatively high, but farmer perceptions of soil erosion as a problem varied, partly due to differences in farmers' knowledge, but mainly because of objective differences in soil and farming conditions (e.g. fallow systems versus continuous cropping). Perceptions of erosion clearly had an effect on adoption behaviour.

Perceptions of, and attitudes to, the recommended technologies suggested that farmers were well informed, from direct or indirect acquisition of site-specific knowledge over several years. Adopters and non-adopters shared perceptions regarding the high labour requirements for establishing and maintaining the technologies, the loss of plantable area and the delay in obtaining benefits, as well as the undesirable side-effects of some forms of hedgerow on weeds, pests and diseases. Nevertheless, contour hedgerow technology in particular was widely seen (at least *within* project villages) as useful and necessary, easy to learn and easy to acquire (though acquisition of planting materials was clearly a problem in some cases).

Thus differences in perceptions as such were not a major factor explaining differences in adoption within project villages; rather, it was the appropriateness or relevance of the technologies to the farmer's specific circumstances (the farm factors) which was the key, emphasizing the need for technologies to be adapted to different sets of farmer circumstances. Beyond project villages, however, there was very little awareness or knowledge of the recommended technologies (particularly regarding methods of implementation), indicating that diffusion does not occur without facilitation.

Farm attributes

There were many farm-specific factors which influenced adoption. Adoption was more likely on larger farms. The physical features of individual fields were also important: adoption of hedgerows in particular was more likely on fields which were larger, steeper, had more erodible soils, were located close to the homestead, had relatively uniform terrain and were oriented down rather than across the slope. In the case of rock walls, the rockiness of a field was an obvious predisposing factor, whereas the prior construction of rock lines discouraged further development. Ownership of lowland fields discouraged investment in upland fields in some contexts.

Land tenure was a major factor but its influence was highly conjunctural. The classification of the land as public or private was only important where farmers occupying public land felt their tenure was insecure *vis-à-vis* the state, in which case stewardship contracts with the government (and the conditions that went with them, including conservation measures) appeared attractive. Otherwise tenancy was the major issue. In general, tenancy discouraged adoption or led farmers to adopt less costly measures such as grass hedgerows, but the crucial factors were the awareness and attitude of the landowner, which varied enormously. Extension strategies may have to involve landlords as important stakeholders where tenancy is a constraint.

The labour requirements of the technologies were an important consideration but the farm-household's labour supply was not a major factor in itself. Rather, it was related to cash flow concerns and the need to use spare labour off-farm in preference to implementing conservation measures. Relatedly, cash income was an important factor promoting adoption, particularly for a subgroup of adopters with high cash income. While all farmers experienced a credit constraint, farmers with higher cash income were less bound by it and therefore able to invest more labour and working capital in their own farms.

Analysis of these farm-specific constraints to adoption can assist in the development and promotion of a wider range of technology options (e.g. low-cost, quick-return options for resource-poor tenant farmers).

Adoption and adaptation of hedgerow intercropping

As mentioned, the main technology promoted and adopted was hedgerow intercropping, in particular the package developed by the Mindanao Baptist Rural Life Centre known as Sloping Agricultural Land Technology (SALT). However, implementation and management of the hedgerow system invariably differed from that followed in research trials and promoted by extension staff (Nelson *et al.*, 1996g; Garcia *et al.*, 1998). Adopters generally established only one plot with hedgerows on one parcel of their land, though typically this plot represented a sizeable proportion (40–100%) of the parcel. Other adaptations involved: (i) reduced hedgerow density due to wider alleys, single rather than double hedgerows and lower planting density within hedgerows; (ii) reduced trimming of hedgerows; (iii) lack of or infrequent mulching of the alleys with hedgerow trimmings; (iv) substitution of recommended hedgerow species with alternatives; (v) incorporating additional components in hedgerows or using them for other purposes; and (vi) changing the cropping pattern.

The factors underlying these adaptations were related to farmers' goals and circumstances in the following ways (Garcia *et al.*, 1998):

1. Technical difficulty. The time and difficulty involved in pegging out contour lines using an A-frame led some farmers to estimate the contour by eye or to plant in a straight line across the slope, resulting in hedgerows which were off the contour.
2. Small landholding. In the standard SALT package, contour hedgerows occupy 27% of the land area – even more on steeper slopes. Farmers with limited land often established hedgerows at wider intervals than recommended and/or with only a single row, to minimize reduction of the cultivable area.
3. Labour constraint. The high labour requirements for establishing and maintaining hedgerows, particularly on steeper slopes, led farmers to make adaptations such as wider alleys, single hedgerows, wider within-row spacing, lower trimming frequency, lack of mulching and replacement of shrub legumes with grasses or natural vegetation.
4. Access to planting materials. The availability and the characteristics of hedgerow planting materials influenced the choice of hedgerow species. Most of the recommended hedgerow species were not indigenous and hence had to be procured by farmers or provided by the project. Limited supply of these planting materials sometimes resulted in the use of alternative species, the planting of single-row hedgerows, wider alleys and wider within-row spacing. Likewise, the lack of planting materials for recommended new crops limited the adoption of crop rotation within the alleys as specified in the SALT system.
5. Market opportunities. Farmers frequently attributed lack of adoption of contour hedgerows to the lack of immediate economic benefits. Consequently, many farmers added components to the hedgerow system from which saleable by-products could be derived and/or changed the relative importance of the system components to emphasize outputs with higher income generating potential (e.g. hedgerow seed, timber, charcoal, fodder, food crops). These adaptations were, of course, dependent on local markets.
6. Hedgerow characteristics. Some of the recommended hedgerow species had undesirable side-effects. For example, *Leucaena* hedgerows on acid soils did not grow well, competing with crops in the alleys and failing to produce substantial biomass for green manuring. Napier grass often proliferated into the alleys, creating a weed problem. Such problems prompted farmers to change the hedgerow species or to remove the hedgerows altogether.
7. Existing farming system. In many cases, adaptations of the contour hedgerows resulted from the need to fit better with the existing farming system. For example, in Domang, where a short-fallow system was practised, *Gliricidia sepium* hedgerows developed big roots and encroached into the alleys during the fallow period, subsequently obstructing cultivation. In Pananag, *Flemingia macrophylla* and *Desmodium rensonii* hedgerows proved better adapted to the short-fallow system practised in the area; the

fallow period also allowed the hedgerows to seed, giving the farmers the opportunity to harvest and sell the seed.

In general, many of the adaptations sought to improve the short-term profitability of hedgerow intercropping, by reducing the up-front costs (primarily, labour for establishment and the opportunity cost of forgone cropping area) and adding to short-term benefits in the form of additional outputs, particularly those with a market value. The longer-term implications of these adaptations for soil conservation, hence crop productivity, are explored by Garcia *et al.* (1998). In brief, it is likely that the adaptations reduced the conservation benefits relative to the recommended versions of the technology, but still made a substantial contribution to soil conservation in comparison with current practice.

Conclusions

We have described two complementary approaches which are necessary to understand constraints to adoption of soil conservation technology at the farm level. The first approach asks: In a relatively unconstrained setting, is the technology profitable? The second asks: Given the range of farm-specific constraints, is the technology adoptable?

Our research has shown that cost–benefit analysis can be used to provide insights into the economic incentives for farmers to adopt conservation farming. Production budgets can be used to quantify the physical production characteristics of the farming system, and to derive an economic value for farm inputs and produce. Erosion–productivity models can be used to predict the benefits of conserving soil in terms of higher sustained crop yields in the long term. Although not explored in this research, cost–benefit analysis can be expanded to investigate the off-site and whole-farm benefits of conservation farming.

The cost–benefit analyses indicated that hedgerow intercropping has the potential to sustain maize production relative to traditional methods of open-field farming, by controlling erosion and contributing nitrogen to the cropping alleys. In the short term, however, both continuous and fallow open-field farming provide higher economic returns to farmers despite high rates of soil loss. This is because the value of sustained maize yields from hedgerow intercropping is not realized rapidly enough to compensate farmers for establishment and maintenance costs. In the long term, the ability of hedgerow intercropping to conserve soil and contribute nitrogen provides an economically attractive method of sustaining cropping relative to continuous open-field farming.

The research project also found that a range of techniques was required to investigate the adoptability of conservation technologies. A case study approach was necessary to bring out the complexity of specific farming environments and project interventions. Within this framework, formal survey methods, while having well-known limitations, provided an effective way of ascertaining the full spectrum of farmers' circumstances and responses to recommended technology. Participatory or group methods were a useful complement to household surveys, particularly with regard to relatively uncontroversial topics (such as cropping calendars), but were not always the best way to investigate contentious issues (such as why a project failed to have an impact on all members of a village). An eclectic approach, drawing on different methods to suit particular requirements, proved capable of yielding useful insights.

The process of adoption of conservation farming technologies in the case study villages was complex and highly variable, and could not be separated from the process of intervention in the villages by government and non-government agencies implementing upland development projects. The case study sites represented a sample of the better-resourced and more successful projects, yet in general, adoption rates were low and diffusion beyond the project site was almost non-existent. The implications for the vast areas of the uplands not in proximity to the 'nodes of diffusion' created by well-resourced upland projects are not encouraging.

Yet recommended soil conservation technologies, particularly hedgerows, were widely seen by farmers who were aware of them as useful and even necessary (consistent with the long term findings of the cost–benefit analysis). Adoption was largely constrained by farmers' specific circumstances (the farm attributes) rather than their personal attributes and perceptions. Moreover, even those who adopted the technologies found it necessary to modify them in various ways. There was thus a trade-off between long-term profitability and adoptability. Farm-level constraints either prevented the adoption of an otherwise profitable technology or led farmers to adapt the technology in order to reduce costs and increase benefits in the short term, possibly at the expense of longer-term profitability.

Were the recommended soil conservation technologies profitable? The answer suggested by erosion–productivity modelling and cost–benefit analysis is that hedgerow intercropping can indeed be profitable, given a sufficiently long-term perspective. Were the technologies adoptable? The answer suggested by the case studies is: in part; by some. Given the range of farmers' goals and circumstances it may be better to focus less on constraints to adoption of a predetermined technology (such as SALT) and more on promoting a wider range of technological options which farmers may find useful in developing their diverse farming systems.

In conclusion, the main 'constraint to adoption' may be the 'transfer-of-technology' approach itself, which unduly privileges the technology being promoted and tends to denigrate farmers' experiments as 'poor adoption' or 'non-adoption', rather than the rational pursuit of diverse livelihood strategies. Based on the findings of the research project, we see a need to invest in a major long-term programme of adaptive research and extension in the Philippine uplands which better accommodates the range of farmers' goals and circumstances, and which actively supports and utilizes farmers' own experiments. Norman and Douglas (1994) provide guidelines for the kind of programme required, incorporating soil conservation into a process of farming systems development. The experience gained in a related project (ACIAR

Project 9220), which utilized a farming systems development approach, is particularly useful in this respect. However, as argued elsewhere (Cramb, 1997), the macro-level, politico-economic constraints to implementing such a programme throughout the Philippines are formidable.

References

Anderson, J.R. (1993) The economics of new technology adaptation and adoption. *Review of Marketing and Agricultural Economics* 61, 301–310.

Anderson, J.R. and Hardaker, J.B. (1979) Economic analysis in design of new technologies for small farmers. In: Valdes, A., Scobie, G.M. and Dillon, J.L. (eds) *Economics and the Design of Small-Farmer Technology.* Iowa State University Press, Ames, pp. 11–26.

Anderson, J.R. and Thampapillai, J. (1991) *Soil Conservation in Developing Countries: Project and Policy Intervention. Policy and Research Series 8.* The World Bank, Washington DC.

Biggs, S.D. (1990) A multiple source of innovation model of agricultural research and technology promotion. *World Development* 18, 1481–1499.

Biggs, S.D. and Clay, E.J. (1981) Sources of innovation in agricultural technology. *World Development* 9, 321–336.

Biggs, S.D. and Smith, G. (1998) Beyond methodologies: coalition-building for participatory technology development. *World Development* (in press).

Biot, Y., Blaikie, P., Jackson, C. and Palmer-Jones, R. (1995) *Rethinking Research on Land Degradation in Developing Countries. World Bank Discussion Papers 289.* The World Bank, Washington DC.

Chamala, S. and Keith, K. (eds) (1995) *Participative Approaches for Landcare.* Academic Press, Brisbane.

Chamala, S. and Mortiss, P. (1990) *Working Together for Landcare: Group Management Skills and Strategies.* Academic Press, Brisbane.

Chambers, R. (1983) *Rural Development: Putting the Last First.* Intermediate Technology Publications, London.

Chambers, R. (1997) *Whose Reality Counts? Putting the First Last.* Intermediate Technology Publications, London.

Chambers, R., Pacey, A. and Thrupp, L.A. (eds) (1989) *Farmer First: Farmer Innovation and Agricultural Research.* Intermediate Technology Publications, London.

Cramb, R.A. (1997) Environment and development in the Philippine uplands: the problem of agricultural land degradation. Paper presented at *Annual Conference of Development Studies Association, University of East Anglia, Norwich, England, 11–13 September.*

Cramb, R.A. (ed.) (1998) *Conservation Farming in the Philippine Uplands: Socio-Economic Perspectives.* ACIAR Proceedings. Australian Centre for International Agricultural Research, Canberra (in press).

Cramb, R.A. and Saguiguit, G.C. (1994) *Socio-Economic Evaluation of Soil Conservation Technologies for Upland Farming Systems in the Philippines: Project Overview. Working Paper No. 1,* SEARCA-UQ Upland Research Project, Los Baños, the Philippines.

Current, D., Lutz, E. and Scherr, S.J. (1995) The costs and benefits of agroforestry to farmers. *World Bank Research Observer* 10, 151–180.

Ellis, F. (1993) *Peasant Economics: Farm Households and Agrarian Development,* 2nd edn. Cambridge University Press, Cambridge.

Ellis, F. (1997) Household strategies and rural livelihood diversification. Paper presented at *Annual Conference of Development Studies Association, University of East Anglia, Norwich, England, 11–13 September.*

Feder, G., Just, R.E. and Zilberman, D. (1985) Adoption of agricultural innovations in developing countries: a survey. *Economic Development and Cultural Change* 33, 255–298.

Frank, B.R. and Chamala, S. (1992) Effectiveness of extension strategies. In: Lawrence, G., Vanclay, F. and Furze, B. (eds) *Agriculture, Environment and Society: Contemporary Issues for Australia.* Macmillan, Melbourne.

Freebairn, D.M., Silburn, D.M. and Loch, R.J. (1989) Evaluation of three soil erosion models for clay soils. *Australian Journal of Soil Research* 27, 199–211.

Garcia, J.N.M., Gerrits, R.V., Cramb, R.A. and Saguiguit, G.C. (1995–1997) *Socio-Economic Survey Reports 1–7.* SEARCA-UQ Upland Research Project, Los Baños, the Philippines.

Garcia, J.N.M., Gerrits, R.V., Cramb, R.A. and Saguiguit, G.C. (1998) Implementation of conservation technologies in the case study sites. In: Cramb, R.A. (ed.) *Conservation Farming in the Philippine Uplands: Socio-Economic Perspectives.* ACIAR Proceedings. Australian Centre for International Agricultural Research, Canberra, (in press).

Garcia, Y.T. (1997) Analysis of decision models for upland soil conservation in Argao, Cebu. PhD dissertation, University of the Philippines, Los Baños.

Gardner, E.A., Freebairn, D.M., Doherty, J.E., Shaw, R.J., Littleboy, M., Titmarsh, G., Silburn, D.M., Connolly, R.D., and Begbie, D. (1992) Hydrological tools to assist decision making in integrated catchment management. Platform Papers Vol. 2: *Technical Papers. 15th Federal Convention, Australian Water and Wastewater Association (AWWA), Gold Coast, 18–23 April.*

Hammer, G. and White, D. (1992) From reality to research and back again by computer model. *Agricultural Science* 5(1), 35–39.

Jackson, C. (1995) Environmental reproduction and gender in the Third World. In: Morse, S. and Stocking, M. (eds) *People and Environment.* UCL Press, London, pp. 109–130.

Jones, G.E. (1967) The adoption and diffusion of agricultural practices. *World Agricultural Economics and Rural Sociology Abstracts* 9(3), 1–34.

Lindner, R.K. (1987) Adoption and diffusion of technology: an overview. In: Champ, B.R., Highley, E. and Remenyi, J.V. (eds) *Technological Change in Postharvest Handling and Transportation of Grains in the Humid Tropics. ACIAR Proceedings No. 19.* Australian Centre for International Agricultural Research, Canberra, pp. 144–151.

Lindner, R.K., Pardey, P.G. and Jarrett, F.G. (1982) Distance to information source and the time lag to early adoption of trace element fertilisers. *Australian Journal of Agricultural Economics* 26, 98–113.

Littleboy, M., Freebairn, D.M., Hammer, G.L. and Silburn, D.M. (1992a) Impact of soil erosion on production in cropping systems. II. Simulation of production and erosion risks for a wheat cropping system. *Australian Journal of Soil Research* 30, 775–788.

Littleboy, M., Coughlan, K.J., Silburn, D.M. and Boaler, L.J. (1992b) Modelling erosion productivity relationships. In: Hamilton, G.J., Howes, K.M. and Attwater, R. (eds) *Erosion/Productivity and Erosion Prediction Workshop. Proceedings of the 5th Australian Soil Conservation Conference,* Vol. 3. Western Australia Department of Agriculture, Perth.

Long, N. and Van der Ploeg, J.D. (1989) Demythologizing planned intervention: an actor perspective. *Sociologia Ruralis* 29, 226–249.

Lovejoy, S.B. and Napier, T.L. (eds) (1986) *Conserving Soil: Insights from Socio-Economic Research*. Soil Conservation Society of America, Ankeny.

Lutz, E., Pagiola, S. and Reiche, C. (1994) Lessons from economic and institutional analyses of soil conservation projects in Central America and the Caribbean. In: Lutz, E., Pagiola, S. and Reiche, C. (eds) *Economic and Institutional Analyses of Soil Conservation Projects in Central America and the Caribbean, World Bank Environment Paper No. 8.* The World Bank, Washington, DC.

McCown, R.L. (1989) Adapting farming systems research concepts to Australian research needs. *Proceedings of the 5th Australian Agronomy Conference*, Perth, September.

McCown, R.L. (1991) Research in a farming systems framework. In: Squires, V. and Taw, P. (eds) *Dryland Farming: a Systems Approach. An Analysis of Dryland Agriculture in Australia.* Sydney University Press, Sydney.

McCown, R.L., Hammer, G.L., Hargreaves, J.N.G., Holzworth, D.P. and Freebairn, D.M. (1996) APSIM: a novel software system for model development, model testing, and simulation in agricultural systems research. *Agricultural Systems* 50, 255–271.

Mikkelsen, B. (1995) *Methods for Development Work and Research: A Guide for Practitioners.* Sage, New Delhi.

Nelson, R.A. (1996) Bioeconomic analysis of hedgerow intercropping in the Philippine uplands. PhD thesis, Department of Agriculture, University of Queensland, Brisbane.

Nelson, R.A., Grist, P.G., Menz, K.M., Paningbatan, E.P. and Mamicpic, M.A. (1996a) *A cost–benefit analysis of hedgerow intercropping in the Philippine uplands using SCUAF. Working Paper No. 7.* SEARCA-UQ Uplands Research Project, Los Baños, the Philippines.

Nelson, R.A., Cramb, R.A., Menz, K.M. and Mamicpic, M.A. (1996b) *Bioeconomic modelling of alternative forms of hedgerow intercropping in the Philippine uplands using SCUAF. Working Paper No. 10.* SEARCA-UQ Uplands Research Project, Los Baños: Los Baños, the Philippines.

Nelson, R.A., Dimes, J.P., Silburn, D.M. and Carberry, P.S. (1996c) *Erosion/productivity modelling of maize farming in the Philippine uplands. Part I: A simple description of the Agricultural Production Systems Simulator. Working Paper No. 12.* SEARCA-UQ Uplands Research Project, Los Baños, the Philippines.

Nelson, R.A., Dimes, J.P., Silburn, D.M. and Carberry, P.S. (1996d) *Erosion/productivity modelling of maize farming in the Philippine uplands. Part II: Parameterising the Agricultural Production Systems Simulator. Working Paper No. 13.* SEARCA-UQ Uplands Research Project, Los Baños, the Philippines.

Nelson, R.A., Dimes, J.P., Silburn, D.M., Paningbatan, E.P. and Cramb, R.A. (1996e) *Erosion/productivity modelling of maize farming in the Philippine uplands. Part III: Simulation of alternative farming methods. Working Paper No. 14.* SEARCA-UQ Uplands Research Project, Los Baños, the Philippines.

Nelson, R.A., Cramb, R.A. and Mamicpic, M.A. (1996f) *Erosion/productivity modelling of maize farming in the Philippine uplands. Part IV: Economic analysis of alternative farming methods. Working Paper No. 15.* SEARCA-UQ Uplands Research Project, Los Baños, the Philippines.

Nelson, R.A., Cramb, R.A. and Mamicpic, M.A. (1996g) *Costs and returns of hedgerow intercropping and open-field maize farming in the Philippine uplands. Working Paper No. 11.* SEARCA-UQ Uplands Research Project, Los Baños, the Philippines.

Norman, D. and Douglas, M. (1994) *Farming Systems Development and Soil*

Conservation. FAO Farm Systems Management Series 7. Food and Agriculture Organization of the United Nations, Rome.

Pagiola, S. (1994) Cost–benefit analysis of soil conservation. In: Lutz, E., Pagiola, S. and Reiche, C. (eds) *Economic and Institutional Analyses of Soil Conservation Projects in Central America and the Caribbean. World Bank Environment Paper No. 8.* The World Bank, Washington, DC.

Paningbatan, E.P (1995) *ACIAR 9201 Sustainable cropping systems in tropical steeplands. Country report: Philippines.* ACIAR 9201 final review, April, Subic Bay.

Parton, K. (1992) Socioeconomic modelling of decision making with respect to choice of technology by resource-poor farmers. In: Probert, M. (ed.) *A Search for Strategies for Sustainable Dryland Cropping in Semi-Arid Eastern Kenya. ACIAR Proceedings 41.* Australian Centre for International Agricultural Research, Canberra, pp. 110–118.

Pretty, J.N. and Shah, P. (1994) *Soil and Water Conservation in the Twentieth Century: A History of Coercion and Control. Research Series No. 1,* Rural History Centre, University of Reading, UK.

Rogers, E.M. (1995) *The Diffusion of Innovations,* 4th edn Free Press, New York.

Scherr, S.J. (1995) Economic factors in farmer adoption of agroforestry: patterns observed in Western Kenya. *World Development* 23, 787–804.

Scherr, S.J. and Muller, E.U. (1991) Technology impact evaluation in agroforestry projects. *Agroforestry Systems* 13, 235–257.

Shively, G.E. (1996) *Assets, attitudes, beliefs and behaviour: explaining patterns of soil conservation adoption among low-income farmers. Working Paper No. 19.* SEARCA-UQ Uplands Research Project, Los Baños, the Philippines.

Silburn, D.M. and Loch, R.J. (1992) Present capabilities and constraints in modelling soil erosion. In: Hamilton, G.J., Howes, K.M. and Attwater, R. (eds) *Erosion/Productivity and Erosion Prediction Workshop. Proceedings of the 5th Australian Soil Conservation Conference,* Vol. 3.

Sinden, J.A. and King, D. (1990) Adoption of soil conservation measures in Manilla Shire, New South Wales. *Review of Marketing and Agricultural Economics* 58, 179–192.

Smit, B. and Smithers, J. (1992) Adoption of soil conservation practices: an empirical analysis in Ontario, Canada. *Land Degradation and Rehabilitation* 3, 1–14.

Stocking, M. (1984) *Erosion and Soil Productivity: a Review. Consultant's Working Paper No. 1.* Food and Agriculture Organization of the United Nations, Rome.

Sumberg, J. and Okali, C. (1997) *Farmers' Experiments: Creating Local Knowledge.* Lynne Rienner, Boulder.

Young, A. and Muraya, P. (1990) *Soil Changes Under Agroforestry. Computer Program with User's Handbook, Version 2.* International Center for Research into Agroforestry, Nairobi.

8

CIAT's Strategic Research for Sustainable Land Management on the Steep Hillsides of Latin America

E. Amézquita, J. Ashby, E.K. Knapp, R. Thomas, K. Müller-Sämann, H. Ravnborg, J. Beltran, J.I. Sanz, I.M. Rao and E. Barrios

Centro Internacional de Agricultura Tropical (CIAT), A.A. 6713, Cali, Colombia

The hillside agroecosystem in Latin America is a major contributor to food supply and is the basis of the livelihood of a large population of the rural poor. A large part of this agroecosystem is inhabited by poor farmers who derive their sustenance from cultivating small parcels of land on steep to very steep slopes, with or without conservation practices. Under these conditions, soil erosion is increasing over time, resulting in: (i) the abandonment of degraded areas, (ii) the opening of new areas which are often less suited to agriculture, and (iii) the start of an endless cycle of resource degradation. As population density increases, less and less area is available for agricultural production. To understand both the biophysical and socio-economic processes that lead to soil erosion, CIAT is using a multifaceted, multidisciplinary research approach which focuses on: (i) the quantification and evaluation of soil erosion parameters, and calibration of the universal soil loss equation (USLE); (ii) agronomic evaluation of prototype production systems designed with farmer participation; (iii) priority setting based on farmer participatory research methods in an interinstitutional framework through a local consortium of farmer organizations and NGOs; and (iv) diagnosis of constraints and extrapolation of results based on geological information system (GIS) methods. The overall goal of this effort is to understand how to manage a watershed to improve the living standards of the farmers while avoiding the degradation of the natural resources. In each of the mentioned areas, important concepts, parameters and methods of adoption of resource-conserving practices have been identified. It is expected that these approaches will allow a better conceptualization of the important aspects that must be taken into account for productive and resource-conserving use of hillside watersheds. This chapter describes CIAT's multidisciplinary research for sustainable agricultural use on the Latin America hillsides.

 Soil Erosion at Multiple Scales (eds F.W.T. Penning de Vries, F. Agus and J. Kerr)

Introduction

The hillside agroecosystem of tropical America constitutes an area of 96 million hectares, excluding Brazil which also has a very large area that would correspond to the hillside environment (Jones, 1993). Of the total area, about 25 million hectares are already highly degraded and 53 million are estimated to be experiencing rapid rates of degradation. Almost 50% of the total area of Colombia suffers from erosion problems, while 83% of the Andean zone exhibits different states of erosion. A similar picture could be painted for neighbouring Andean countries such as Ecuador, Peru and Bolivia, as well as Costa Rica, Nicaragua, Honduras and El Salvador. The hillside agroecosystem is characterized by land use patterns of intensive coffee production, pastures and the cultivation of annual crops (maize, beans and cassava). Population pressure and poverty in the small farm sector often leads to the deforestation of steeply sloping lands that are highly vulnerable to degradation but only marginally suited to production of annual semisubsistence crops (Pachico *et al.*, 1994). Specific environmental characteristics are: 800–2000 m above sea level, acid soils (pH < 5.3), mean temperature between 18 and 23.5°C during the growing season and mean rainfall (1500–1800 mm year^{-1}) exceeding 60% of potential evapotranspiration during 6–9 months annually. Strictly defined, this agroecosystem covers 30×10^6 ha (Pachico *et al.*, 1994). The hillsides of Central America and the foothills of the Andes through to Brazil have extensive areas that are essentially identical.

This chapter presents an overview of the strategic research activities that a multidisciplinary group of CIAT is developing in the hillside area of Cauca Department in the South Andean region of Colombia.

Research Activities

Quantification and evaluation of 'USLE' parameters

In response to the serious problems of soil erosion and land degradation in the Andean region of Colombia, a Gesellschaft für Technische Zusammenarbeit (GTZ)-sponsored project was established at CIAT in 1987. This project has made commendable contributions in understanding: (i) the basic processes governing soil erosion in Andean inceptisols, especially with reference to their erosivity and erodibility (Ruppenthal *et al.*, 1996); (ii) the development of resource conservation-oriented production systems (Muhr *et al.*, 1995a,b; Tscherning *et al.*, 1995); and (iii) the establishment of a network of on-farm experiments that lay down the foundation for transfer of technology (Lal, 1992).

Field trials in plots of 22 × 11 m (standard plot) and 16 × 8 m were initiated in early 1987, on slopes of 7–20% gradient at two locations in Santander de Quilichao and Mondomo in Cauca, Colombia, to determine the individual factors of the universal soil loss equation (USLE). Soils are acid and infertile, but present favourable physical conditions such as high structural stability and high infiltration rates (Reining, 1992).

The kinetic energy of the rainfall was 21.3 J m^{-2} in Santander and 20.4 J m^{-2} in Mondomo. Most of the rain falls with an intensity of 10 mm h^{-1}. The erosivity indices that correlated better with soil loss were: the rainfall amount, the kinetic energy *E*, the maximum 30-min intensity (I_{30}), 15-min intensity (I_{15}) and 7.5 min intensity ($I_{7.5}$) of erosive rains and products of two or more of these indices (Reining, 1992).

In standard plots free of vegetation, soil losses were up to 311 t $ha^{-1}year^{-1}$ (dry soil). The extent of soil loss under cropping systems was up to 68 t $ha^{-1}year^{-1}$. Other treatments of minimum tillage, contour ridges and contour strips of grass (Mondomo) showed the best erosion control and resulted in soil losses of 1–5 t $ha^{-1}year^{-1}$, within the limits of tolerable loss of soil for the studied area (Reining, 1992). Table 8.1 shows soil losses of different cassava-based cropping systems in Santander and Mondomo. The greatest soil losses were found in the clean-tilled fallow systems (control plots). In the cropping systems, the greatest soil losses were measured at the beginning of the growing season. This was especially evident in plots with cassava planted on ridges down slope. Results showed that conservation measures must be directed towards the reduction of surface runoff during the first months after planting (Reining, 1992). In this context those cropping systems which reduced the quantity and velocity of runoff by physical barriers were the most efficient in erosion control.

Ruppenthal (1995) found rainfall erosivity (*R*) values between 9042 and 11,078 for Santander de Quilichao and between 5142 to 7548 MJ $h^{-1}mm^{-1}ha^{-1}year^{-1}$ for Mondomo and concluded in a preliminary approach that the USLE *R*-factor is a promising index. Correlation coefficients with soil losses for permanent bare plots ranged from 0.77 to 0.82 in Quilichao and from 0.85 to 0.90 in Mondomo. Based on these data, it was concluded that the potential of soil erosion in the South Colombian Andes is high.

Soil erodibility was found to be a dynamic property (Ruppenthal *et al.*, 1996). It is lowest after clearing of grassland or bush fallow but changes with time of soil use. Five years of permanent bare fallow led to a steady increase of the soil erodibility (*K*) factor from 0.004 to 0.015 in Quilichao and from 0.003 to 0.012 in Mondomo (Table 8.2). The USLE monograph proved not to

Table 8.1. Dry soil losses (t ha^{-1} $year^{-1}$) in Santander de Quilichao and Mondomo in different cassava-based cropping systems during 1990–1992 (Lal, 1992).

Cropping systems	Santander	Mondomo
Fallow	152.5	183.2
Contour ridges	3.0	2.0
Flat	7.8	16.4
Kudzu	9.7	n.d.
Zornia	14.2	3.8
Vetiver	1.3	2.6
Centro	7.0	2.1
Elephant grass	3.8	0.7

Table 8.2. Kinetic energy and erodibility as calculated from the USLE and measured in erosion plots at Santander de Quilichao and Mondomo.

	Santander		Mondomo	
	Kinetic energy MJ ha^{-1} year^{-1}	Erodibility	Kinetic energy MJ ha^{-1} year^{-1}	Erodibility
Computed[a]				
1987		0.080		0.080
1992		0.110		0.102
Determined in the field				
in 1987/88[b]	311	0.004	304	0.003
in 1988/89	512	0.013	488	0.010
in 1990/91	325	0.015	273	0.011
in 1991/92	359	0.018	256	0.012

[a] Equation: $K = 2.77 \times M^{1.14} \times 10^{-6} \times (12 - OM) + 0.042\,(A - 2) + 0.032\,(D - 3)$
OM, organic matter; *M*, soil texture index; *A*, aggregate class; *D*, permeability class.
[b] Plots established in 1986.
Source: Ruppenthal (1995), Ruppenthal *et al.* (1996).

be applicable to estimate the erodibility of inceptisols such as those of the Colombian Andes, because it greatly underestimated *K*-values (Ruppenthal, 1995; Ruppenthal *et al.*, 1996).

Agronomic evaluation of production systems in relation to crop productivity and sustainability

In 1994 CIAT initiated a strategic research experiment to generate principles for improved crop–livestock production and resource conservation. Treatments (systems) are based on the 'rooting ability' of components and range from short-cycle shallow rooted to deep-rooted perennial components (beans, cassava, forage and trees). Treatments were established on steep slopes (25–43%) using contour strips for crops (15–25 m wide), bordered by runoff channels protected by live productive barriers. Different soil-cover treatments are being used: monocrop (maize, bean, cassava), maize–legume, cassava–legume, cassava–maize–legume, pasture, legume trees, bare soil and soil covered with permanent mulch (Fig. 8.1). All treatments were fertilized with chicken manure and were planted either under fallow or continuous maize–bean cropping.

Initial results show that under previously cropped soil, maize production and nutrient uptake were greater than those of the previously fallowed soil, due to the prior heavy manure application (Table 8.3). Maize yields were also reduced when sown with a legume cover crop (*Arachis pintoi*) due to possible competition for water and nutrients.

Contrary to the common belief, a land fallowed for several years is not necessarily in better condition than one continuously cropped, but rather

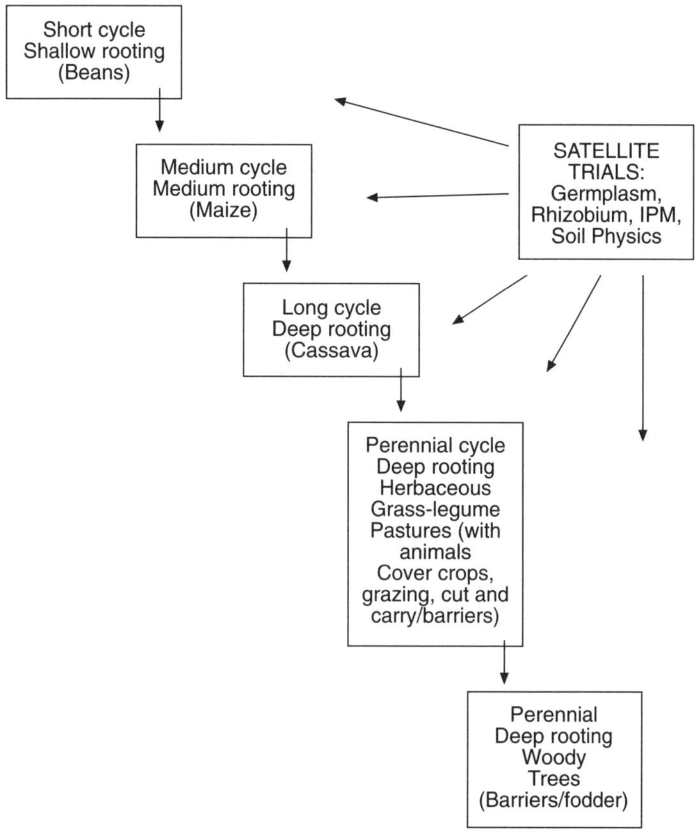

Fig. 8.1. Field prototype studies: crop type (Sanz, 1994).

depends on the management of the cropped land and on the quality of the biomass that cycles nutrients in the fallow. In Pescador, the continuously cropped land was fertilized with chicken manure for several years prior to the trial and received at the establishment of the trial a new application in the form recommended by the farmers in the region, namely 6 t ha^{-1}. The fallowed land had a plant cover fully dominated by the grass species *Melinis minutiflora* for several years, and received the same manure application as the non-fallowed land for our trial. The initial soil analysis showed greater nutrient availability in the non-fallowed land (Sanz, 1995).

It is clear that 6 t ha^{-1} of chicken manure, commonly used in the region, exceeds by far the total amounts of nutrients taken up by the above ground biomass and by even more the amount exported from the field. This adds to the abundance of available nutrients in the non-fallow land before fertilizing and results in a tremendous waste of nutrients and farmer's capital. Even in the fallow land, there were still nutrients in adequate amounts before fertilizing, particularly calcium. In both the non-fallowed and the fallowed land, the

Table 8.3. Nutrient balances in the first season of prototype systems on hillsides, in Pescador, Cauca, Colombia, in 1995.

		After no fallow							After fallow			
		Maize–legumes (yield = 2.7 t ha^{-1})				Maize (3.3 t ha^{-1})			Maize–legumes (1.7 t ha^{-1})			
	Fertilization:		Nutrient uptake[a]			Nutrient uptake				Nutrient uptake		
Nutrients	Chicken manure (6 t ha^{-1})	Initial soil	Total	% export[b]	Final soil	Total	% export	Final soil	Initial soil	Total	% export	Final soil
OM (%)	–	12	–	–	–	–	–	–	8	–	–	–
N (kg ha^{-1})	298	–	101	(47)	–	116	(51)	–	–	50	(45)	–
P (kg ha^{-1})	117	14	20	(69)	10	22	(74)	10	6	14	(71)	14
K (kg ha^{-1})	136	393	118	(20)	163	131	(19)	155	67	76	(22)	141
Ca (kg ha^{-1})	272	565	22	(7)	514	19	(7.5)	543	344	15	(5.2)	531
Mg (kg ha^{-1})	29	334	17	(32)	186	21	(44)	183	113	10	(33)	174
Zn (kg ha^{-1})	1.3	5.2	0.3	(46)	4.7	0.2	(50)	4.4	8.7	0.2	(42)	4.6
B (kg ha^{-1})	0.2	0.6	0.05	(20)	0.5	0.04	(25)	0.5	0.3	0.03	(0.0)	0.6

[a] Nutrient uptake by total above ground biomass.
[b] Percentage of taken up nutrients exported from the field in grain + cob + stover.

applied P which is not used by the crop becomes fixed in the soil and is not available as shown by the soil analysis after the harvest. The available cations K, Ca and Mg, particularly in the non-fallow, are lost in large amounts from the system, either by leaching or by erosion or both (Sanz, 1995). On the other hand, in the fallow the applied cations are found in the final soil analysis. Surprisingly, the uptake of nutrients by the maize crop in the fallow is lower than in the non-fallow when their availability is not actually a limiting factor. Perhaps the 4.1% difference in quantity of organic matter explains the grain yield difference in the maize crop between the two fallows (Sanz, 1995).

Soil losses in this experiment are being determined by the use of a micro-relief meter (Amézquita *et al.*, 1997). With this apparatus it is possible to determine the amount or depth of soil that is lost by measuring in selected places the changes of soil micro-relief between successive rainfall events. Figure 8.2 shows losses or gains of soil under different systems. The higher losses (up to 25 mm or 135 Mg ha^{-1}) occurred under beans and maize–legumes and the lower losses under grasses and bushes legumes. At the time of evaluation, plots with elephant grass were accumulating soil.

Fallowing with native vegetation species is a traditional land management practice throughout the tropics for restoration of soil fertility which is lost during cropping. The successful restoration of soil fertility normally requires a long fallow period for sufficient regeneration of the native vegetation. Increasing population pressure on limited agricultural land, however, has resulted in a reduction in length of fallows or increased use of fertilizers.

When purchasing power is low, one alternative to traditional fallows is managed fallows with plants that replenish soil nutrient stocks faster than plants in the natural succession. Short-duration fallows with planted trees and

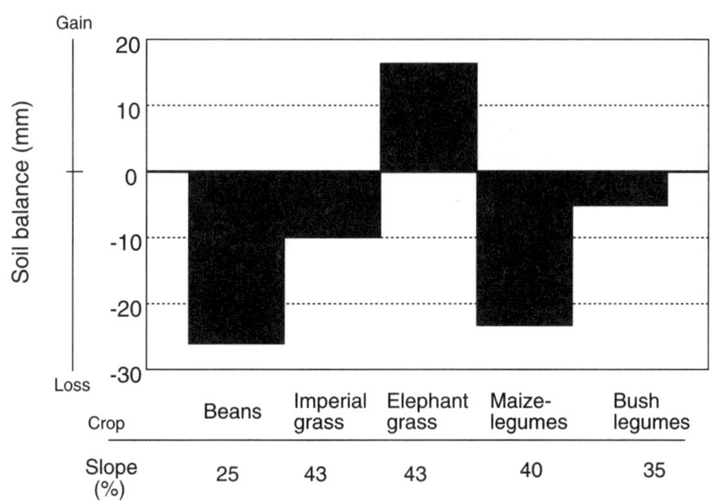

Fig. 8.2. Soil balance in various hillside production systems as determined by a relief meter between March and September 1995 in Pescador, Cauca, Colombia.

shrubs have the potential to restore soil fertility in N- and/or P-limited soils and upon harvest also provide a source of firewood. Two leguminous trees (*Indigofera constricta* and *Calliandra calothyrsus*) and a tropical shrub (*Tithonia diversifolia*) are being compared with natural fallows, in on-farm trials, for their potential to regenerate soil fertility faster than the native flora.

Rooting characteristics and root competition between crop and forage components could have important effects on yields and soil loss in hillside crop–livestock systems (Muhr *et al.*, 1995a,b, Tscherning *et al.*, 1995; Zhiping *et al.*, 1998). Root biomass distribution by soil depth (0–80 cm) was 44% greater in the top 0–10 cm of soil under cassava intercropped with cover legumes (*Arachis pintoi* and *Centrosema acutifolium*) than under the cassava monocrop (Table 8.4). However, root length density values were similar between the two systems indicating that roots of cover legumes were coarser than those of the cassava crop. Fodder grasses (elephant grass and imperial grass) growing as live barriers for 26 months presented marked differences in root growth and distribution (Table 8.4). The root biomass of elephant grass was 2.2-fold greater than that of imperial grass and provided greater protection to the soil against erosion. These differences in root growth could explain why it is possible to obtain more fodder cuttings per year with

Table 8.4. Differences in root growth and distribution among crop–forage systems grown in a volcanic ash soil of the Andean hillsides.

Soil depth (cm)	Grass	Root length density (cm cm^{-3})	Root biomass (kg ha^{-1})
0–10	Cassava monocrop	0.98	410
	Cassava intercrop	0.96	180
	Elephant grass	23.5	3561
	Imperial grass	5.04	1122
10–20	Cassava monocrop	0.13	260
	Cassava intercrop	0.36	280
	Elephant grass	14.8	2770
	Imperial grass	4.23	1331
20–40	Cassava monocrop	0.28	190
	Cassava intercrop	0.28	310
	Elephant grass	5.74	1610
	Imperial grass	2.26	1212
40–80	Cassava monocrop	0.09	140
	Cassava intercrop	0.09	140
	Elephant grass	2.97	1360
	Imperial grass	0.65	589
0–80	Cassava monocrop	0.31	1060
	Cassava intercrop	0.42	1510
	Elephant grass	11.6	9301
	Imperial grass	3.04	4253

elephant grass, which can explore greater volumes of soil for acquisition of nutrients (Zhiping *et al.*, 1998). Thus elephant grass can be used as an effective grass fodder barrier to reduce soil erosion in Andean hillsides.

Farmer participatory research approach

Agricultural research not only needs to be farmer oriented but farmer driven in order to ensure its continuous relevance to farmers' needs. Farmer driven implies that farmers should be involved in the decision-making process with respect to establishing research priorities, planning and conducting trials and evaluating results (Ravnborg, 1994). Local agricultural research committees (CIALs, Spanish acronym) can take responsibility for carrying out adaptive research trials which are statistically valid and address priorities defined by their community. CIALs can reduce the cost to the public sector of doing adaptive research (Ashby, 1994).

Based on the premise that local-level organization plays an integral part in successful watershed management, a study of the process of organizing for local-level management was carried out in a watershed (Rio Cabuyal) of 6500 ha (Ravnborg and Ashby, 1996). The experience suggests that despite the many features which complicate the process of watershed management, potential does exist for local people to take responsibility in this area if the existing structure of local power is controlled by identifying all stakeholders and ensuring their representation. Stakeholder analysis was a good tool to identify internal conflicts (Ravnborg and Guerrero, 1997).

Six functions for local-level watershed management organizations were identified as important by Ravnborg and Ashby (1995) based on the experiences of the Rio Cabuyal watershed:

1. Identifying stakeholders and ensuring their representation in management efforts.
2. Providing a forum for analysis and negotiation of diverse interests.
3. Defining rules and norms for the use of resources within the watershed.
4. Initiating a process of local-level resource-monitoring research.
5. Formulating and exerting demand for services from external institutions in support of local management efforts.
6. Negotiating internal versus external watershed interests.

Strengthening interinstitutional linkages

To alleviate poverty and to practise ecologically sound management of natural resources, communities have to establish agreements with individuals in order to avoid conflicts. Consortia should conduct an equitable negotiation between individuals and the needs of the community. CIPASLA (Spanish acronym for Interinstitutional Consortium for a Sustainable Agriculture in Hillsides) is a good example of such a consortium. Its mission is to develop collaborative activities with communities and other institutions to try to reach a better level of life for the inhabitants of the Rio Cabuyal microwatershed through a series of organized actions in a logical chain of sustainability (organization, training, research, production, conservation of natural resources, commercialization and transformation of products) by respecting the local and cultural values of the community (Beltran, *et al.*, 1996).

CIPASLA integrates seven government, six non-government and two international institutions which work with local organizations for a more integrated and productive management of the Rio Cabuyal microwatershed. Some of the actions that have been taken are listed below:

- Participation in planning
- Elaboration of tridimensional models of the watershed
- Study of land use changes
- Development of crop prototype systems for better utilization of hillsides
- Development of technologies for soil fertility management
- Development of local committees for research
- Reforestation
- Establishment of live barriers with farmers' participation
- Recuperation of small deposits of water
- Development of proposals for environmental curricular education
- Floristic characterization
- Establishment of rural agroindustries
- Marketing studies for agricultural products

GIS support systems

Geographical Information Systems (GIS) can deliver a wealth of complex land-use information in forms that are relatively easy to use and understand. Some of these products are being designed to give local communities and institutions a quantitative basis for their decisions about resource management. Others help influential people at the national and higher level to study resource issues and appraise the options for government decision. Through the use of GIS systems, it has been possible to produce thematic maps for different objectives, to bring maps up to date and to develop strategies for research. Today it is possible to have on a computer screen a detailed map of small areas showing land configuration for planning conservation practices.

Concluding Remarks

The steady degradation of tropical hillside soils is robbing small-scale farmers of their best hope for giving themselves and their children a better future. Crops and livestock may survive on these eroded steep slopes, but they cannot thrive. CIAT scientists are exploring various options that enable farmers to intensify agricultural production on tropical hillsides, while preserving fragile soils. Crop–pasture systems, particularly those including forage legumes and fodder grasses, increase crop and livestock production while improving soil quality through enhanced biological activity and physical structure. Developing sustainable production systems is ultimately a local challenge, requiring that researchers and farmers tailor new options to specific environments. To help them do this more efficiently, CIAT scientists act locally in collaboration with partners, while thinking globally. Through strategic research and farmer participatory studies at the watershed/community level, CIAT scientists are identifying basic principles that can guide the development of improved production systems and are helping to predict whether or not they will be sustainable under different biophysical and socio-economic scenarios. This is being achieved through an effective integration of GIS tools to encourage collective action.

References

Amézquita, E., Chavez, L.F. and Alvarez, A. (1997) Diseño y construcción de un microrelievémetro para evaluar la dinámica de la erosión en árez de ladera. *Revista Suelos Ecuatoriales* 27, 145–150.

Ashby, J. (1994) The CIAL method. In: *CIAT, Hillsides Program Annual Report 1993–1994.* Cali, Colombia, pp. 131–152.

Beltran, J.A., Hurtado, M. and Morales, M.E. (1996) *Consorcio internacional para una agricultura sostenible en laderas (CIPASLA). CIAT, Hillsides Program.* Cali, Colombia, 20 pp.

Jones, P. (1993) *Hillsides Definition and Classification.* CIAT, Cali, Colombia, 7 pp.

Lal, R. (1992) *Soil conservation in small-holder hillsides farming of the Andean inceptisols through cassava–forage legume intercropping.* Internal report on GTZ/CIAT cooperative project, 39 pp.

Muhr, L., Leihner, D.E., Hilger, T.H. and Müller-Sümann, K. (1995a) Intercropping of cassava with herbaceous legumes. I. Rooting patterns and their potential importance for below-ground competition. *Angewandte Botanik* 69, 17–21.

Muhr, L., Leihner, D.E., Hilger, T.H. and Müller-Sümann, K. (1995b) Intercropping of cassava with herbaceous legumes. II. Yields as affected by below-ground competition. *Angewandte Botanik* 69, 22–26.

Pachico, D., Ashby, J. and Sanint, L.R. (1994) Natural resource and agricultural prospects for hillsides of Latin America. Paper prepared for *IFPRI 2020-Vision Workshop Washington, 7–10 November.* In: *CIAT, Hillsides Program Annual Report 1993–1994.* Cali, Colombia, pp. 283–321.

Ravnborg, H. (1994) Organizing watershed user associations for farmer-drive research in the Andean hillsides. In: *CIAT Hillsides Program Annual Report 1993–1994.* Cali, Colombia, pp. 121–130.

Ravnborg, H. and Ashby, J. (1995) Organizing for local–level watershed management. Lessons from Rio Cabuyal watershed, Colombia. In: *CIAT Hillsides Annual Report 1994–1995.* Cali, Colombia, pp. 131–152.

Ravnborg, H. and Ashby, J. (1996) Organizing for local–level watershed management: lessons from Rio Cabuyal watershed, Colombia. *ODI, Agricultural Research and Extension Network. Network paper No 65,* 14 pp.

Ravnborg, H. and Guerrero, M.P. (1997) Stakeholder analysis in natural resource management. *ILEIA Newsletter* 13(1), 12–13.

Reining, L. (1992) *Erosion in Andean hillside farming. Characterization and reduction of soil erosion by water in small scale cassava cropping systems in the southern Central Cordillera of Colombia.* Center for Agriculture in the Tropics and Subtropics, University of Hohenheim, Germany, 219 pp.

Ruppenthal, M. (1995) *Soil conservation in Andean cropping systems: Soil erosion and crop productivity in traditional and forage–legume based cassava cropping systems in the south Colombian Andes.* Center for Agriculture in the Tropics and Subtropics, University of Hohenheim, Germany. 110 pp.

Ruppenthal, M., Leihner, D.E., Hilger, T.H. and Castillo, J.A. (1996) Rainfall erosivity and erodibility of inceptisols in the southwest Colombian Andes. *Experimental Agriculture* 32, 91–101.

Sanz, J.I. (1994) Prototype systems for ecologically sound intensification of production in the hillsides (Interprogram project). In: *CIAT, Hillsides Program Annual Report 1993–1994.* Cali, Colombia, pp. 105–110.

Sanz, J.I (1995) Prototype systems for ecologically sound intensification of production in the Hillsides (Interprogram project). In: *CIAT, Hillsides Program Annual Report 1994–1995.* Cali, Colombia, pp. 89–99.

Tscherning, K., Leihner, D.E., Hilger, T.H., Müller-Sümann, K.M. and Sharkawy, M.E. (1995) Grass barriers in cassava hillside cultivation: rooting patterns and root growth dynamics. *Field Crops Research* 43, 131–140.

Zhiping, Q., Rao, I., Ricaurte, J., Amezquita, E., Sanz, J. and Kerridge, P. (1998) Root distribution effects on nutrient uptake and soil erosion in crop forage systems on Andean hillsides. *Proceedings of the 16th International Soil Science Congress, Montpellier Cedex 5, France.*

9

Measuring Erosion as a Component of Sustainability

D.E. Swete Kelly[1] and A.A. Gomez[2]
[1]*Maroochy Research Station, Queensland Horticultural Institute, Nambour, Queensland, Australia;* [2]*Department of Agronomy, University of the Philippines Los Baños, College, Laguna, the Philippines*

Over the last four years, work in the Philippines and Australia has developed and evaluated sustainable farming systems for some important steep-land industries and communities. Two difficult issues in this work were: first, how to define what a sustainable farming system was, and second, how then could the sustainability of the system be measured. This issue of sustainability and its indicators has absorbed the talents and time of many professionals over the years and, although at times the arguments have been convoluted and confusing, some significant progress has been made.

Obviously, minimal erosion is one key ingredient in long-term sustainability. It is important for us, therefore, to monitor effectively the levels of erosion occurring and then define these levels in terms of their impact on long-term sustainability. However, when we came to identify simple yet effective methods to measure erosion on farm, very few seemed to be available. Techniques are well established to study erosion processes on small erosion plots with carefully controlled land management practices. However, no well-defined techniques were available which could be used cheaply and easily on farm. For our work, therefore, we used a range of techniques which seemed to best fit the situation at hand.

The techniques used included social approaches such as surveying of attitudes to and appreciation of erosion, as well as biophysical assessments. In this chapter we will first describe the techniques used to measure erosion on farm. Second, we will describe briefly the process used to interpret the figures obtained and incorporate these into an overall model of farming system sustainability.

 Soil Erosion at Multiple Scales (eds F.W.T. Penning de Vries, F. Agus and J. Kerr)

Introduction

Over the last four years, the Australian Centre for International Agricultural Research (ACIAR) has been supporting work to integrate resource conservation strategies into improved and hopefully more sustainable farming systems. This chapter draws on progress made within three linked projects:

- The major SEARCA & DPI project on 'Development and evaluation of sustainable production systems for steep lands' (ACIAR PN LWR-9220);
- the IBSRAM-led project 'Assessment of efficiency of farming systems on vertisols and sloping lands using a FESLM' (ACIAR 93/772); and
- the IBSRAM-led project 'Sustainability of farming systems in Fiji and Papua New Guinea' (ACIAR 95/126).

These projects aim to develop a technique which evaluates the sustainability of current and introduced farming systems. Being systems focused the evaluation needs to be broad based and take into account production, social and resource conservation impacts. To guide this process we use the evolving Framework for the Evaluation of Sustainable Land Management (FESLM), which identifies pillars of productivity, conservation, risk, viability and socio-cultural acceptability (Smyth *et al.*, 1993). Work on the FESLM has progressed on many fronts. The approach taken in this chapter is based on the definition that: 'a farming system is said to be sustainable at the farm level if it satisfies the farm manager's needs (over time) while conserving the natural resource' (Gomez *et al.*, 1996)

On this basis the FESLM pillars of productivity, viability, stability and socio-cultural acceptability are clearly linked as attributes which relate to the overall satisfaction of the farmer (or manager of the system). As such these pillars must always be defined, measured and interpreted relative to their impacts on the manager's expectations. This leaves resource conservation separate and able to be identified in more absolute terms.

Erosion in its many forms constitutes the major and most widespread form of resource degradation. Hence there is an obvious need to understand and monitor erosion. However, erosion is a complex and ongoing natural process. As such its occurrence and its impacts are site and scale specific. This workshop is all about defining the available techniques to measure erosion at these various scales. One important outcome should be to provide workers with a suite of possible erosion monitoring techniques which meet their needs for appropriate scale and accuracy. Unless appropriate techniques are defined then either much expense, time and equipment will be wasted collecting superfluous data, or insufficient information will be available to draw conclusions.

In our projects on farming system development it is not necessary either to understand the intricacies of the erosion process or to measure soil loss accurately. The need to minimize erosion is only one of the many important components of farming systems development. What then does farming system work need to know about erosion? First, we need to understand the relative level of risk that erosion poses for the target farming systems; second, we must ensure protective measures are integrated into the farming system at a level likely to provide protection given this risk; and third, we must monitor the impact of the improved system to ensure it is delivering the desired erosion outcomes. Therefore, what we need to know then are the risks, appropriate practices and impacts.

In relation to the theme 'assessing erosion' we require simple indicators, first of *risk* and second of *impact* at the farm level. To assess the level of risk we need to identify the key factors contributing to erosion; whereas to assess impact, we need to know the degrees to which erosion risk is affected with the implementation of new farming systems. In both cases we are not interested in a major study of the erosion process.

Simple factors such as rainfall volume and intensity, ground cover, soil type, slope, slope length, cultivation practices, visible erosion features and farmer experience all combine to provide a rough estimate of the level of erosion threatening traditional farming systems. With this information, we are able to arrive at an estimate of erosion risk – severe, medium, or low – that allows us to select the appropriate conservation practices to incorporate into improved farming systems. However, it would have been more useful to have a simple field tool that would provide risk assessment in a more standardized fashion.

Once the farming systems were designed and implemented, we needed simple tools to monitor their impact on erosion relative to the traditional practices. Techniques are well established to study erosion processes on small erosion plots with carefully controlled land management practices. However, no standardized techniques were available which were cheap and easy to use on farm. For our work, therefore, we used a range of indicators which seemed to best fit the situation at hand. These are listed in Table 9.1.

Table 9.1. Erosion indicators used in farming system studies.

Indicator	Unit of measurement	Related indicators
Rainfall (quantity and intensity)		Bed load
Runoff (quantity and intensity)		Suspended sediment
Surface cover	% ground cover (to 10 cm impeding runoff)	Depth of top soil (cm)
Slope length	m	Erosion breaks
Water-stable aggregates (WSA)	% grain size distribution < 0.125 mm (modified bottom withdrawal tube)	Erosion features: rills, gullies, etc
Water-dispersible aggregates (WDA)	% grain size distribution (wet sieving)	Riser height
Depositability	% soil aggregates with a settling velocity < 0.01 m s^{-1})	
Slope	%	

In monitoring the impact of improved systems on the level of erosion we used two techniques. For the Australian sites we followed the 'process technique' (Ciesiolka *et al.*, 1995). The approach aims to record a range of data related to erosion events and analyse results simultaneously in order to understand accurately the processes involved. For the Philippine sites we required much simpler techniques. The sites were quite isolated and we aimed at having up to ten grower cooperators at each site. The process technique would have been too expensive to install and maintain and too difficult to manage in this situation. Additionally, experience has shown that equipment was frequently stolen from the isolated on-farm sites. Instead we used a range of erosion indicators (referred to in this chapter as the 'indicator technique').

As work progressed, we were also interested in understanding which resource characteristics the communities we worked with saw as being important. Hence for one study in East New Britain (Papua New Guinea) we worked with the communities to define resource indicators in a participative manner.

Methodology

Process technique

This is an intensive monitoring technique whereby runoff from delineated treatments is channelled into steel or plastic troughs which collect soil bed load. Where steel troughs are used, runoff is channelled through a tipping bucket which measures the total runoff volume and rate at 1-min intervals. A small splitter sample is collected to determine suspended sediment. Added to this information is rainfall volume and intensity collected via a standard pluviometer. All data are logged on to data chips in the field and regularly downloaded. Bed load and suspended sediment samples are collected after every major erosion event. All downloaded data are analysed using the programs DATALOG and DATASUM. This method was used in two cropping situations in Australia. One involved testing minimum tillage practices in annual crops (beans), the other tested improved layouts and introduced groundcover plants in plantation crops (bananas).

Soil loss was calculated from both the amount of soil captured in troughs (bed load) and that suspended in a known volume of runoff water (suspended sediment). Results from the Australian sites indicate that soil loss from the traditional practice was 30 to 200% higher than the introduced practice. Enrichment concentrations of nitrogen, phosphorus and calcium were consistently higher in eroded soil compared with the top 10 cm of the profile.

The technique has the potential to provide detailed and accurate data that are useful in comparing alternative management systems, as shown in Table 9.2 for the banana site in Australia.

Our experience with this equipment has highlighted a number of issues:

1. The technique works well where trials are conducted on research managed plots or for comparing the performance of farming systems that differ in

Table 9.2. Impact of best practice treatments on a range of factors during summer 1996 (bananas, Mullumbimby)

	Conventional	Best practice
Runoff	1.03 Ml ha^{-1}	0.12 Ml ha^{-1}
Runoff as % of rainfall	6	0.7
Soil loss	15 t ha^{-1} +	0.2 t ha^{-1}
Nitrogen lost in runoff	10.4 kg N ha^{-1}	0.7 kg N ha^{-1}
Phosphorus lost in runoff	24.5 kg P ha^{-1}	1.7 kg P ha^{-1}
Nitrogen lost in eroded soil	3.2 kg N ha^{-1}	0.02 kg N ha^{-1}
Phosphorus lost in eroded soil	4.3 kg P ha^{-1}	0.04 kg P ha^{-1}

a few simple components. It worked well for the bean industry and banana industry trials in Australia as the cropping systems are simple and standardized. As the farming system becomes more complex (due to increased crop options or planting times) and as testing is conducted more participatively on farm, then many of the data collected are difficult to extrapolate because of the uniqueness of each system studied.

2. Farming system development work will often involve a significant number of farmer cooperators (six to nine at each of our Philippine sites) and the cost of installation and management can be prohibitive. Even in Australia, where each cropping system has three sites, the cost of installation, management and analysis using this technique was the largest budget item for the project. Given that erosion is only one component of the assessment of the improved farming system, a cheaper approach would have been preferable.

3. The technique has the ability to produce extensive data sets. These often become difficult to manage and interpret without skilled assistance. A lot of time is required in validating each data set before further analysis as simple errors due to logger or operator malfunction commonly occurred.

4. The use of the technique in short-term annual cropping patterns (e.g. beans in Australia) required the sites to be set up each year. This was costly and the short 10- to 12-week growing season often did not coincide with significant rainfall events, limiting data considerably. The technique was much more useful in perennial cropping systems.

5. An adaptation of Murphy's law applies to the data logging equipment – 'If the logger is down it will rain'.

6. The technique provides very accurate and integrated data for the plot studied. However, extrapolation to other similar farms is risky. Our work may have benefited from a parallel but simpler technique whereby less accurate data from more farms could have been collected.

7. Many of the data sets collected are not directly applicable to farming system development. It will be very useful to use these for erosion modelling work, but the impact of the improved systems on reducing erosion could have been obtained more simply.

8. The technique has provided very good data on nutrient loss in runoff.

These data have been very useful in helping these two Australian industries identify potential off-farm impacts.

Indicator technique

For the Philippine sites we decided to measure a range of simple erosion indicators. Although each measurement is less accurate, we were able to take a large number of measurements at many more sites, thus greatly improving the reliability and precision of the computed averages. The indicators used included:

- rills and gullies;
- erosion breaks and riser height;
- soil surface level;
- soil physical characteristics; and
- farmer-selected indicators.

The main Philippine sites studied are at Halang (Luzon) and Ned (Mindanao). Halang has a shallow degraded soil and is subject to a distinct dry season. Ned has a deeper top soil and a relatively even rainfall. At both these sites the traditional system uses few conservation practices. The improved systems introduced in these areas are shown in Table 9.3. The two upland Cebu sites have a range of conservation practices in place and offered the opportunity to rate the impact of these practices over a longer time period.

Table 9.3. Characteristics of the Philippine steep-land farming systems studied.

Name	Slope class (%)	Cropping pattern	Conservation treatment
Cropping pattern 1 (CP 1)	35–70% with permanent ground cover	Perennial tree crops	Legume cover crop Vetiver bunds around trees
Cropping pattern 2 (CP2)	20–35%	Fruit trees in hedgerow Traditional annual cropping during fruit tree establishment	Hedgerow (legume or grass) Mulch from pruning Contour cultivation Reduced tillage
Cropping pattern 3 (CP3)	< 20%	High-value annual alley cropping in rotation with traditional crops	Hedgerow (legume or grass) Contour cultivation Drainage for bedded crops Integrated water management Reduced tillage

Results

Rills and gullies

Rills were observed in all conventional plots and treatments using annual crops and hedgerows. Rills were not recorded where cover crops or ground covers were used (CP 1, Table 9.4). Most rills occurred earlier in the project. Reasons for this include the following:

1. As barriers establish, the opportunity for erosion rills obviously declines. The project specifications required the replanting of all hedgerow gaps to ensure a continuous barrier.
2. Terrace formation between hedgerows reduces slope over time resulting in less rill erosion.

Grass hedgerows had consistently lower numbers of rills developing than shrub legume hedgerows. One newly planted site at Ned recorded 14 rills in the grass barrier and 74 in the adjacent legume hedgerow. Additionally, the rills in the legume hedgerow system were on average four times larger than rills which developed below the grass hedgerow.

Rill erosion was also greater when the contour was inaccurately laid out or where the distance between contour barriers was large. A deviation from the contour of more than 2% was sufficient to allow surface water flow to break through barriers and cause rill development. This occurred in all slope categories at both sites.

Hedgerow spacing in annual cropping systems ranged from 6 to 8 m. These spacings reflect the district/farmer consensus on minimum sizes suitable for efficient management. Wider spacing on steeper slopes led to more rill erosion. Once hedgerow barriers were fully established, the impact of contour accuracy and alley width was reduced. Our experience in measuring rills and gullies suggests the following.

1. They need to be measured after each event. This may take time and in some cases data were lost when the farmer had already cultivated the field.
2. The impact of rills on erosion is difficult to quantify in terms of soil loss, but still they were useful as a guide. The work of Karl Herwig (1996) was useful in streamlining the technique.

Table 9.4. Averaged rill data for three cropping patterns at two sites in the Philippines.

Site	No. of rills	Width (cm)	Depth (cm)
Halang	35	5.35	2.6
Ned	57	6.85	3.1
CP 1	0	0	0
CP 2	39	4.2	1.75
CP 3	53	8	4

Although the results are highly variable, the average over several farmers conforms with our expectations, namely higher ratings for Ned than Halang and for CP 3 (annuals) than CP 2 (annuals and perennials).

Erosion breaks and riser height

Where vegetative contour barriers were installed, observations were made on the riser height and the number of erosion breaks. An erosion break was defined to have occurred when a rill or gully cut through the vegetative barrier. Recordings were made after each significant rainfall event (> 25 mm in any 24-h period). Data on erosion breaks are shown in Table 9.5.

During development of improved farming systems in the Philippines, risers developed below each contour barrier. This was primarily aided by cultivation, but erosion also had an impact. The heights of the risers formed over a 2-year period are shown in Tables 9.5 and 9.6. This was expected and confirms the suitability of contour barrier technology in annual cropping systems on sloping land.

The differences in the rate of riser development reflects site characteristics. Ned has a deep friable soil. Halang is eroded, with exposed subsoil and parent material. Riser development shows the relative ease with which the Ned soil is moved through erosion and cultivation.

The rapid soil deposition associated with the development of risers may impose limitations on many of the hedgerow species currently used. This

Table 9.5. Hedgerow performance to 24 months after establishment.

Vegetative barrier species	Site	Height of risers (cm)	Erosion breaks (per 100 m of hedgerow)
Legume	Halang	19	1.02
	Ned	75	0.24
Grass	Halang	11	0
	Ned	53	0

Table 9.6. Average height of risers for three cropping practices at two Philippine sites after 2 years.

Site	Cropping practice	Average height (cm)
Halang	CP1	13
	CP2	22
	CP3	17
Ned	CP1	70
	CP2	80
	CP3	55

applies particularly to some shrub legumes which are already stressed by a regime of continual pruning. The grass species in the trial are more resistant to these high levels of soil build up. The stoloniferous *Arachis glabrata* shows potential as a species for situations where riser development is occurring rapidly.

Our experience in measuring erosion breaks and riser height suggests the same issues as those for rills and gullies. While the data are variable and are not accurate enough to be translated into soil loss, they are, nevertheless, able to distinguish the comparative magnitude of soil movement across two sites and three farming systems.

Soil surface level

Erosion pins were established on plots at Halang and Ned in the Philippines. Each plot ranged in size from 500 to 1200 m^2 and was relatively uniform in both slope and soil type. Pins made of 10-mm steel reinforcing rod were cut to 1.8 m and hammered into the soil. More than 1 m was left exposed above ground and a mark was etched at the 1-m level. Plot treatments were farmer managed and included a range of grass or legume contour barriers, complete living mulch cover and traditional unprotected cultivation. Pins were placed at about 10-m spacings across the slope and at 5- to 10-m spacing down the slope (Fig. 9.1). Slopes ranged from 25 to 50%. Where contour structures occurred, pins were placed in the contour strip.

To enable measurement, a nylon string was stretched from a pin at the top of the slope to a pin at the bottom of the slope and then connected to each pin in between. In each case the nylon string was attached at the original 1-m marking above the ground. Measurements were made between the string

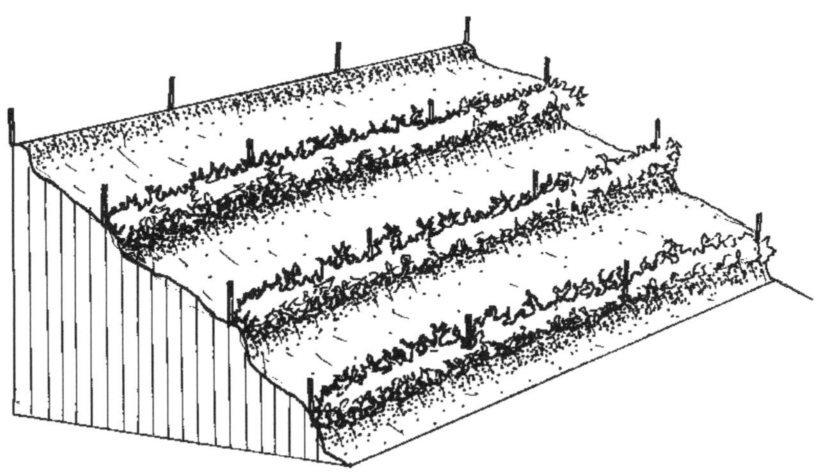

Fig. 9.1. Placement of erosion pins.

and the soil at each pin and then 1-m uphill and 1-m downhill from the pin, thus avoiding any impact the pin itself would have on erosion or cultivation. Measurements were made at the end of the cropping cycle to reduce the impact of cultivation or soil disturbance.

This method measures soil movement and not necessarily soil erosion. It is important to understand how sensitive the procedure is in detecting changes in soil height. Table 9.7 shows the number of measurements that must be made to detect a given level of movement (measured in centimetre increase or decrease in soil depth). The change in soil depth which can be detected is also affected by the number of plot replications.

As can be seen from Table 9.7, the greatest sensitivity achieved for the technique was 0.8 cm of change in soil surface level. If all this soil surface was lost due to erosion (which is doubtful), this would equate to 80 t of lost surface soil per hectare.

Overall the use of pins was not sufficiently sensitive to measure soil movement due to erosion in the short term. In the 2 years of measurement so far, movement due to cultivation was the major cause of change in soil surface level. It was impossible to identify what contribution erosion may have had. The technique is easily and cheaply installed and monitoring and maintenance is not a difficult task. As the pins are established in the contour barrier they should not impede cultivation or be subject to disturbance. Pins are most useful in plotting the development of terraces in fields where permanent contour barriers have been established.

Farmer-selected indicators

In the majority of our studies we have come into a situation and assessed the characteristics of the farming system which we, as researchers, considered relevant to long-term sustainability. It became apparent that the farming communities themselves not only were valued suppliers of information but might have some insights into the criteria they considered essential to what constitutes a 'good' farm. We therefore decided to ask the communities themselves what characteristics they considered important. This study was conducted in

Table 9.7. Change in soil depth (in cm) detected by pin measurements over a given number of replicates and samples per replicate.

	No. of plot replications				
No. of pins	3	4	5	6	8
10	1.72	1.49	1.32	1.22	1.06
20	1.46	1.26	1.13	1.03	0.89
30	1.36	1.18	1.05	0.96	0.83
40	1.31	1.21	1.01	0.93	0.80

Computed from estimates of experimental error = 50.88; sampling error = 3.86.

three villages in East New Britain. The approach did not prompt the farmers with erosion, production or any other characteristic. Farmers instead supplied characteristics they considered important to a 'good' or 'bad' farm. The process is described in Swete Kelly (1997). The characteristics identified added another dimension to the assessment of overall farming system sustainability. Men's and women's responses were sought separately. Indicative outcomes for one village are shown in Table 9.8.

Erosion was not mentioned by any of the communities even though two of these (Malasait and Vunamami) had gardens on some very steep land. However, factors related to erosion were raised. Issues such as 'soil colour', 'soil texture', 'slope' and 'stoniness' may all relate to erosion, but they were not mentioned in that context. For example, heavy clay soils were identified as poor garden areas and often related to the gardens which had been cropped without fallow over a long period. However, the main reason for raising the issue of heavy clay related to its impact on workability. The point here is that erosion impacts are primarily important for their effects on management. Possibly, in future, our focus should be to emphasize these management factors rather than pushing a somewhat abstract and incipient process we call erosion.

The factors identified by these communities were added to the researcher-defined indicators and assessed on farm.

Erosion, sustainability and indicators

For farming system development work, erosion is one component of a multifaceted evaluation. The key objective for erosion is minimization, which contributes to the conservation of the soil and to the overall sustainability of the system. In this section we introduce the concept and procedure for measuring sustainability and relate this process to that of soil erosion.

Indicators are measurable characteristics that mirror the behaviour of the attribute being evaluated. For sustainability, the possible indicators are far too

Table 9.8. Farm characteristics mentioned by farmers as important features of a 'good' farm.

Men (n =15)		Women (n = 12)	
Farm characteristics	Rating (%)	Farm characteristics	Rating (%)
Planting time	34	Soil colour and texture	52
Soil colour and texture	33	Distance	49
Garden size	31	Fallow length	42
Drainage	26	Slope	36
Mixed crops	25	Stoniness	21
Soil fertility	25		
Weediness	24		
Slope	22		

many and too tedious to measure completely. Much effort can be saved if a few can be selected to represent the larger set. In selecting the representative indicators we used the following guidelines.

1. They must represent the major components of sustainability, namely chemical, physical and management-related properties of the natural resource and viability, productivity and stability for farmer satisfaction.
2. They should be sufficiently sensitive to changes in the management and environmental factors as shown by the size of their variance.
3. They should have a high correlation with indicators in the same component and a low correlation with those in other components.
4. They should be easy and cheap to measure.

To test the selection process at the farm level, we used 21 indicators for resource conservation and 12 indicators of farmer satisfaction. Shown in Table 9.9 are the variance and correlation ratings of the selected representative set for resource conservation and farmer satisfaction.

While additional data may be needed to establish the general applicability of these selected indicators, it is interesting to note that the sustainability rating of the representative set coincides very closely with that using the complete set of 33 indicators. On this basis, we will use this selected set to

Table 9.9. Indicators for sustainability and their corresponding variance and correlation ratings.

Indicators	Coefficient of variation	Correlation index	
		Within group	With others
For resource conservation			
Chemical properties			
Organic carbon (%)	20	19	1
Cation exchange capacity	14	13	2
Physical properties			
Soil depth	24	1	1
Moisture content at 1/3 bar	9	8	3
Water-dispersible aggregates (%)	8	2	9
Management-related properties			
Permanent ground cover (%)	49	0	6
For farmer satisfaction			
Productivity			
Gross return (US$ ha^{-1})	79	12	6
Viability			
Labour cost (US$ ha^{-1})	73	12	5
Material cost (US$ ha^{-1})	83	12	3
Net return (US$ ha^{-1})	160	11	7
Stability			
Cropping diversity (No.)	33	11	2

demonstrate the process by which the sustainability of farming systems can be represented and compared.

Measuring sustainability

To measure the overall sustainability of a farming system a threshold must be determined by which the current practice and potential changes can be compared. We looked at two alternative approaches for specifying the threshold level. First, that a threshold is an absolute and must be constant over all locations. This seemed sensible for biological or physical characteristics (e.g. organic carbon or soil depth), but it was not at all appropriate for social or economic factors. Second, that the threshold is location specific and must be different for each resource or community. We felt that the second is the more logical as farmers generally judge their satisfaction upon their relative position in the local community (i.e. they compare themselves to their neighbours). Additionally, base resource condition varies and the impact of time or changed practices must be gauged relative to the original resource endowment. Thus we opted to use the average of similar farms in the community as the threshold for all indicators.

The procedure for computing the sustainability of a given farming system for a particular farm is as follows.

Step 1. Measure the selected indicators for farmer satisfaction and resource conservation.
Step 2. Compute the average of each indicator across all farms in the community and designate this average as the threshold level, that is, the level whereby the farm manager judges his success in managing his farm and protecting the available resource. Summarized resource conservation data for farms at Ned are shown in Table 9.10.
Step 3. Convert all values of the indicators measured in step 1 into threshold units by dividing each value with the threshold level (Table 9.11). In the case of negative indicators, such as water-dispersible aggregates (WDA) (where an increase in level is considered detrimental), the threshold should be divided by the actual value.
Step 4. Compute the index for resource conservation by averaging the threshold units (last column of Table 9.11).

At the end of this process one should have threshold units for each of the selected indicators and an overall index for farmer satisfaction and resource conservation. If either of these indices is less than one, then the sustainability of the system is questioned and remedial action should be required. If both are above one, then the systems are more sustainable than the community average. In both situations the trend over time is also important. In the case shown in Table 9.11, the improved farming system introduced at Ned shows a marginal improvement for resource conservation in the 2 years the project has been conducted. However, the process remains transparent allowing workers to identify the areas of prime improvement or loss.

Table 9.10. Resource conservation indicator values for improved farming system and the community average (Ned, Mindanao).

	Resource conservation					
	Organic carbon	1/3 Bar	Water-dispersible aggregate	Soil depth	Ground cover	Cation exchange capacity
Improved farming system	2.96	20.95	21.6	17.8	18	12.91
Community average	3.73	23.37	20.37	16.89	9.75	16

Table 9.11. Indices of sustainability represented in threshold units.

	Resource conservation						
	Organic carbon	1/3 Bar	Water-dispersible aggregate	Soil depth	Ground cover	Cation exchange capacity	Average
Improved farming system	0.79	0.89	0.94	1.06	1.85	0.8	1.06
Community average	1	1	1	1	1	1	1

These results can also be represented graphically as shown in Fig. 9.2. This graph is a good tool to identify immediately the specific component indicators that contribute to sustainability. Individual workers can easily see the relative contribution of each index and draw their own conclusions based on their perception of the importance of each.

Conclusions

Obviously, minimal erosion is one key ingredient in long-term sustainability. It is important for us, therefore, to monitor effectively the levels of erosion occurring and then define these levels in terms of their impact on long-term sustainability. Although techniques are well established for studying erosion processes on small erosion plots with carefully controlled land management practices, we could not find standard techniques that are cheap and simple enough for use in on-farm trials involving several farmer cooperators. For our work, therefore, we used a range of techniques which seemed to best fit the situation at hand. Although these techniques are not accurate enough to measure the amount of soil loss due to erosion, we found them sufficient for farming system development work where an indicative measure is all that is usually required.

The procedure for measuring sustainability of farming systems has been developed from a definition of sustainability at the farm level based on

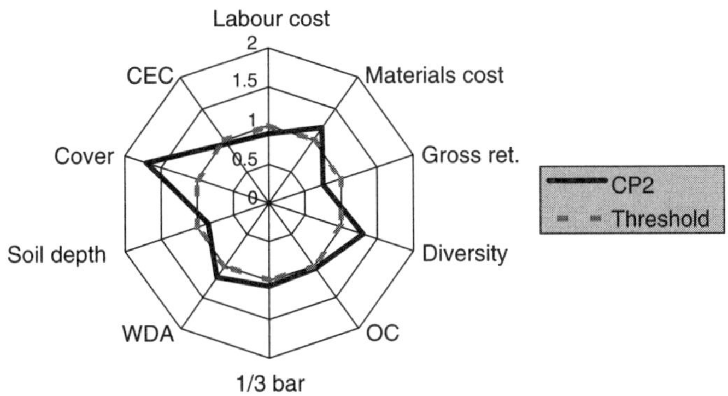

Fig. 9.2. Radar graph showing the threshold or trigger line and the comparatively sustainable situation in the improved cropping pattern (CP2). CEC, cation exchange capacity; OC, organic carbon; WDA, water-dispersible aggregate.

farmer satisfaction and resource conservation. Erosion is obviously an important component of resource conservation, but its importance must be kept in perspective when considering the overall sustainability of the farming system. Unless all aspects of resource conservation and farmer satisfaction are considered, overall long-term sustainability cannot be assured. A process to screen potential indicators for sustainability using variance and correlation analysis has shown initial promise but needs further validation.

Acknowledgements

This chapter draws on work completed as part of three ACIAR-funded projects. These include two projects led by IBSRAM. The input of Keith Syers and Tony Dowling as leaders of these projects is acknowledged.

The assistance of Cyril Ciesiolka (Department of Natural Resources, Queensland) in the design, implementation and interpretation of erosion monitoring layouts used in Australia is greatly appreciated.

References

Ciesiolka, C.A., Coughlan, K.J., Rose, C.W., Escalante, M.C., Hashim, G.M., Paningbatan, Sombatpanit, S. (1995) Methodology for a multi-country study of soil erosion management. *Soil Technology* 8, 179–192.

Gomez, A.A., Swete Kelly, D.E., Syers, J.K. and Coughlan, K.J. (1996) Measuring sustainability of agricultural systems at the farm level. In: Doran, J.W. and Jones, A.J. (eds) *Methods for Assessing Soil Quality. SSSA Special Publication No 49.* Soil Science Society of America, Madison, Wisconsin, pp. 401–410.

Herwig, K. (1996) *Field Manual for Assessment of Current Erosion Damage.* Soil Conservation Research Program, Ethiopia and Centre for Development and Environment, University of Berne, Switzerland, 69 pp.

Smyth, A.J., Dumanski, J., Spendjian, G., Swift, M.J. and Thornton, P.K. (1993) *FESLM:*

An International Framework for Evaluating Sustainable Land Management. FAO Discussion Paper, FAO, Rome.

Swete Kelly, D.E. (1997) *A Process to Identify Indicators of Sustainability of Importance to the Farmer.* ACIAR 95/126 Working Paper, Canberra, Australia.

10

Agronomic Consequences of Soil Erosion

R. Lal

School of Natural Resources, The Ohio State University, Columbus, OH 43210, USA

Agronomic effects of soil erosion, impact on actual and potential productivity, are complex and governed by numerous interacting factors that may be: on-site where erosion occurs or off-site where runoff and sediment enter another field, direct due to damage to crops/plants or indirect due to other factors that affect plant growth, short term that alter present productivity or long term that affect potential productivity. Agronomic impact assessment of erosion may also involve measurements to be made at different temporal or spatial scales. In terms of temporal scales, agronomic productivity may be assessed as a result of past, present or future erosion rates. In terms of spatial scales, erosion impact may be assessed at soilscape, landscape, farm, watershed, geographical, regional or national scales. Methods of assessment of agronomic productivity differ depending on the specific nature of the impact to be assessed and the scale at which measurements are to be made. In addition, erosional impact on productivity is governed by soil profile characteristics, weather during the growing season, and management. Establishing a reference state, uneroded soil, is important. The research data generated on erosional impact on productivity must, therefore, be supported by the information on soil quality, weather conditions, farming/cropping systems, scale of measurement, and the nature of the impact assessed (e.g. on-site versus off-site, direct versus indirect, the time horizon, and spatial scale). Unless the objectives and methods used are clearly stated, results obtained using different approaches are not comparable and lead to conflicting views and debatable conclusions. There is a strong need to standardize methods, establish long-term field experiments on benchmark soils in principal ecoregions, and develop scaling procedures to extrapolate the data from soilscape to regional and global scales.

 Soil Erosion at Multiple Scales (eds F.W.T. Penning de Vries, F. Agus and J. Kerr)

Introduction

The relative magnitude of on-site versus off-site effects of soil erosion at different scales (e.g. local, regional and global) has become a debatable issue. There are two schools of thought. One school affirms that the on-site effects are of a little consequence and the principal adverse effects of accelerated soil erosion are off-site (Crosson, 1998). The proponents of this school argue that on-site effects of accelerated soil erosion are easily compensated by soil and crop management options and additional inputs. Further, the land manager or farmer has vested interests in looking after his/her soil resources and in managing soil erosion. The second school argues that on-site effects of soil erosion are extremely severe and a major economic concern at all scales (Brown, 1995; Pimmentel *et al.*, 1995). The severity of adverse on-site effects is attributed to the global shortage of prime arable land, use of marginal or fragile soil resources, harsh climate, and non-availability or non-affordability of the essential inputs needed. Resolution of this debate is important to the development of a sound policy for sustainable management of soil and water resources. An objective resolution requires a thorough knowledge of the magnitude of on-site versus off-site effects, the cause–effect relationship, and underlying principles of methods used in assessing these effects.

The objectives of this chapter are to: (i) explain on-site versus off-site effects, (ii) describe methods of assessing the on-site and off-site impacts on agronomic productivity, (iii) explain possible sources of error and confusion, and (iv) identify researchable topics.

On-site Versus Off-site Impacts

On-site effects are those that happen at the site where erosion occurs. These effects include two primary consequences of accelerated erosion: (i) short-term/immediate or direct effects on plant growth, such as seed/plant uprooting and washing away and loss of water and fertilizer in runoff and sediments; and (ii) long-term/potential or indirect effects due to adverse changes in soil quality (Fig. 10.1). The immediate or direct effects usually involve damage to the crop on the ground. The long-term effects involve changes in soil quality that impact growth and productivity of crops to be grown in the future. Adverse on-site effects on soil quality include: (i) decrease in effective rooting depth, (ii) reduction of plant-available water capacity (AWC), (iii) decrease in soil organic carbon (SOC) content, and (iv) exposure of the relatively infertile subsoil with unfavourable soil physical properties.

Off-site effects are those that occur when runoff and sediments from one field, watershed or waterway enter another. The off-site effects are complex and involve at least three categories: (i) damage to present and future crop growth, (ii) adverse changes in environment, and (iii) damage to civil structures and other industries (Fig. 10.2). The off-site effects on the crop/plants growing at the time may be due to inundation, and siltation or deposition, and on the growth and productivity of future crops due to long-term changes in soil quality. The latter may be positive or negative.

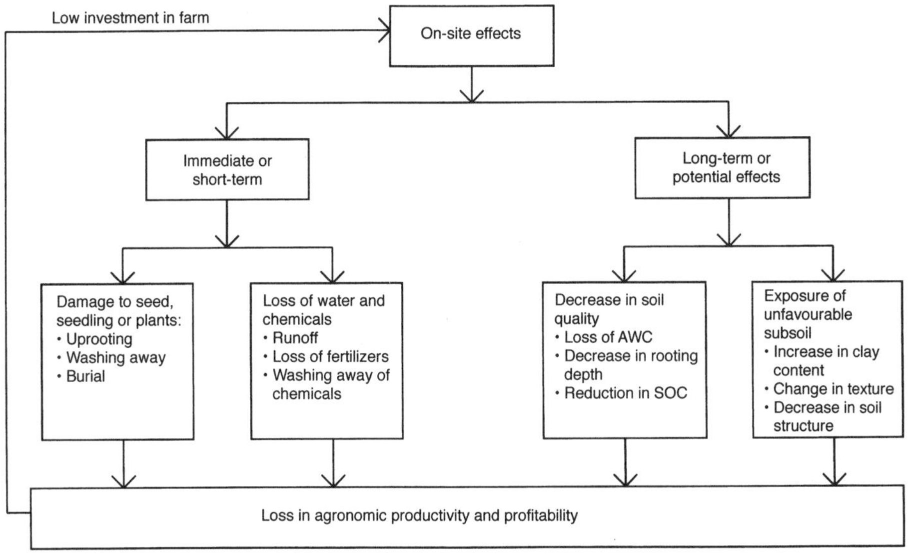

Fig. 10.1. Types of on-site effects on agronomic productivity. AWC, available water capacity; SOC, soil organic carbon.

The agronomic effects on productivity, therefore, involve two distinct components. One component is due to the on-site effects as described in Fig. 10.1. The second component is the off-site agronomic effects as described in the first module of Fig. 10.2. While some research information exists with regard to the on-site agronomic effects for a few soil types, there is little information on the off-site agronomic effects of soil erosion.

A principal cause of the debate and confusion is the lack of understanding of the processes involved and scarcity of quantitative information of all aspects of on-site and off-site effects identified in Figs 10.1 and 10.2. Credible information is particularly lacking on the on-site and off-site effects on present and potential productivity for different soils, crops and cropping systems and management scenarios.

Possible Sources of Error

There are numerous sources of error and/or missing information that accentuate the confusion about the magnitude of the agronomic effects of erosion. Principal among these are briefly described below.

Magnitude of soil erosion

The cause–effect relationship between soil erosion and agronomic productivity can only be established by obtaining reliable information on both dependent (crop yield and income) and independent (soil erosion) variables. Similar to

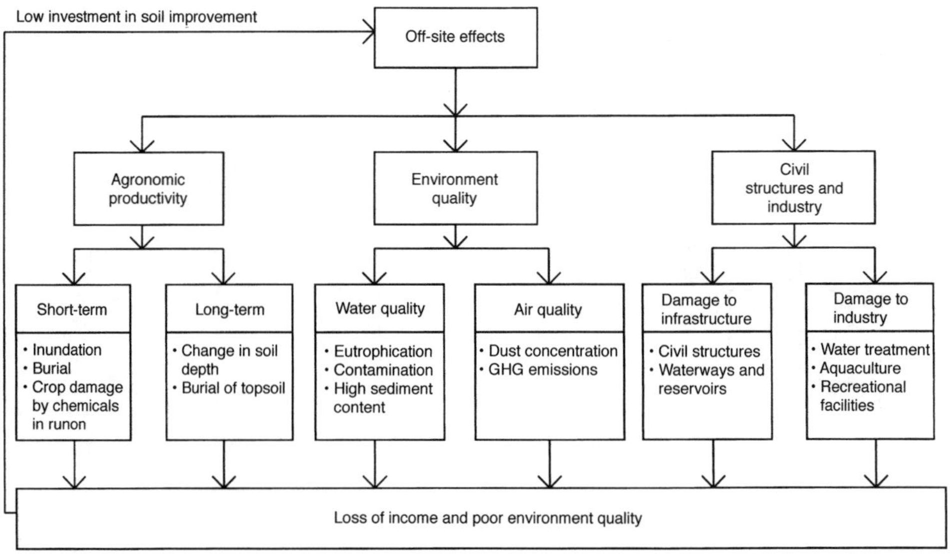

Fig. 10.2. Off-site effects of soil erosion on agronomic productivity, environment quality and civil structures and industry.

the lack of quantitative data on agronomic productivity, the magnitude of soil erosion for principal soils and ecoregions is also not known. The available information on the magnitude or severity of soil erosion, voluminous and often replete with rhetoric, is confusing, qualitative, incomplete and unreliable (Fig. 10.3). It is confusing because of the diverse terminology used, such as soil loss, erosion, sediment yield, and denudation rate. These terms need to be defined, the evaluation procedures of the underlying processes standardized, and their comparability refined by providing supporting information on delivery ratio. Information on delivery ratio is scanty, and difficult to obtain especially for soils of the tropics. Further, the available data on soil erosion are often qualitative, descriptive, highly subjective and usually based on casual observations rather than systematic experimentation. Research information on soil erosion is incomplete unless it has supporting information on soil profile and landscape characteristics, climate and its erosivity, land use and cropping/farming systems along with inputs used. The information is also unreliable because it is often obtained by unstandardized methods, such as runoff plots and collection/monitoring systems of varying shapes, sizes and of different degrees of automation, and of low reliability because of the poor design.

The information on soil erosion is also erratic because of the lack of scaling procedures. It is difficult to aggregate the data from point or field scale to landscape, watershed, ecoregional and global scales.

The confusion regarding the agronomic impact is exacerbated by the lack

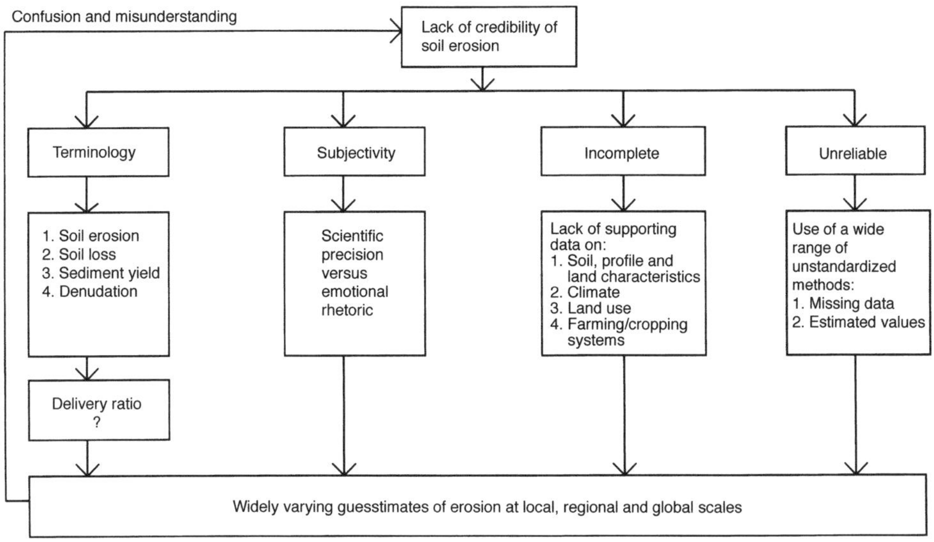

Fig. 10.3. Reasons for the lack of credibility of available data on the magnitude of soil erosion.

of distinction among different temporal scales of assessing soil erosion. Soil erosion has three distinct temporal scales: past, present and future. While assessing the on-site agronomic impacts, it is important to make distinction between the past cumulative soil erosion, present soil erosion rate and the future projected erosion rate. Methods of assessment of soil erosion for different temporal periods are different (Table 10.1). Use of these methods requires careful analyses of the objectives. Soil erosion data are often reported without clearly stating the objectives of measurements, the methods used and the temporal scales involved.

Soil loss tolerance

The information on agronomic effects of soil erosion is also incomplete without the supporting data on the soil loss tolerance. The information on soil loss tolerance is not available, especially for the most fragile soils in harsh environments and in the ecologically fragile ecoregions. The widely used figure of 5 t acre^{-1} year^{-1} or 11.2 Mg ha^{-1} year^{-1} is too high, especially for many intensively used but shallow soils of the tropics and subtropics. Lal (1984, 1985) observed that for some shallow alfisols, with root-restrictive layers at 20 to 30 cm depth, soil loss tolerance was as low as 1 Mg ha^{-1} year^{-1}. In contrast, soil loss tolerance for deep and fertile soils (e.g. soils derived from loess, alluvium or volcanic ash parent materials) may be even higher than 11.2 Mg ha^{-1} year^{-1}. The magnitude of soil loss tolerance depends on numerous interacting factors (Fig. 10.4). Important among these are: (i) rate of new

Table 10.1. Methods of soil erosion assessment in relation to temporal attributes to quantify the on-site agronomic impacts.

Past cumulative soil erosion	Present erosion rate	Future erosion rate
Topsoil depth	Runoff plots	Predictive models
Depth to root-restrictive layer	Field methods, e.g. buried nail, rill volume	Assess soil life
Soil quality attributes	Paired watershed monitoring techniques	
Tracer techniques, e.g. ^{137}Cs, Pb, As, flyash	Rainfall simulation	
Reservoir sedimentation	Predictive models	
Tree roots exposure, fence post burial		
Models		

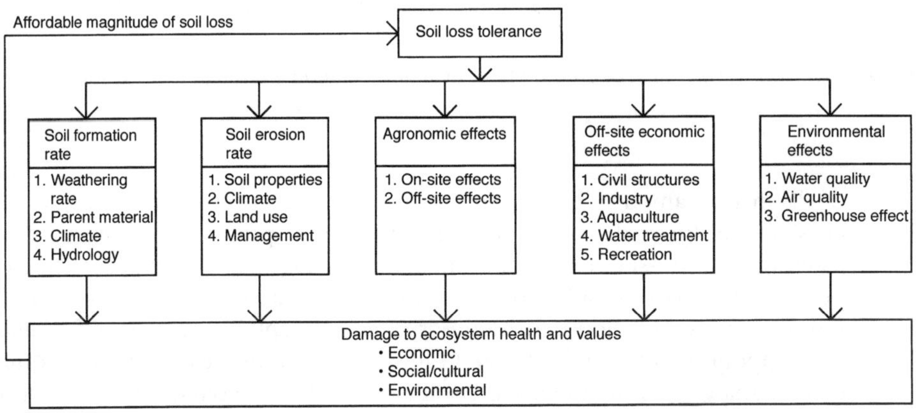

Fig. 10.4. Factors affecting soil loss tolerance.

soil formation, (ii) rate of soil erosion, (iii) on-site and off-site agronomic effects, (iv) off-site economic effects, and (v) environmental effects on water and air quality and the greenhouse effect. Quantitative and credible information on soil loss tolerance is virtually non-existent for any soil and environment.

The masking effect of improved agricultural technology

One of the reasons for the controversy about the magnitude of on-site versus off-site effects is the masking effects of improved agricultural technology on increase in crop yields. Crop yields are increasing worldwide despite the increasing severity of soil erosion. Increases in crop yields are due to the widespread use of improved genotypes, tillage methods, fertilizers, water management systems, past control measures and harvesting techniques. Consequently, the adverse effects of soil erosion and decline in soil quality are partly compensated by use of high-yielding technologies. However, gains in productivity would have been much larger if there was no or slight rather than severe soil erosion. The magnitude of the masking effect generally increases with increasing use of the input and improved technology and depends on several factors (Table 10.2). The most drastic effects of soil erosion on crop yield are due to loss of soil fertility, loss of water, decrease in rooting depth and reduction in AWC. Therefore, the masking effect is generally in the order of fertilizers and organic manures ≥ irrigation and water management > tillage methods and residue management > crop varieties > pest control measures.

Adoption of conservation-effective measures to control soil erosion, although without causing any increase in crop yield, would lead to several indirect benefits. These savings include: (i) low inputs of fertilizer, (ii) saving in irrigation water, (iii) reduction in the frequency or intensity of tillage, (iv) saving in seed used, and (v) saving in land that can be used for other purposes, such as wildlife reserves, wetland restoration and environmental buffers.

Methods of assessment of on-site agronomic effects

There is a wide range of methods used for assessment of the on-site impacts on crop yield and agronomic productivity. Results obtained by different methods are not comparable, and doing so exacerbates the confusion and accentuates the debate. There are numerous ways to group the methods used into different categories. One of the categorization procedures is based on the scales, which range from point scale to watershed scale (Fig. 10.5). While the data obtained from large scales are more representative of the natural

Table 10.2. Factors affecting the magnitude of the masking effect of improved technology.

Parameter	Masking technology
Shallow topsoil depth	Fertilizer use, irrigation, deep-rooting crops
Nutrient and SOC stratification	Manuring, fertilizer use, deep placement of soil amendment in the topsoil layer
Soil compaction, crusting and structural decline	Subsoiling, chiselling, use of soil conditioners and amendments
Drought-prone soils in dry climates	Irrigation, water harvesting, *in situ* water conservation, short-duration and deep-rooting cultivars
Inundation	Surface and subsurface drainage
Soil variability	Site-specific management, compensatory effects of depositional sites

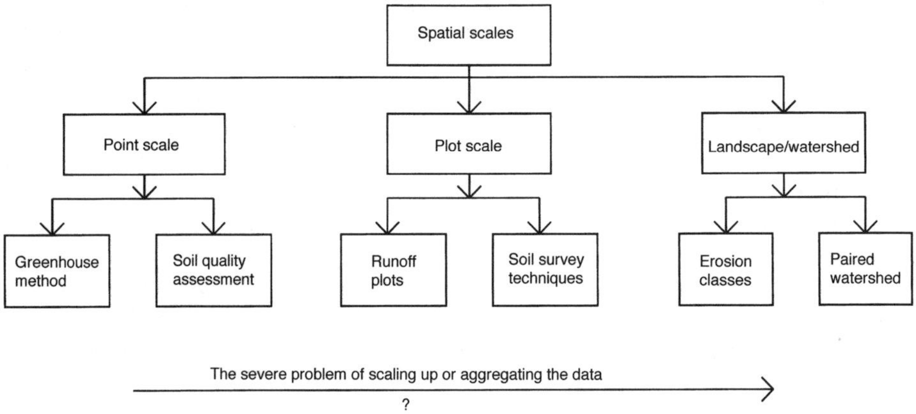

Fig. 10.5. Spatial scales of methods used to assess the on-site agronomic effects of soil erosion.

conditions, development of appropriate scaling procedures can lead to use of small-scale techniques which are cost-effective and easy to use.

The second method of characterization of procedures used is based on the time period involved, such as effect of past, present or future erosion. Methods used to characterize on-site effects on productivity for different time periods are listed in Table 10.3. Commonly used methods to assess the impact of past soil erosion include field techniques based on soil survey methods. An important consideration is to establish the direct relationship between soil erosion and the topsoil depth or depth to the root-restrictive layer. Knowledge surveys have been used to assess qualitatively the impact of past erosion on agronomic productivity (Dregne, 1998). In addition to soil erosion, changes in topsoil depth can occur as a result of numerous pedogenic factors. The effect of present rate of soil erosion on agronomic yield can be assessed by using runoff plots and monitored watersheds whereby seasonal or annual rates of soil erosion and water loss can be related to crop growth. The impact of projected soil erosion in the future on crop growth may be determined empirically by a topsoil removal and addition technique and use of predictive models. The topsoil removal and addition technique is an erosion simulation method, and wholesale/gross soil removal by scalping is not the same as the preferential removal of colloids (SOC and clay) by the natural process. It is also difficult to predict yield effects under field conditions on the basis of relative seedling growth measured on surface soil versus subsoil under the greenhouse conditions.

The third method of characterizing the techniques used is based on experimental versus modelling approaches. Numerous experimental techniques used are outlined in Table 10.3 and Fig. 10.5. There is also a wide range of modelling techniques, which are basically of two types: biophysical models and economic models (Fig. 10.6). There are at least five subcategories

Table 10.3. Methods used to assess on-site effects of soil erosion for different temporal scales.

Temporal scale	Method used
Past soil erosion	Topsoil depth, depth to root-restrictive layer, soil erosion phases within a landscape, soil quality and pedotransfer functions, predictive models, knowledge surveys
Present erosion rate	Runoff plots, paired watershed technique
Future erosion	Topsoil removal, greenhouse experiments on crop growth in topsoil versus subsoil, soil quality and pedotransfer functions, predictive models

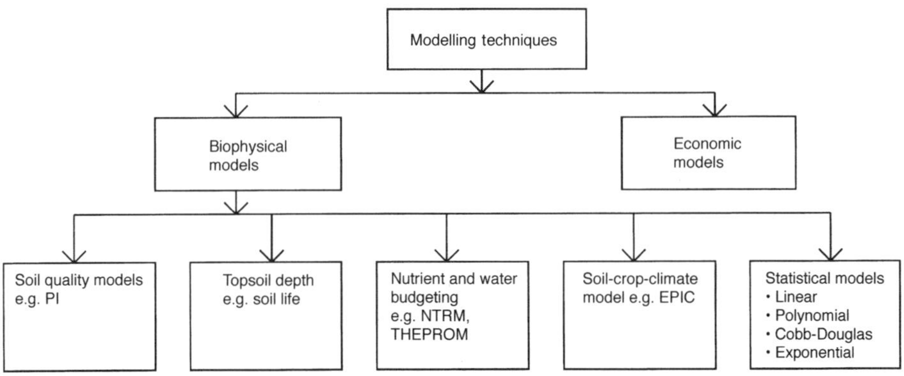

Fig. 10.6. Modelling techniques used to assess erosional impacts on agronomic productivity.

of biophysical models based on different conceptual approaches. The most commonly used models are based on erosion-induced changes in soil quality (e.g. PI; Kiniry *et al.*, 1983; Pierce *et al.*, 1984), changes in topsoil depth, water and nutrient budgeting, soil–climate–plant interaction (EPIC; Williams *et al.*, 1983; Williams and Renard, 1985), and simple statistical procedures based on regression analysis. The applicability of different methods depends on the availability of the input data and on the validation/calibration of the model for site-specific situations.

Methods of assessment of off-site agronomic effects

Off-site agronomic effects of soil erosion can be severe, and it is difficult to quantify these effects. These effects may be positive or negative depending on the landscape and quality of the sediment brought in by runon. The positive effects are due to increase in topsoil depth by deposition of good quality sediments and improvements in soil's nutrient reserves and AWC. The negative impacts on agronomic productivity are due to inundation, anaerobiosis due to impeded drainage, runon of agricultural chemicals and biocides with adverse impacts on crop growth, and burial of good quality soil with sediments of poor quality.

The off-site agronomic impact may be assessed by evaluating the crop damage by floods. The depositional impact on crop yield may also be assessed by evaluating the weighted average yield on a landscape (Salchow and Lal, 1997).

General Discussion and Conclusions

Agronomic effects of erosion on crop yield have not been adequately assessed. The available information is confusing and erroneous because of the confounding effects of numerous interacting factors, such as the masking effect of improved technology, lack of reliable data on the magnitude of past and present erosion, incomplete data on soil profile and land use, wide range of unstandardized methods used, and the complexity caused by prevailing weather conditions. Over and above the effect of erosion, rainfall amount and its distribution during the growing season affect crop yield. In seasons with normal rains, agronomic yields are usually less from eroded soils with shallow topsoil depth. In seasons with below normal rains, crop yields may be more on eroded than uneroded soils (Ebeid *et al.*, 1995). Use of improved technology has strong masking effects on agronomic yield, and the magnitude of this effect cannot be easily assessed.

There are numerous techniques of evaluating the on-site agronomic effect. Results obtained from these techniques are not comparable because of differences in spatial and temporal scales and the lack of appropriate scaling procedures for aggregating the data.

A major cause of controversy and confusion about the agronomic impact of erosion is due to the weak, incomplete and unreliable data on soil erosion and its impact on productivity. The data and information base can be vastly improved by obtaining credible information for benchmark soils and predominant ecoregions for the following:

- soil erosion rate and its impact on soil quality;
- erosion-induced temporal changes in soil quality under different land uses and farming/cropping systems;
- on-site and off-site agronomic effects of erosion; and
- supporting data on soil profile, landscape, land use and climatic characteristics.

A coordinated effort is needed to establish networks/task forces for conducting experiments with standardized methods to obtain the following research data:

- the magnitude and severity of past soil erosion and present erosion rates using the soil survey methods and modern techniques such as GIS and remote sensing;
- cause–effect relationships between erosion-induced changes in soil quality and crop yield;
- losses of nutrients, water and agricultural chemicals by soil erosion; and
- the magnitude of the off-site agronomic impacts due to inundation, burial, deposition and anaerobiosis.

There is also a strong need to develop and standardize methods with regards to the following:

- scaling methods of aggregating the data from point or plot scale to regional and global scale;
- methods to evaluate the magnitude of the masking effect of improved technology; and
- methods to assess the relative magnitude of agronomic versus other impacts of soil erosion.

References

Brown, L.R. (1995) *Who will Feed China? Wake Up Call for a Small Plant.* The Worldwatch Institute, Norton, New York, 166 pp.

Crosson, P. (1998) The on-farm economic costs of soil erosion. In: Lal, R., Blum, W., Valentine, C. and Stewart, B.A. (eds) *Methods of Assessment of Soil Degradation.* CRC Press, Boca Raton, Florida.

Dregne, H.E. (1998) Desertification assessment. In: Lal, R., Blum, W., Valentine, C. and Stewart, B.A. (eds) *Methods of Assessment of Soil Degradation.* CRC Press, Boca Raton, Florida.

Ebeid, M.M., Lal, R., Hall, G.F. and Miller, E. (1995) Erosion effects on soil properties and soybean yield of a Miamian soil in western Ohio in a season with below normal rainfall. *Soil Technology* 8, 97–108.

Kiniry, L.N., Scrivner, C.L. and Keener, M.E. (1983) A soil productivity index based upon predicted water depletion and root growth. *Research Bulletin 1051, Missouri Agricultural Experiment Station Bulletin.* Columbia, Missouri.

Lal, R. (1984) Soil erosion from tropical arable lands and its control. *Advances in Agronomy* 37, 183–248.

Lal, R. (1985) Soil erosion and crop productivity relationship for a tropical soil. In: El-Swaify, S.A., Moldenhauer, W.C. and Lo, A. (eds) *Soil Erosion and Conservation.* Soil Conservation Society of America, Ankeny, Iowa, pp. 237–257.

Pierce, F.J., Larson, W.E., Dowdy, R.H. and Graham, W.A.P. (1984) Productivity of soils: assessing long-term changes due to erosion. *Journal of Soil and Water Conservation* 38, 39–44.

Pimmentel, D.C., Harvey, P., Resosudarmo, K. *et al.* (1995) Environmental and economic costs of soil erosion and conservation benefits. *Science* 267, 1117–1120.

Salchow, E. and Lal, R. (1997) Erosional effects on crop yields on a complex landscape in west-central Ohio. *52nd Annual Conference of the Soil and Water Conservation Society,* Toronto, Canada, 26–27 July, *Journal of Soil and Water Conservation* (in press).

Williams, J.R., Renard, K.G. and Dyke, P.T. (1983) EPIC: A new method for assessing erosion's effect on soil productivity. *Journal of Soil and Water Conservation* 38, 381–383.

Williams, J.R. and Renard, K.G. (1985) Assessment of soil erosion and crop productivity with process models (EPIC). In: Follett, R.F. and Stewart, B.A. (eds) *Soil Erosion and Crop Productivity.* Soil Science Society of America, Madison, Wisconsin, pp. 67–103.

11

Catchment Approach to Combating Soil Erosion in Asia – the Managing Soil Erosion Consortium

E.T. Craswell, C. Niamskul and F.W.T. Penning de Vries

International Board for Soil Research and Management, PO Box 9-109, Jatujak, Bangkok 10900, Thailand

Population growth and the exploitation of tropical timber resources have caused widespread intensification of agriculture and deforestation across the extensive hilly and mountainous areas of South and South-East Asia. Sediment outflows from the region's river systems to the ocean are massive, indicating that soil erosion is a serious problem in many areas. However, the relative importance of natural and human-induced impacts on stream flows and sediment loads is not clear. In contrast to farmers in high-potential lowlands, cultivators in upper catchments are commonly poor, isolated from the mainstream of development, and do not have secure land tenure. They need research support to develop and adopt sustainable land management systems that are productive, acceptable and combat erosion. Furthermore, there is a need for better understanding of the on- and off-site impacts of sloping land management to help decision-makers better assess policies and other measures needed to address these problems.

In 1997, national agricultural research and extension systems (NARES) in eight Asian countries formed a consortium with international agricultural research centres (IARC) and advanced research organizations (ARO) in Europe, Australia and North America to tackle the problem of managing soil erosion. Instead of addressing erosion narrowly as a soil- and climate-related problem, the consortium adopted a catchment approach, in which multidisciplinarity and stakeholder participation are pivotal concepts. The consortium will build the capacity of involved institutions, while drawing on the strengths of each. Model catchment studies will be selected in each country to provide the focus for the collaborative research. The main outputs of the consortium will be: acceptable and sustainable community-based land management systems; and reliable information and guidelines for decision-makers on on- and off-site agricultural, environmental and social impacts of managing soil erosion.

Introduction

Land degradation is a major concern to many nations and to the international community. The Earth Summit in Rio de Janeiro, 1992, and subsequent meetings, reviewed the problem and called for intensified efforts to arrest the widespread degradation of land and water resources. Despite this concern and considerable investments in research and development activities, examples of successful programmes on soil and water conservation in tropical areas are still rare. Reviewing these problems, Greenland *et al.* (1994) called for the adoption of a new agenda to expand global research on soil, water and nutrient management. They advocated a new paradigm for research that takes into account the social and economic dimensions as well as the biological and physical aspects of soil management. Sustainable land management requires research that addresses the needs of the land user, utilizes the participatory mode and takes into account the policy and institutional dimensions. A central tenet of the new paradigm is the catchment approach, which was strongly endorsed by the Technical Advisory Committee of the Consultative Group on International Agricultural Research (TAC, 1996).

This chapter reviews the case of land and water resources in the hilly and mountainous areas of tropical Asia where dynamic changes in population pressure and economic development are significantly affecting land use, with consequent impacts on soil erosion and landscape hydrology. The chapter describes the organization and research approach of a new consortium on managing soil erosion (MSEC) that brings together scientists from developed countries and eight Asian developing countries with the common goal of improved integrated catchment management. The MSEC programme is part of a CGIAR system-wide programme on soil, water, and nutrient management, which is coconvened by CIAT and IBSRAM. MSEC also comes under the umbrella of the CGIAR ecoregional programme for the warm, humid tropics and subtropics of Asia, which is convened by IRRI.

Physiographically and agriculturally, tropical Asia can be generally divided into the uplands which are hilly and mountainous and the lowlands where paddy rice is cultivated. The steep slopes and young landscapes of the uplands of Asia are naturally erosion prone and deliver vast quantities of sediment to the surrounding oceans (Fig. 11.1).

The growth of the population in Asia has slowed in recent years, but the population of the region is projected to grow from 3.4×10^9 in 1995 to 5.4×10^9 in the year 2050 (Fischer and Heilig, 1997). Since the high-potential lowlands are already heavily used for food production, the upland areas of Asia will increasingly be developed for agriculture. This trend will be exacerbated by greater urbanization that will reduce land availability for peri-urban agriculture. The resultant pressure on fragile sloping lands will seriously affect the hydrology and sediment outflows from upper catchments unless efforts to promote sustainable land management in the uplands are successful.

The Green Revolution of the 1970s was successful in increasing food production to meet the demands of a rapidly growing population. However, farming practices employed in many areas of Asia have become environmentally

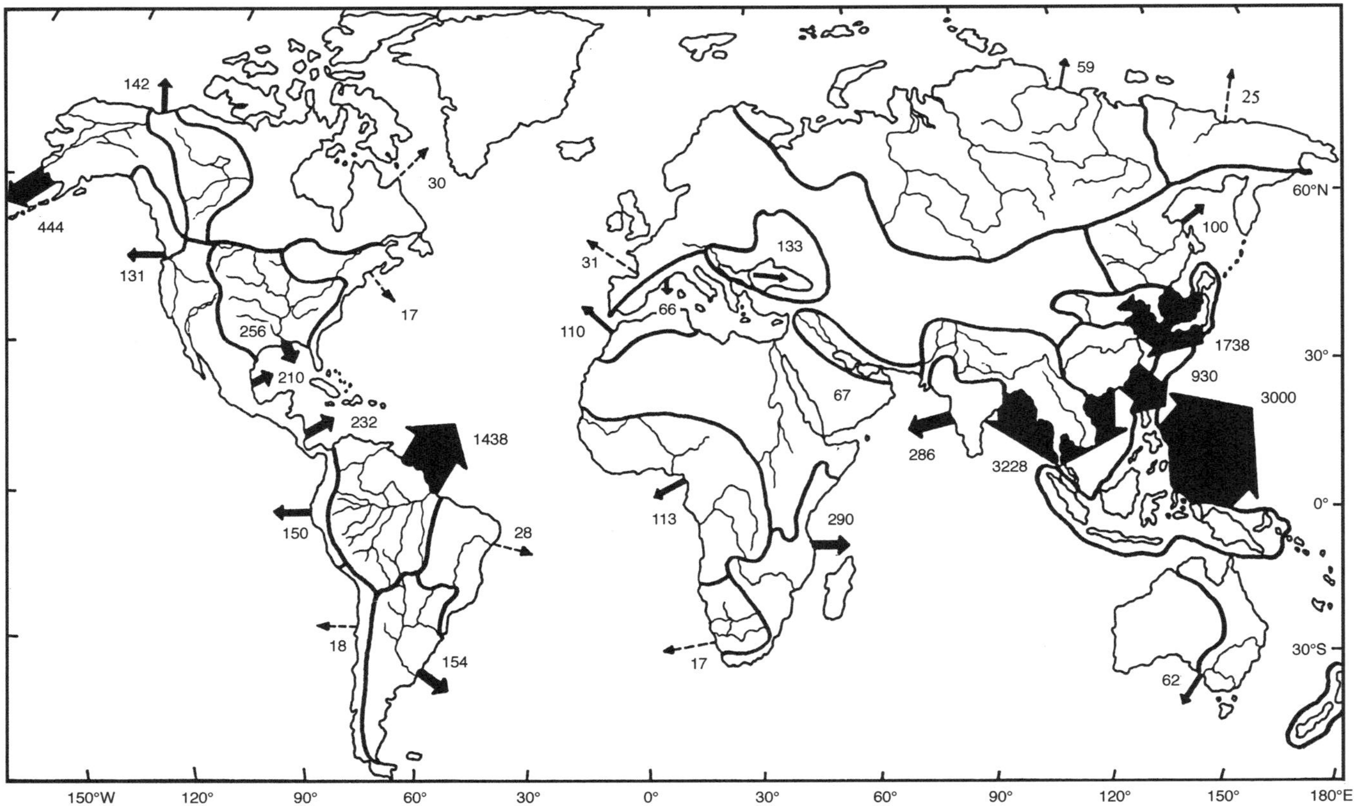

Fig. 11.1. Annual discharge of suspended sediment from various drainage basins of the world. Width of the arrows corresponds to relative discharge; numbers refer to average annual input in millions of tons. Source: Milliman and Meade (1983).

unsustainable, especially in the uplands where the rural poor have been bypassed by the Green Revolution. In these uplands, soil erosion is a widespread problem. Coupled with continuing deforestation and urbanization, the resultant land degradation is reducing the area of land suitable for cultivation and the quantity of water available for rural, industrial and urban use. Thus, Asian nations face the almost insurmountable challenge to produce more food with less water and cultivated land available, without degrading the natural resource base. Poor upland farmers are in greatest need of new, sustainable and remunerative farming systems.

To address these concerns, the International Board for Soil Research and Management (IBSRAM) spearheaded the formation of a consortium with institutions in and outside Asia to pool their resources and adopt a common strategy focused on tackling the soil erosion problem. The Managing Soil Erosion Consortium (MSEC) and its research approach was planned at a series of meetings: in March 1995 in Thailand, in June 1995 in Germany, in February 1996 in the Philippines, and in February 1997 in Thailand, while a meeting in Vietnam in May 1998 has been planned. The consortium members discussed and agreed upon a project that will address the erosion problem in Asia through a series of catchment studies. The scope of the problem and the research approach developed are described below.

Scope of the Erosion Problem

Soil erosion is particularly serious in tropical Asia because the climate is monsoonal and much of the land is hilly and mountainous. In Vietnam, for example, 75% of the area is classified as sloping land. A recent IBSRAM survey of seven Asian countries showed that substantial areas of land have been seriously affected by soil erosion resulting from overintensive land use and the inappropriate cultivation of steep and sloping lands. In the Philippines, 5×10^6 ha of uplands, or 17% of the land area, are estimated to be severely eroded. The seriousness of the problem in the region is illustrated ultimately by the discharge of suspended sediment from rivers (Table 11.1). The total annual sediment outflow from Asian countries to the tropical oceans is more than 7×10^9 t, which represents an enormous net loss of fertile topsoil and nutrients. It is also higher than any other region in the world. The sediment outflow is highest from the short, fast flowing rivers in the many islands in the region – for example the entire basin of the Cilutung River in Indonesia has an average sediment yield as high as 120 t $ha^{-1}year^{-1}$. Further evidence that the soil erosion problem in tropical areas of Asia is severe comes from a recent assessment by ISRIC: high to very high soil erosion by water is by far the most widespread cause of land degradation (Van Lynden and Oldeman, 1997). In Thailand, the Department of Land Development estimates that 30% of the land area is moderately to severely eroded (Chisiri *et al.*, 1990).

On-site soil loss reduces the chemical fertility of the soil by nutrient and organic-matter depletion, and in some cases by the exposure of acid subsoils. Erosion also damages the physical fertility by removing surface soil and reducing soil depth and water-holding capacity. These soil changes will,

Table 11.1. Sample data on sediment outflows to the ocean from rivers in tropical Asia (Milliman and Syvitski, 1992).

River	Drainage basin area (10^6 km^2)	Sediment load (10^6 t year^{-1})	Sediment yield (t km^{-2} year^{-1})
Irrawaddy (Burma)	0.43	260	620
Mekong (Vietnam)	0.79	160	200
Angat (Philippines)	0.00057	4.6	8000
Cimatur (Indonesia)	0.00058	1.9	3000
Cilutung (Indonesia)	0.00060	7.2	12,000
Agno (Philippines)	0.0012	5.0	4350
Citanday (Indonesia)	0.0025	9.5	3700
Cimanuk (Indonesia)	0.0032	25	7800
Solo (Indonesia)	0.16	19	1200
Hungho (Vietnam)	0.12	130	1100

slowly, reduce crop yields and hence farm income and household nutrition, and require increasing amounts of inputs to achieve certain yield levels. In a region like tropical Asia where the population will grow by 1.6 billion over the next quarter century, and arable land per person in most countries is already lower than 0.15 ha, loss of land productivity through erosion is an especially serious problem.

The off-site effects of erosion on the quality and seasonal availability of water can be very serious. The many off-site impacts of soil erosion, important and highly location specific, are not well analysed and documented (see also Enters, Chapter 1, this volume). Major social, economic and environmental effects include:

- increased peak surface runoff, often resulting in flooding, which displaces people in low lying areas and damages infrastructure;
- increased sediment, nutrient and pollutant loads in streams, which degrade the quality of household water supplies and increase the risk to health;
- siltation of dams and irrigation channels, resulting in reduced security of water supply for irrigated crops and shorter lifetimes of reservoirs; and
- sediment deposition in offshore fisheries, reducing the availability of aquatic food supplies and ecotourism.

There is a lack of principles and techniques for practical catchment-oriented approaches to evaluate resource-use systems and management practices in terms of soil erosion risk and on- and off-site effects. The information and decision support needs of the full range of stakeholders, from resource users to policy-makers, on resource use management and planning at the catchment scale need to be assessed over all stages of the planning processes.

Existing approaches are limited in that they primarily focus on the biophysical impacts of soil erosion management for on-site productivity of the land resource at a particular location, and do not address the environmental, economic and social implications of resource use and management at the catchment scale (see also Bricquet and Claude, Chapter 12, this volume). For reasons of efficacy and efficiency, there is a need for: (i) the incorporation of indigenous knowledge bases relating to land use and management into resource-use evaluation systems; (ii) the capability to assess the implications of these indigenous practices for soil erosion and its on- and off-site effects; and (iii) the establishment of practical indicators of sustainability that relate to both state/condition and process.

Targets and Research Needs

The MSEC objectives are:

- to develop sustainable and acceptable community-based land management systems that are suitable for entire catchments;
- to quantify biophysical, environmental and socio-economic on- and off-site impacts of soil erosion;
- to prepare reliable information and guidelines for decision-makers; and
- to improve training on soil erosion control.

The challenge for MSEC participants is to recognize that land degradation, including soil erosion, is driven largely by socio-economic factors (Kerr, Chapter 2, this volume). There is an urgent need to change how decisions are made regarding land use at all levels, from the individual land user to national and international policy-makers, taking into account the development of improved land-use practices. A new paradigm for research is required which ensures that the whole range of stakeholders, from land users to policy-makers, is involved in the generation and promotion of improved land-use practices.

The first step to include sociological and economic aspects for improved catchment management is to make full use of the indigenous knowledge. Researchers must heed the views of farm households (men and women) on what is desirable and possible, because they are the best and most relevant integrators of their own biophysical and socio-economic environment. Research of the *ASIALAND* sloping lands network in seven Asian countries, coordinated by IBSRAM, has shown that technologies such as alley cropping can significantly reduce erosion: in the Philippines, soil loss over 3 years on plots with alley cropping was 26 t ha^{-1}, whereas losses totalled 135 t ha^{-1} with farmers' practice. Although the alley cropping gives high economic returns, costs of labour in the first year reduce profits too strongly and the rates of adoption were low (Craswell *et al.* 1997). A participatory approach would involve stakeholders (farmers and other decision-makers) in the process, harnessing indigenous knowledge for the identification of real bottlenecks and suggestions for solutions, and encouraging the ownership of research results.

Also, people living far from steep-lands are affected by land management, and are stakeholders. There is a serious knowledge gap about off-site impacts of erosion in the upper watersheds on sediment and streamflows in catchments. In addition, few data differentiate human-induced impacts from natural erosion. Not enough is known about the erosion and deposition processes in landscapes of the humid tropics, and even less is known about the agronomic and economic consequences of these processes within catchment systems. These reasons provided the main motivation for this workshop.

MSEC uses a water catchment approach because erosion and sedimentation have strong interactions at this level, and because small communities decide in an integral manner on the management of the resources in the catchments. MSEC focuses on catchments of 5–50 km^2 with different forms of land use ('farm types', rangeland, forestry) practised by 10–100 farmers, and at microcatchments of 1–5 km^2 for detailed observations. The catchment approach will ensure that all stakeholder groups in the landscape affected by soil erosion, including farmers and policy-makers, benefit from the knowledge generated, recognize the scope and severity of the problem, and make appropriate decisions about investments and land-use policy in the sloping land areas. Some research and development projects in the Asian region have adopted these new approaches, but few studies are truly comprehensive, particularly with respect to the participatory and multidisciplinary aspects of the catchment-level research paradigm.

MSEC will provide hard data on how soil and water fluxes vary in response to changes in agricultural and forestry management on slopes. This information will facilitate the participatory development of improved soil management technologies with farmers. An output of the MSEC benchmark sites will be data on hydrological, soil management, economic, sociological and policy aspects that guide the management of sloping lands for the benefit of all the stakeholders. The main hypothesis to be tested is that a better understanding of the interaction of people, land and water resources across the landscape will produce more appropriate technologies and policy options and, ultimately, expanded use of appropriate land husbandry practices to control soil erosion.

Specific outputs of the project will be:

1. Improved techniques and tools that facilitate a negotiated approach to the development of appropriate resource-based catchment plans and the implementation of appropriate catchment-oriented practices for managing soil erosion.
2. Enhancement of the planning capacity of the full range of stakeholders related to sustainable resource use, through participatory approaches to R&D on catchment management.
3. Data quantifying biophysical and socio-economic, on- and off-site cost of soil erosion.
4. Established sites for catchment research for the next 5–15 years.

Major clients for the outputs of MSEC will be bilateral and multilateral development agencies that invest considerable grants and loans in projects designed to protect watersheds in the region. The target beneficiaries will gain from the project as follows:

1. Upland farmers will benefit from improved policies and investments, and the higher income from more sustainable and profitable land management practices.
2. Lowland farmers will benefit from improved water supplies for irrigation, less flash flooding, and less damage to fisheries by sediments.
3. Policy-makers will benefit from improved information about on- and off-site impacts of soil erosion on which to base land-use and other policies to combat the problem.
4. The environment will benefit through reduced sedimentation in waterways, more regular flow of water from upper catchments, a higher water quality, and non-declining or even increased biodiversity due to reduced land clearing for permanent agriculture.
5. Financial institutions will benefit from the expanded and improved knowledge base about soil erosion and from improved methods for integrated catchment management, which can be utilized in the many watershed management projects that they fund.

MSEC Consortium and Research Activities

The Asian members of the consortium include: (i) Center for Soil and Agroclimate Research in Indonesia; (ii) Department of Agriculture and Extension in Laos; (iii) National Agricultural Research and Services Center in Nepal; (iv) Philippine Council for Agriculture, Forestry and Natural Resources Research and Development in the Philippines; (v) Royal Forestry Department and Department of Land Development in Thailand; (vi) National Institute for Soils and Fertilizer and the Forestry Department in Vietnam; (vii) Forest Department, Ministry of Forestry and Environment in Sri Lanka; and (viii) University of Hangzhou, and Fujian Academy of Agricultural Sciences, China.

Other consortium members are international centres including International Rice Research Institute (IRRI), International Centre for Integrated Mountain Development (ICIMOD), International Centre for Research on Agroforestry (ICRAF) and International Food Policy Research Institute (IFPRI), and advanced research organizations such as Institut Français de recherche scientifique pour le développement en cooperation (ORSTOM), the University of Bayreuth, the Australian National University, and Chulalongkorn University.

The MSEC studies focus on the effects of participatory land management on water/element inputs and outputs of meso- and/or microcatchments, and process studies to quantify water and element fluxes between different ecosystem components, in selected model catchments in eight participating Asian countries, namely China, Indonesia, Laos, Nepal, the Philippines, Sri Lanka, Thailand and Vietnam. MSEC activities at the consortium level (Table 11.2) are organized to exploit to the maximum synergies between and amongst NARES and AROs.

To assess the significance of sustainability factors on steep-lands, mid- to long-term experiments on a catchment basis are essential. Such studies will allow the quantification of the on- as well as the off-site effects between the

Table 11.2. Framework of MSEC activities at the consortium level.

1. Evaluate indigenous land-use systems and management practices
2. Establish standard methodologies for the selection of the representative catchments
3. Identify technologies and methodologies to be tested and validated on project sites
4. Hold training courses on standard research methodology for MSEC. The training will include all aspects of socio-economic, biophysical and policy issues
5. Disseminate information on validated technologies, on- and off-site effects, and impact assessment by publication of scientific papers, a newsletter and other materials
6. Organize consortium assemblies for the dissemination and interchange of research results and developments among all stakeholders, i.e. NARES, AROs, IARCs and donors.
7. Ensure, through periodic technical workshops, that the research results are made available to both representatives of the respective NARES and to other NARES in the region for which the research results would be relevant

different compartments of the ecosystems. Supplementary plot experiments will be carried out to study special questions that may arise due to the participatory approach. Catchment experiments require significant investments for appropriate equipment. The present proposal builds on ongoing studies to reduce the costs. The sequence of activities at selected sites, from establishment to presentation of a decision support system for policy-makers, is described briefly below (see also Table 11.3).

A weir at every catchment will allow us to measure what is happening both on and off site with water, sediments and nutrients. Installation of the equipment for monitoring water and element balances at different levels (mesocatchment, microcatchment, plot; see also Bricquet and Claude, Chapter 12, this volume), including an automatic water level recorder (WLR), meteorological station (MET), suction cups, tensiometers, automatic rain gauge recorders (RGR), time domain reflectometry (TDR), and automatic water and sediment collectors, will be done shortly thereafter. At selected sites, such as in Thailand, suction cups and tensiometers will be used to quantify water and nutrient fluxes in the soil.

Understanding crucial balances (water, soil, nutrients) and extended analyses (runoff and erosion) requires the use of models (see also Zech *et al.*, Chapter 13, this volume). The project will only involve the simplest models possible. Inputs for the models will be measured (to drive the models). Outputs will be modelled and some of them will also be measured, so that validation of these models becomes feasible, and, if necessary, to allow for calibration. Modelling is an indispensable tool to establish the consistency of analyses ('no crucial processes omitted'), to achieve a certain level of generality (to allow appropriate adaptation of the techniques that are best in one catchment to the different conditions of other catchments) and to enable exploration of the sustainability of the proposed techniques under a wider range of conditions than is possible through experimentation (e.g. a wider range of weather conditions; effects of different land-use practices). Special

Table 11.3. Framework of MSEC activities at catchment sites.

1. Select model catchments using the political, socio-economic and biophysical criteria developed in Cebu, the Philippines, February 1996
2. Through a participatory approach, elicit knowledge on indigenous land-use systems and management practices
3. Identify project staff, including extension workers and GO/NGO agencies who are stakeholders, and encourage their collaborative input
4. Conduct biophysical inventories of the meso- and microcatchments
5. Conduct socio-economic inventory in meso- and microcatchments
6. Calibrate identified catchments
7. Calibrate microcatchment
8. Carry out participatory monitoring activities for regular assessment (6–12 months). Emphasis on the process of learning experiences and collaborative efforts put into the implementation activities
9. Assess existing policies, programmes and institutional management relevant to natural resource management, agricultural development, land rights and community-based participation.
10. Examine the on- and off-site effects of different land-use practices at the meso- and microcatchment level and extrapolate by modelling, taking into consideration the results of the participatory activities.

attention will be given to the effect of scale (see also van Noordwijk *et al.*, Chapter 15, this volume).

From a socio-economic perspective, catchment experiments carried out by the biophysical science team members will have an impact on people at both on- and off-site locations, directly and indirectly. People need to be informed and involved in activities undertaken. This is especially important if the outcomes of the findings are to lead to a move towards more sustainable catchment resource management planning and implementation.

The project experiments will be carried out at different levels of intensity in different countries. The country representatives have agreed during the consortium assembly in Cebu, and in Nan, about the phasing of activities. All countries will start together with socio-economic, policy, institutional and biophysical inventories. Installation of equipment and calibration will be started in Thailand, Indonesia and the Philippines followed by Nepal, Laos and Vietnam. In Sri Lanka, sites established under the recent Asian Development Bank (ADB)-supported watershed project will participate.

MSEC Site Selection

A very crucial activity in the evolution of MSEC is selection of the sites, where for many years observations will be made and trials implemented. Based on the recommendations of the MSEC meeting in Cebu in February 1996, a selection team was formed to evaluate candidate MSEC sites as follows (IBSRAM, 1997):

Step 1. Interviews and discussions with the local chiefs of relevant agencies or research stations, their personnel and villagers, mainly based on questionnaires, open-ended interviews and the examination of available reports.
Step 2. Field assessment and examination of:

- existing equipment;
- land-use systems;
- soil-conservation practices of the farmers;
- off-site effects;
- related R&D activities;
- socio-economic conditions of on- and off-site communities; potential for their participation.

Step 3. A summarizing discussion involving all team members to evaluate the following issues:

- representiveness of the catchment with respect to people, land use, hydrology, soil erosion, nutrients, and pollution;
- potential activities to be recommended;
- main problems and options.

The MSEC site selection mission recommended sites in Thailand, Indonesia and the Philippines that were endorsed at the Nan Consortium Assembly in February 1997 (see IBSRAM, 1997, for details). The mission also identified a number of key principles relating to land and water resource use in Thailand, Indonesia and the Philippines:

1. The pattern of land use and management within a catchment was in general a function of the interaction of biophysical, economic, sociocultural, policy and institutional factors operating at any one period of time or locality. The way in which these factors interact influence: (i) the nature and severity of the soil erosion and degradation problems present, and (ii) the apparent success of community acceptance and implementation of sustainable soil- and water-conservation practices. As agricultural land use is an important part of each country's national economy, increasing population pressure and continuing and increasing land degradation on a broad catchment/regional scale is perceived to threaten living standards, and national and regional development opportunities, as well as the protection of biodiversity and genetic resources.
2. Decision-making and action relating to land resource use and management occurs largely at the individual landholder level (e.g. farmer, forestry industry, commercial crop plantation owner). Collectively, such decisions and actions at a catchment scale may have significant environmental, economic and social implications to other parts of a catchment system and/or communities, as well as to the current and potential on-site resource uses and functions. The spatial and temporal dislocation of the cause-and-effects of these natural resource issues means that community-based action for change is required for effective resolution. However, a major issue in achieving such action is that, although at the catchment perspective the practices of resource use and management may need to change, from the individual resource user's perspective it may not be economical or socially acceptable to change. The key problem

being that the use of techniques or methodological approaches to improve resource use and management, rather than increase economic performance, are borne by the individual resource user, while the broader benefits of these improvements may accrue to society as a whole.

3. A land-use system or management practice that may be assessed as sustainable at one scale (e.g. in terms of production at the field or plot scale) may not be sustainable, however, at another scale due to, for example, the cumulative effects at a catchment level (such as downstream sedimentation and increased flooding). Moreover what is sustainable in the short term may not be so in the longer term without considerable additional on-site inputs which may have economic consequences to farmers/resource users and/or undesirable off-site environmental and social impacts.

4. The diversity of stakeholders of R&D on soil-erosion management, including farmers (both upstream and downstream), the forestry and agricultural industries, local and regional governments, extension and advisory agencies, as well as policy-makers, means that frequently there are conflicting and/or competing interests involved. The costs and benefits of soil erosion are complex and involve environmental, economic and sociocultural considerations. Soil-erosion impacts may be considered beneficial by some stakeholders in some contexts while at the same time distinctly disadvantageous by others, especially in the long term. This requires a negotiated participatory approach to the implementation and communication of the R&D for effective long-term outcomes to be achieved.

The MSEC programme has intensified activities in 1998. The planning phase has been protracted but necessary, considering the ambitious and complex nature of the consortium. The methodological guidelines and other outcomes of this Bogor workshop will make a significant contribution to MSEC, by enhancing the scientific rigour of the methods and helping to avoid duplication of effort. Past and future MSEC progress is built on the dedication and hard work of many individuals. We acknowledge their inputs and look forward to greater collaboration in the future.

Conclusions

In order to experience land use with minimal soil erosion in Asian uplands, research must have a number of new characteristics: it must be participatory (meaning that stakeholders, such as farmers and local planners, contribute indigenous knowledge), interdisciplinary (particularly integrating biophysics with socio-economics) and catchment based (as this is a unit of management both from the natural resources and from the communities point of view). These features are all present in the new Management of Soil Erosion Consortium (MSEC) that is coordinated by IBSRAM. MSEC already includes NARES from eight Asian countries, several IARCs and also some AROs as active members, and has already nearly 20 catchment sites selected for its research.

References

Chisiri, S., Palasuan, N., Inthapan, P. and Srikumde, B. (1990) Conservation project design – Thailand Country Review. In: *ASOCON Report 3 Regional Action Learning Programme on Conservation Project Design.* ASOCON, Jakarta, Indonesia.

Craswell, E.T., Sajjapongse, A., Howlett, D.J.B. and Dowling, A.J. (1997) Agroforestry systems in the management of sloping lands in Asia and the Pacific. *Agroforestry Systems* 38, 121–137.

Fischer, G. and Heilig, G.K. (1997) Population momentum and the demand on land and water resources. *Philosophical Transactions of the Royal Society of London* B352, 862–869.

Greenland, D.J., Bowen, G., Eswaran, H., Rhoades, R. and Valentin, C. (1994) *Soil Water and Nutrient Management Research – A New Agenda.* IBSRAM position paper. IBSRAM, Bangkok, Thailand.

IBSRAM (1997) *Model Catchment Selection for the Management of Soil Erosion Consortium (MSEC) of IBSRAM.* International Board for Soil Research and Management, Bangkok. Thailand, 109 pp.

Milliman, J.D. and Meade, R.H. (1983) World-wide delivery of river sediment to the oceans. *Journal of Geology* 91, 751–762.

Milliman, J.D. and Syvidski, J.P.M. (1992) Geomorphic/tectonic control of sediment discharge to the ocean: The importance of small mountainous rivers. *Journal of Geology* 100, 525–544.

TAC (1996) Technical Advisory Committee of the Consultative Group on International Agricultural Research. *TAC's framework for priority setting, CGIAR. Milestones of Renewal,* October. FAO, Rome.

van Lynden, G.W.T. and Oldeman, L.R. (1997) *The Assessment of the Status of Human-induced Soil Degradation in South and Southeast Asia.* UNEP (Nairobi, Kenia), FAO (Rome), and ISRIC (Wageningen, Netherlands).

12

Latest Developments in the Design of Hydrological Studies of Watersheds

J.P. Bricquet and J. Claude
Laboratoire d'hydrologie, ORSTOM, BP 5045, 34032 Montpellier cedex, France

The use of catchments to study hydrological processes has evolved much during the last 15 years. The hydrological response of catchments is controlled by a complex function of ecological, climatic and geomorphic processes, especially in the tropical environment. However, for such knowledge to have wide application, the hydrological response needs to be understood in terms of the processes driving it, such as climate, vegetation, drainage network, soils and land use.

New tools and concepts have been developed that permit a different analysis of processes (remote sensing, mathematical models) and allow catchments to be considered as entire units. Hydrological characteristics can be generated, such as by use of one of several runoff models, or they may be gathered and examined without the use of a model. This is often the case for flood frequencies, annual water yields or erosion rates. In either case, high quality, long-term hydrological data are required to define adequately the hydrological response of a catchment. This hydrological response then needs to be related back to the specific ecological, climatic and geomorphic conditions.

Predicting the hydrological response of catchments can be accomplished using regionalization techniques, assuming that new catchments behave similarly to other, gauged catchments. Because there are many different approaches to regionalization, and because these approaches are often region specific, collaboration between hydrology and remote sensing for modelling is especially important.

Thus, through the range of disciplines involved, watershed research has become important for the true understanding of the environment and its conservation.

Introduction

Traditionally, hydrology, soil science and geochemistry have been separate sciences, taught in different departments with different purposes and aims. This separation has stood the test of time in that each discipline seemed to work adequately within its sphere. However, when we consider environmental effects of human activities on terrestrial ecosystems, this division becomes a liability, as it denies proper understanding at the system level. As a result, the concept of watershed has rapidly become essential in the study of natural systems. The use of a watershed as a unit of research in the environmental sciences has been fully developed for about 15 years, and considerable progress has been made (e.g. mathematical models, hydrology of the slope). These studies recognize that water is simultaneously a resource, a vector for other elements and an agent in many processes – indeed, the centre of life. The water cycle, therefore, conditions the working of the entire natural environment.

The great physical and biological overlap and the great temporal and spatial variability demand an interdisciplinary approach. Until recently, studies on watersheds had a finality practice (phase 'project' works) and studied parameters that were directly linked to objectives. For some years, and supported with new tools (remote sensing, RS and geographical information systems, GIS) and new equipment (electronic recording, automation, transmission by satellite), studies have taken a new direction. This chapter describes some of these developments and directions in the hydrology of watersheds.

Hydrological Research in Watersheds

The endless recirculation of water in the atmosphere–hydrosphere–lithosphere is known as the hydrological cycle. The cycle can be studied according to the particular scale, such as the global or basin scale. From the hydrological point of view, the basin scale cycle is emphasized. The basin-scale cycle is considered to be a continuous circulation of water from water vapour to precipitation, stream flow, lakes, reservoirs, soil moisture, groundwater and out of basin transfer, to the return to water vapour through evaporation and transpiration (Fig. 12.1). Within a basin, the dynamics of the hydrological processes are governed partially by the temporal and spatial characteristics of inputs and outputs and the land use/land cover conditions. Currently, little is known about the accuracy of the estimation of hydrological parameters in a given area, primarily because of the wide variation of the distribution in space and time, and the lack of information concerning the optimization of the conventional measurements.

The concept of a watershed as a unit for research

The basin is indeed a functional and basic unit for the study of the water cycle, and for the coupling of water flows with other elements. Therefore, it is also

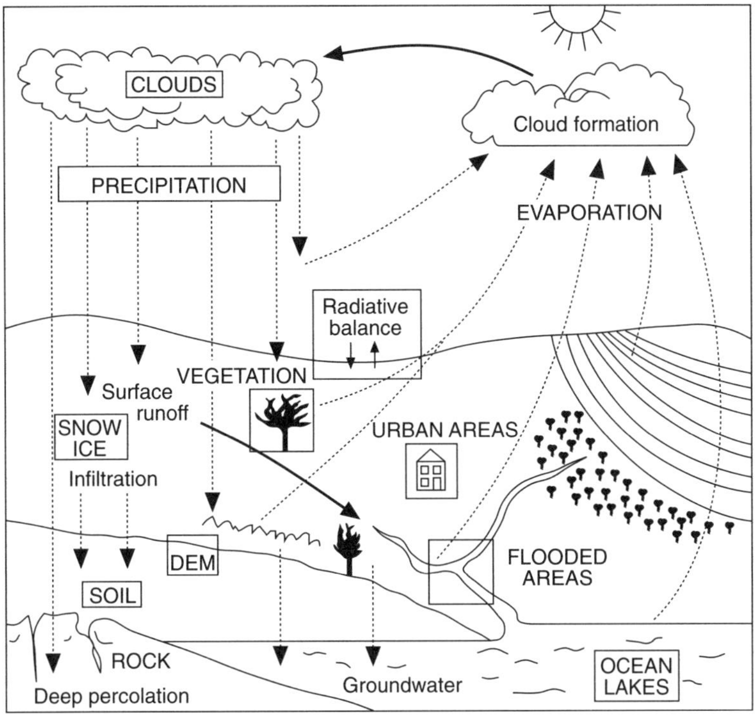

Fig. 12.1. The hydrological cycle. DEM, digital elevation model.

a basic unit for management and planning of water resources. The watershed is an integrator concept that can be applied to all spatial scales. A research watershed is therefore a catchment whose physiographical characteristics and functioning are considered as typical or representative of a certain hydrological region. Results obtained on these basins are transposable to the whole region or to similar basins. The research watershed is therefore a specially selected area, equipped for the detailed study of the water cycle and coupled flows, as well as consequences of artificial or natural evolution.

Research watersheds are also used to investigate many problems, including water as a resource, vector or agent: to specify the effect of management and human perturbation (vegetable shelter change, occupation of the ground, land use); to improve hydrochemical balances, to understand cycles of elements in the ground–plants–water system, and the dynamics of the ground. The scientific approach to the hydrological cycle includes the quantification and the modelling of flows and assessments, the identification of process and factors, and the modelling of its dynamics (Likens *et al.*, 1977). A research watershed, as a tool for many thematic studies, comprises four distinct but complementary functions:

- a field laboratory for interdisciplinary research, such as detailed balance establishment, study of water and soil movements, *in situ* experimentation;
- an environmental observatory to measure, in the long term, flows and parameters that control them, to divulge evolution linked to human perturbations or to climatic changes;
- a site for method development and tool validation, to test and calibrate collectors, to validate several models and to provide ground truth for remote sensing;
- training to and by the research, permanent training.

The research watershed in tropical environments

The big differences between tropical environments and temperate regions are in the intensity of the processes of the water cycle (evaporation, precipitation) and the nature of the receiving body of water (soil, vegetation, land use). About 400 watersheds have been studied by ORSTOM since 1951 for periods of a few months (one rainy season) to more than 10 years (Dubreuil, 1986; Rodier and Ribstein, 1988). This collection of catchments concerns mainly Africa, but also Latin America and Oceania. From the 1970s, several research basins have been exploited in these regions in the humid savannah zone (Ivory Coast for example; Chevallier, 1990) and in the forest zone (French Guyana; Fritsch, 1993, 1994).

Strong evapotranspiration

The tropical regions present a strong potential evapotranspiration. It varies between 1000 mm year^{-1} in the equatorial forest regions (Abidjan, Brazzaville or Manaus) to more than 2000 mm year^{-1} in the semi-arid regions of the Brazilian Nordeste and the Sahel. In one location, it reaches the near maximal value of the 3000 mm year^{-1} in subdesert regions.

Volume, form, and spatial and temporal distribution of precipitation

Precipitation in the tropical zone is very variable in space and in time. It goes from the equatorial zone where the annual precipitation often exceeds several metres distributed throughout the year, to the subdesert zone where one observes only some precipitation during one short season of rains (Fig. 12.2).

Shape and soil type of the watershed

The receiving body in the tropical environment is fragile. The natural vegetation possesses a huge complexity, but is in peril as a result of recent, disorganized and often destructive exploitation. The tropical soils formed themselves in conditions of heat and humidity, under natural processes and human activities. For example, one observes a great diversity of circulation mechanisms of water in watersheds where, according to circumstances, one recognizes various 'ways of water', such as total infiltration, surface runoff, rapid subsurface flow, infiltration of depth of surface runoff, runoff on saturated surface, and exchanges between groundwater and river (Fig. 12.3).

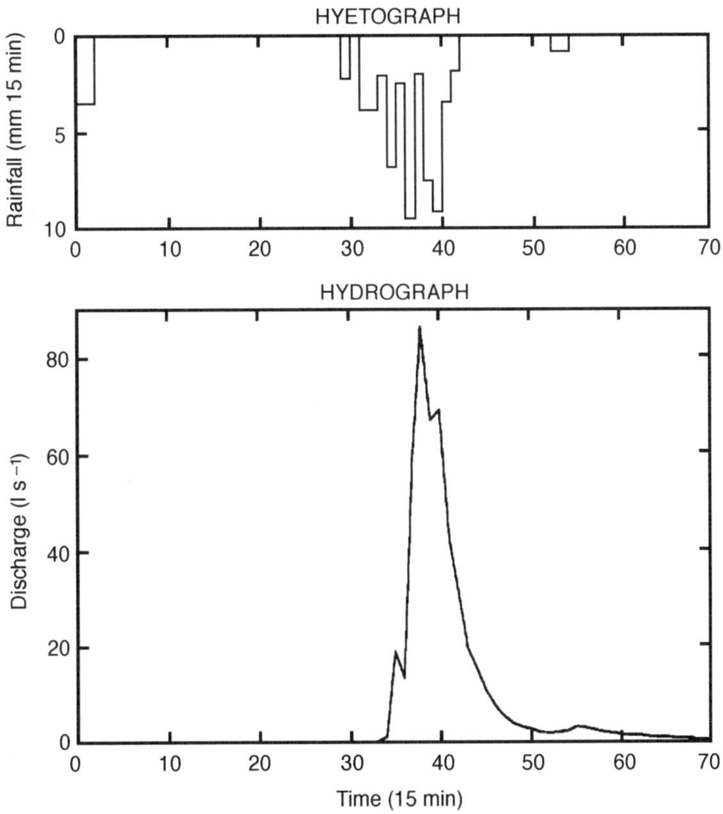

Fig. 12.2. Hyetograph and hydrograph in ECEREX catchment (French Guyana).

Review of Hydrological Tools

Prior to the advent of the unit hydrograph, by Sherman (1932), hydrological modelling was mostly empirical and based on limited data. In those days, graphs, tables and simple analytical solutions were the standard models, and hand calculations in conjunction with a slide rule reflected the computing prowess. In the decades that followed, perhaps up to the end of the 1950s, simple analytical models received a great deal of attention, the range of such models was limited and data processing was easily handled manually. In the 1960s, a new frontier in hydrological modelling unfolded on account of the digital revolution. For the first time it became possible to synthesize the entire hydrological cycle and process a large quantity of data (Crawford and Linsley, 1966). In subsequent years and decades, computing capability grew in leaps and bounds. Today's computing capability is virtually limitless. In the hydrological arena, synthesis of the entire hydrological cycle became popular. Its computer models are too many to count (Singh and Fiorentino, 1996). In the

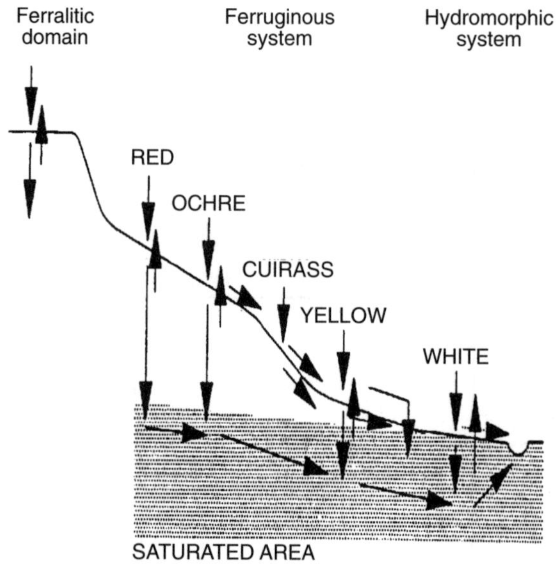

Fig. 12.3. Schematic description of water ways on a typical slope of the Booro Borotou basin (Ivory Coast, Hyperbav team, 1990).

meantime, growing environmental and ecological concerns demanded greater information from these models. Thus, these models were expanded in their scope to encompass simulation of erosion and sediment transport, chemical transport and the effect of land-use changes due to human activities. Clearly, the input data needed for such models multiplied, and so did the need for data processing. Models continued to expand, incorporating statistical analysis, risk and reliability, and graphics, and more and more demands were imposed on these models. Today, the models are becoming more and more widely distributed and some of them are being integrated with global circulation models. The net result is that the data needs have become enormous and so has the need for processing them. The old ways of handling and managing the data are no longer adequate. Furthermore, technology exists that can handle and process massive quantities of data.

Hydrologic modelling with GIS

Hydrologic models that lack a spatial component do not use GIS. However, concern for resource management and environmental quality requires application of distributed models. One characteristic of distributed-parameter models is that they are data intensive. Multiple data types such as hydrometeorology, topography, land use, soils, geology and streamflows are commonly required. Many of these data are often used to identify a hydrological unit, because application of these models requires partitioning of the watershed into homogeneous units. This often proves to be cumbersome. Therefore, a spatial data analysis and manipulation tool is desirable, and that tool could be GIS. Furthermore, the use of computers in hydrological modelling has become so widespread that a marriage between hydrology and GIS is a logical step.

Zhang *et al.* (1990) and Schultz (1993) presented overviews of hydrological modelling with GIS, and De Vantier and Feldman (1993) reviewed GIS applications in hydrological modelling, with particular reference to rainfall–runoff models, flood-plain management and forecasting, erosion prediction and control, water quality prediction and control, and drainage utility implementation. The integration of GIS with hydrological models is increasingly facilitating their design, calibration, modification and comparison. A short review of hydrological modelling with GIS is in order.

Hydrometeorological forecasting

Barrett (1993) presented an application of GIS in the Nile hydrometeorological forecast system in Egypt. A customized GIS was developed to show political boundaries, rivers, terrain elevation and gauging network and was employed to develop hydrological parameters for the basin and for multiple display features (maps and graphics).

Watershed modelling

Perhaps the greatest application of GIS in hydrology has been in the area of watershed modelling. This is understandable because the computer models of watershed hydrology are highly data intensive and GIS is a natural technology to process volumes of data. Schumann (1993) discussed development of conceptual semidistributed models and estimation of their parameters with the aid of GIS. Schultz (1996) reported on mesoscale modelling of runoff and water balances using RS and GIS data. By integrating hydrological modelling with GIS, Ross and Tara (1993) obtained increased detail in evaluation, minimized user subjectivity in parameter selection, and reduced costs of analysis as a result of significant time savings. In developing watershed models for two Sierra Nevada basins, Jeton and Smith (1993) found that the GIS was an objective, efficient methodology for characterizing a watershed and for delineating hydrological response units. Mitchell *et al.* (1993) validated the agricultural non-point source (AGNPS) model by use of a GIS for predicting runoff and sediment delivery from small watersheds of mild topography. Neumann and Schultz (1989) quantified the effect of catchment characteristics with the use of a GIS and a hydrological model. The influence of seasonally varying vegetation status on flood flows and the effect of land-use changes over the years on runoff were investigated.

Flood prediction

Chairat and Delleur (1993) coupled a GIS-GRASS with a physically based hydrological model, TOPMODEL (Beven and Kirby, 1979) to predict peak flows and to quantify the effects of resolution in a digital elevation model (DEM) and of contour lengths on the predicted peak flows. Muzik and

Pommery (1990) coupled a GIS with a flood prediction software package, based on the unit hydrograph theory, for prediction of design flood hydrographs for selected watersheds in Alberta's Rocky Mountain foothills.

Hydrological Modelling with Digital Terrain Models and Digital Elevation Models

The spatial distribution of topographical and geomorphological features of drainage basins and of drainage networks is relevant for determining the spatial variability of some hydrological phenomena (Fig. 12.4). Such phenomena are usually analysed with hydrological distributed models, which require input data for some parameters. Among these the most important are: catchment area; topographical form, slope and aspect; topological structure and slope of links for the drainage network. The quantification of these parameters is tedious and time-consuming when this is to be accomplished manually. There is, therefore, a need for automation. This is possible with GIS and digital elevation models (DEM) or digital terrain models (DTM), which together provide an ideal support for hydrological distributed models.

DTMs allow the determination of topographical parameters relevant for many hydrological models. TOPMODEL, developed by Beven and Kirkby (1979), is a physically based, topographically driven flood-forecasting model. It allows the prediction of distributed soil moisture status on the basis of spatial indices which depend on slope and cumulative upslope area derived by flow pathways (Quinn *et al.*, 1991). The Geomorphic Instantaneous Unit

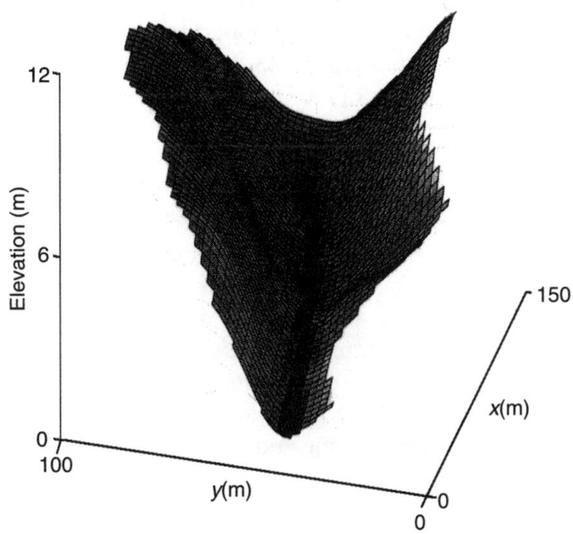

Fig. 12.4. 2 × 2 m DTM of the Seternbekken catchment (Lamb *et al.*, 1997).

Hydrograph (GIUH) (Rodriguez-Iturbe and Valdez, 1979, and others) is a commonly used method for determining the rainfall–runoff response of catchment; it uses Horton's ratios in order to calculate its characteristic parameters. Horton's ratios can be determined from digital channel networks derived from DTM. DTM and GIS were used by Djokic and Maidment (1991) for modelling consequences of urban storms. In particular, they used DTM for calculating parameters of concentration time. Lanza *et al.* (1993) developed a set of automatic procedures for flash-flood forecasting, using a DTM and an hydrologically oriented GIS.

Data sources for generating DTM and DEM

Data for DTM should be observations about the elevation and shape of the terrain surface, with particular attention to surface discontinuities (breaklines) and special locations (passes, pits, peaks, points of change in slope, ridges, stream channels, etc.). These data can be collected using a variety of techniques: ground surveys, photogrammetry using aerial photographs or satellite images, digitizing or scanning existing maps (Weibel and Heller, 1991). The choice of data source is dependent on a combination of several factors, such as the extent of the study area, and the desired precision and accuracy of DTM in relation to specific applications. Accuracy is a measure of how much an estimated value differs from the true value, while precision, in statistical terminology, is a measure of the dispersion of observations about a mean value and is not related to the meaning of the data but to equipment used to collect data. None of the previous collection methods is, however, error free and the extent of error is dependent on the specific technique.

Products derived from DTM: an example

Several algorithms have been developed in the past for demarcating a watershed. Most of them are combined algorithms which delineate both the watershed and drainage network, and are based on elevation matrices. One algorithm that delineates only the watershed has been proposed by Marks *et al.* (1984). Preprocessing of the elevation matrix is needed in order to remove pits as much as possible, because they could influence the basin delineation.

The algorithm delineates the drainage basin using slope and aspect matrices, and an outlet cell interactively selected by the user. In addition, the user has to assign a threshold value of slope for the definition of flat regions. Aspect values are coded using eight intervals each one of 45° (Fig. 12.5). The algorithm uses a 3×3 moving window, initially centred on the outlet selected cell. It marks this cell as belonging to the basin and sequentially checks the adjacent eight neighbours. A neighbour cell is upstream (i.e. within the same basin) if its aspect faces toward the centre cell or if it has a slope lower than the threshold value chosen by the user.

The algorithm is recursive in that, once an upstream neighbour is found, it is marked as belonging to the basin and the 3×3 window is moved and centred on it. At this stage the algorithm analyses neighbour cells and again

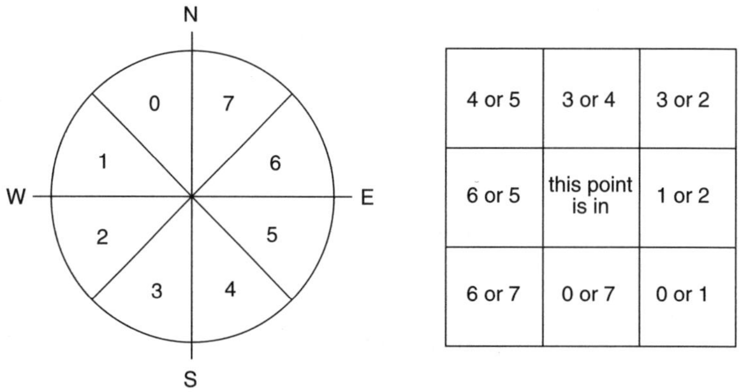

Fig. 12.5. Aspect classes and values of these classes for which a neighbour cell belongs to the same basin of the centre cell

starts checking for upstream cells. The algorithm is repeated until it encounters a basin boundary or an edge of the grid. It then backtracks, repeating the procedure until all cells belonging to the basin have been marked.

This algorithm is simple and has the advantage of analysing only grid cells that are in or just adjacent to the drainage basin and ignoring the rest of the elevation grid. In addition, during the recursive process, cells already marked as upstream are ignored.

Hydrological Modelling and Remote Sensing

In general, the conventional methods used in hydrological parameter estimation are expensive, time-consuming and difficult to update. While the conventional methods continue to be the principal technique, gathering a comprehensive overview of the distribution in a given area can be performed more rapidly and cheaply by remote sensing. However, satellite data have not yet been used widely for managing earth resources. This could be due to the lack of knowledge and understanding of data applicability, availability and accessibility. These matters all imply that the public use of satellite data to manage earth resources is still in its infancy, and that more techniques and programmes for application, data access, publicity and education are urgently needed.

Traditionally, the hydrological data, maps, aerial photography and tabular data describing tracts of land use/land cover have been stored separately. Updating the information has been cumbersome and expensive. Now, computer and remotely sensed data have given us the opportunity to merge data sets, and update the information by a low-cost operation. In particular, the widely used technique of GIS can provide a means of merging spatial and attribute data into the computerized database systems, allowing input, storage, retrieval, overlay, analysis and tabulation of geographically referenced data.

Usually, the information used in hydrological modelling is from ground-based observations, except for some specific atmospheric observations. Catchment data are taken from topographical maps, while conventional data are often gained by specific survey, if gathered at all. With RS techniques, the required input data or parameters are either provided indirectly or directly, giving the opportunity to use area-based large-scale synoptic information.

General information on this is cited in a variety of monographs and papers (e.g. International Symposium on Remote Sensing and Water Resources, 1990; Engman and Gurney, 1991; Rango, 1994; Schultz, 1994a,b). They show the following features to be detectable by RS: cloud cover, precipitation, radiation balance, extent of snow and ice and their properties, soil units, land use and DEM, and geomorphological or morphometrical features. So far, the most widely used application of RS has been the indirect generation of catchment data, needed for traditional hydrological models either as, for example, land use, soil units/properties and geomorphological or morphometrical features. In combination with auxiliary data and GIS, these data are used for calculation of, for example, infiltration capacity of soils, subcatchment slope and length of flood routing reach in models. A delineation of homogeneous subareas can also be done (Allewijn, 1986).

General approaches are directed towards hazard assessment, mitigation and remediation, especially in the field of mass movement such as soil erosion, landslides and mud flows (Claure *et al*., 1994) or geomorphologically and morphometrically derived flash-flood potential (Ashmawy, 1994). Furthermore, the use of land-use classification in soil erosion quantification using the universal soil loss equation (USLE) and other morphometric features and comparison with drainage density (Gonzalez Loyart *et al*., 1990) is a proven RS application. For water-balance models and soil–vegetation–atmosphere models, average land surface parameters can be estimated, yielding canopy resistance, albedo, leaf area index (LAI) and fractional vegetation cover (e.g. Huang *et al*., 1995). Concerning land use, the definition of classes and classification accuracy, including spectral and geometric resolution, are issues to be considered. Kite (1995), for example, determined 10% relative differences when comparing Landsat multispectral scanner (MSS) and National Oceanic and Atmospheric Administration (NOAA) advanced very high resolution radiometer (AVHRR)-derived land cover with 80-m and 1-km geometric resolution, respectively.

New Equipment in Hydrometry

Field hydrometry has evolved, as has the interpretation software. From entirely mechanical and manual materials, we passed progressively to autonomous supplies that can integrate a large amount of data. This evolution started in 1980, when ORSTOM installed the first automatic station on the Senegal river. Since this time, progress has been constant and this type of equipment covers all hydrological activities (hydrometry, meteorology, etc.).

A large number of sensors as well as interfacers and tools have been developed for some years. The sensors perform the actual measurement of the environment, converting the variable to be measured into an equivalent electrical signal. They are intelligent sensors, able to treat and visualize information on site. The data transmission can take place from afar by cable or by satellite. Data are memorized on the site (cartridge memory or electronic memory) and can be captured in portable terminals or computers. Finally, the measurements can be used by several devices that are connected in the network.

A sensor used in a hydrometer: an example in water-level recorders

The traditional system is the float-operated sensor. The water level is measured by means of a float. Via a cable or a perforated metal belt and the float wheel, the float causes the wheel to rotate. The wheel is connected to an encoder via a series of cogwheels (Fig. 12.6). The data are recorded on a paper diagram. Extraction of the data is difficult because the operator must interpolate the diagram to obtain processable values.

Now, with electronic data, we use a pressure sensor. The pressure sensor is an electronic device that is based on the deformation of a piece of silicon. When the water stresses (through a diaphragm) the silicon component, the silicon provides an electrical signal proportional to the hydrostatic pressure of water. There are two main groups of pressure transducers: piezoresistive sensors which give variation in resistance, and capacity sensors which give variation in electric capacity. For a pressure sensor, 0.1% is about the best resolution obtainable, and this is equivalent to 1 cm in a level change of 10 m or 1 mm in 1 m. However, pressure sensors can be installed without stilling wells and so are more adaptable than the float system. When the pressure is measured, the reading must be corrected for temperature and the level of the water.

Elsyde France has developed, with the ORSTOM Laboratory of Hydrology, an electronic and computer device for measurement of the level

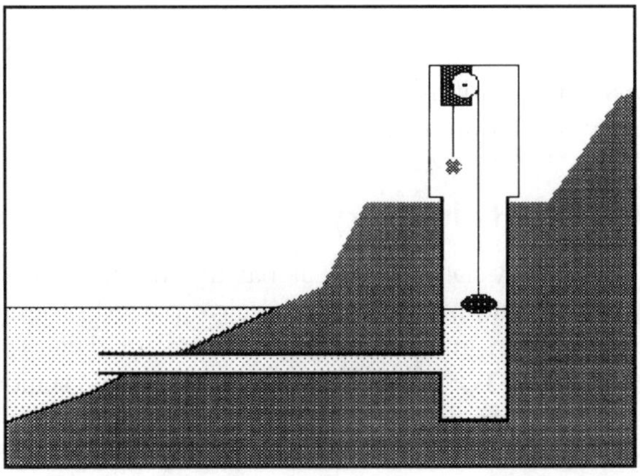

Fig. 12.6. Diagram of traditional water-level recorder.

and temperature of a constant-density liquid. Water levels and temperature are measured by a pressure transducer providing an electrical signal proportional to the pressure of liquid versus atmospheric pressure. This piezoresistive transducer is located in a stainless steel box and isolated from liquid by a diaphragm made of tantal platinum. The SPI probe contains a microprocessor that controls it and carries out transducer acquisitions, scaling and temperature compensation, to deliver true scale results including water level and temperature. This information is given in a binary slow-speed serial message (RS232), opto-isolator coupled and protected from lightning. Each SPI probe is pressure and temperature calibrated; calibration tables are recorded in the microprocessor memory. Zero and scale factor can be remote loaded.

Description of a data logging system

The essential component of these new automatic stations is the data logger system. It is the main part of the station and controls all of the components (Fig. 12.7). The electronics of a modern data logger consist of the interface, the clock, control circuits, memory functions and the power supply. The interface is a special piece of circuitry that handles the signals travelling between the logger and another piece of equipment, such as a sensor or a serial link. The interface converts the logger signals into signals suitable for the other pieces of equipment and vice versa. The interface also controls the speed at which serial data travels between control circuits and external equipment, such as a terminal or a computer. Sensors rarely produce electrical signals that are immediately acceptable by a logger, and some processing is needed to convert the sensor outputs from a low voltage level, resistance, capacitance or contacts closure into a suitable voltage. The ADC (analogue to digital converter) converts the sensor signals from their analogue form into digital form. ADCs are available that can produce 8, 10, 12 or 16 bit resolution; 10 and 12 bits are the norm. Pulse-producing sensors, such as tipping-bucket rain gauges, require no interfacing as they are already digital.

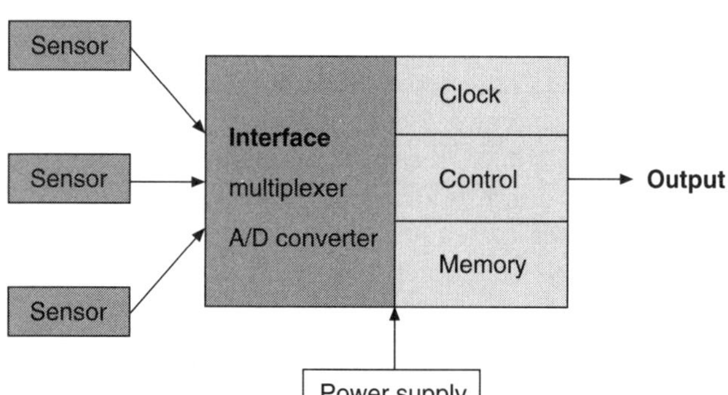

Fig. 12.7. Schematic diagram of a data logger.

The multiplexer sequentially switches the sensor signal to the analogue to digital converter. In the clock, the digital circuit chips do not operate continuously but are powered-up periodically for a few milliseconds to make their measurements before being powered-down again into a sleep mode to save energy. This requires a real-time clock and logic chips to orchestrate the whole process, using a series of pulses generated by a quartz crystal clock. The processing speed of the digital device is measured by the number of these pulses its clock produces per second. In recent loggers, more and more of these functions are performed by one single, self contained, computer chip.

The control unit circuits the way in which a logger function is programmed by the operator, via portable computer software or a terminal. This requires onboard memory in the logger to hold the instructions input to it from the computer (regarding such things as which of its channels are to be used, how often it is to log and what type of sensors are connected). The control unit is the part of the logger that performs all the operations needed to carry out a particular task.

The processed signals are stored in the memory. This is an area inside the logger (or microcomputer) where information is stored. To store data, it can use static RAM with a backup battery to preserve the data should the main battery fail. Some data loggers use memory cards, which are external static RAM with a backup battery made into a credit-card size cartridge.

Loggers are built with low-power consumption chips, which also run for most of the time in a sleep mode (in which only 5 or 10 μamp are consumed), so that they can be operated for long periods without a battery change. It is common practice to use solar panels to charge lead–acid or nickel–cadmium batteries via a specific regulator.

These new sensors and instruments are being improved continually, and are now employed in many thematic explorations in earth sciences. They are used easily, quickly and reliably with field data, that we can use in our models and other programmes.

Conclusions

Watersheds provide unique opportunities for scientific thematic research in hydrology in a broad sense and in the related sciences. Thinking and working at the level of entire watersheds provides a much-needed structure for the hydrological community. Fortunately, catchments' research is also relatively inexpensive. Of course research catchments are not a panacea: the spatial unit 'catchment' is not relevant for research on larger-scale hydrosystems. In research on water cycles and balances, there is much benefit in combining the different tools (catchments, weather stations, experimental plots, permanent and temporary sites) as well as the different approach (*in situ*, laboratory, various levels of modelling). Research catchments constitute preferential sites for interdisciplinary studies, the urgent need for which is increasingly felt in the international scientific community (Christophersen and Neal, 1990; De Coursey, 1991).

References

Allewijn, R. (1986) A methodology for integrating satellite imagery and field observations for hydrological regionalisation in Alpine catchments. In: Damen, M.C.J., Sicco Smit, G. and Verstappen, H.T. (eds) *Remote Sensing for Resources Development and Environmental Management, Proceedings,* Vol. 2. A. A. Balkema, Rotterdam, pp. 693–697.

Ashmawy, M.H. (1994) Assessment of flash flood potential of the Red Sea drainage basins along the Qena–Safaga highway, Eastern Desert, Egypt, *ITC Journal,* 1994–2, 119–128.

Barrett, C.B. (1993) The development of the Nile hydrometeorological forecast system. *Water Resources Bulletin* 29(6), 933–938.

Beven, K.J. and Kirby, M.J. (1979) A physically-based, variable contributoring area model of basin hydrology. *Hydrological Sciences Journal* 24, 43–69.

Chairat, S. and Delleur, J.W. (1993) Effects of the topographic index distribution on predicted runoff using GRASS. *Water Resources Bulletin* 29, 1029–1034.

Chevallier, P. (1990) *Complexité hydrologique du petit bassin versant. Exemple en savane humide: Booro-Borotou* (Côte d'Ivoire). ORSTOM, Paris, coll. Etudes et Thèses, 332 pp.

Christophersen, N. and Neal, C. (1990) Linking hydrological, geochemical and soil chemical processes on the catchment scale: year interplay between modelling and field work. *Water Resources Research* 26, 3087–3100.

Claure, B., Maldonado, J., Vargas, O. and Valenzuela, C.R. (1994) A conceptual approach to evaluating watershed hazards: the Tunari watershed, Cochabamba, Bolivia, *ITC Journal* 1994–3, 283–291.

Crawford, N.H. and Linsley, R.K. (1966) *Digital simulation in hydrology: The Stanford Watershed Simulation Model IV. Technical Report No. 39,* Department of Civil Engineering, Stanford University, Stanford, California.

De Coursey, D.G. (1991) Mathematical models: research tools for experimental watersheds. In: Bowles, D.S. and O'Connel, P.E. (eds) *Recent Advances in the Modelling of Hydrologic Systems, NATO ASI Series C,* Vol. 345. Kluwer Academic Publishers, 591–612.

De Vantier, B.A. and Feldman, A.D. (1993) Review of GIS applications in hydrological modelling. *Journal of Water Resources Planning and Management* 119, 246–261.

Djokic, D. and Maidment, D.R. (1991) Terrain analysis for urban stormwater modelling. *Hydrological Processes* 5.

Dubreuil, P. (1986) Review of relationships between geophysical factors and hydrological characteristics in the tropics. *Journal of Hydrology* 80, 201–222.

Engman, E.T. and Gurney, R.J. (1991) *Remote Sensing in Hydrology.* Chapman & Hall, London, 225 pp.

Fritsch, J.M. (1993) Hydrological effects of deforestation and alternative land uses: a French experiment in the Amazonian rain forest. In: *Environment and Forestry Management.* Asian Productivity Organization, Tokyo, pp. 74–105.

Fritsch, J.M. (1994) The hydrological effects of clearing tropical rain forest and of the implementation of alternative land uses. *Proceedings of IAHS Scientific Assembly, Yokohama, 13–15 July, IAHS Publication No. 216,* pp. 53–66.

Gonzalez Loyart, M.M., Leguizamo, S. and Zano-Rodrigue, B. (1990) A simple method to assess potential soil erosion by using geographically referenced information in digital form. In: *International Symposium on Remote Sensing and Water Resources, Enschede, 20–24 August, Proceedings.* R. van Acken GmbH, Lingen (Ems), pp. 859–870.

Huang, X., Lyons, T.J., Smith, R.C.G. and Hacker, J.M. (1995) Estimation of land surface parameters using satellite data, *Hydrological Processes,* 9, 631–643.

Hyperbav team (1990) *Structure et fonctionnement hydropédologique d'un petit bassin versant de savane humide.* Journées hydropédologiques de l'ORSTOM, 12–13 septembre 1989, ORSTOM, Paris, coll. Etudes et Thèses, 308 pp.

Jeton, A.E. and Smith, J.L. (1993) Development of watershed models for two Sierra Nevada basins using a geographic information system. *Water Resources Bulletin* 29, 923–932.

Kite, G.W. (1995) Scaling of input data for macroscale hydrological modelling, *Water Resources Research* 31, 2769–2781.

Lamb, R., Beven, K. and Myrab, D.S. (1997) Discharge and water table predictions using a generalized TOPMODEL formulation. *Hydrological Processes* 11, 1145–1167.

Lanza, L., Conti, M. and La Barbera, P. (1993) Automated modelling of flashfloods in the Mediterranean area. The September 1992 event over the Liguria region. *Proceedings of the 12th IASTED International Conference, Innsbruck, Austria.*

Likens, G.E., Bormann, F.H., Pierce, R.S., Eaton, J.S. and Johnson, N.M (1977) *Biogeochemistry of a Forested Ecosystem.* Springer Verlag, Berlin, 146 pp.

Marks, D., Dozier, J. and Frew, J. (1984) Automated basin delineation from digital elevation data. *Geoprocessing* 12.

Mitchell, J.K., Engel, B.A., Srinivasan, R. and Wang, S.S.Y. (1993) Validation of AGNPS for small watersheds using an integrated AGNPS/GIS system. *Water Resources Bulletin* 29, 833–842.

Muzik, I. and Pommery, S.J. (1990) A geographic information system for prediction of design flood hydrographs. *Canadian Journal of Civil Engineering* 17, 965–973.

Neumann, P. and Schultz, G.A. (1989) Hydrological effects of catchment characteristics and land use changes determined by satellite imagery and GIS. *IAHS Publication,* No. 186, pp. 169–176.

Quinn, P., Beven, K., Chevallier, P. and Planchon, O. (1991) The prediction of hillslope flow paths for distributed hydrological modelling using digital terrain models. *Hydrological Processes* 5, 59–79

Rango, A. (1994) Applications of remote sensing by satellite, radar and other methods to hydrology. *WMO-No. B04.* Secretariat of the World Meteorological Organization, Geneva, Switzerland, p. 33.

Rodier, J.A. and Ribstein, P. (1988) Estimation des caractéristiques de la crue décennale pour les petits bassins versants couvrant de 1–10 km. ORSTOM, Montpellier, France, 133 pp.

Rodriguez-Iturbe, I. and Valdez, J.B. (1979) The geomorphological structure of hydrological response. *Water Resources Research* 15, 1409–1420.

Ross, M.A. and Tara, P.D. (1993) Integrated hydrological modelling with geographic information systems. *Journal of Water Resources Planning and Management* 119, 129–140.

Schultz, G.A. (1993) Application of GIS and remote sensing in hydrology. *IAHS Publication No. 211,* pp. 127–140.

Schultz, G.A. (1994a) Remote sensing for forecasting of floods. In: Rossi, G., Harmancioglu, N. and Yevjevich, V. (eds) *Coping with Floods, Proceedings of the NATO Advanced Study Institute on Coping with Floods, Erice, Italy, 3–15 November 1992.* Kluwer Academic Publishers, Dordrecht, pp. 459–472.

Schultz, G.A. (1994b) Meso-scale modelling of runoff and water balances using remote sensing and other GIS data, *Hydrological Sciences Journal* 39, 121–142.

Schumann, A.H. (1993) Development of conceptual semi-distributed hydrological models and estimation of their parameters with the aid of GIS. *Hydrological Sciences Journal* 38, 519–528.

Sherman, L.K. (1932) Streamflow from rainfall by unit graph method. *Engineering News Record* 108, 501–505.

Singh, V.P. and Fiorentino, M. (1996) Hydrologic modeling with GIS. In: Singh, V.P. and Fiorentino, M. (eds) *Geographical Information Systems in Hydrology.* Water Science and Technology Library, Vol. No. 26, Kluwer Academic Publisher, Dordrecht, Netherlands, pp. 1–13.

Weibel, R. and Heller, M. (1991) Digital terrain modelling. In: Maguire, D.J., Goodchild, M.F. and Rhind, D.W. (eds) *Geographical Information Systems*, Vol. 1. Longman Scientific Technical.

Zhang, H., Haan, C.T. and Nofziger, D.L. (1990) Hydrologic modelling with GIS: an overview. *Applied Engineering in Agriculture, ASAE* 6, 453–458.

13

Off-site Impacts and Nutrient Dynamics in Catchment Research

W. Zech, K. Kaiser, W. Wilcke, J. Lehmann and B. Huwe
Institute of Soil Science and Geography, University of Bayreuth, 95440 Bayreuth, Germany

To maintain and increase sustainability is a major global challenge. It includes biophysical and socio-economic endeavours. Biophysically seen, the sustainability of a given land-use system under defined site conditions can only be evaluated in a reliable way if we know the impact of land use on water and element fluxes within the different compartments of the ecosystem. A suitable approach to assess human impact on ecosystems is based on the establishment of an ecosystem balance, quantifying input, output and the internal water and element fluxes using, for example, rain collectors, tensiometers and suction cups.

With respect to the catchment approach, different levels of intensity can be realized:

1. Quantifying only water discharge and element output in solid and liquid states may give initial information on off-site effects. As an example, the effects of forest harvesting on stream water pH and NH_4^+ concentration in an small-scale alpine water catchment is presented.
2. Sustainability can be better evaluated if, besides the output, the input is known (e.g. via precipitation). Input and output of solutes of small-scale alpine water catchments are given to illustrate this approach.
3. The most sophisticated design for studying the nutrient dynamics is based on a combination of plot experiments measuring water and element fluxes in the plant cover and soil and the catchment approach. Exemplary results of two plot experiments testing different land-use practices in Brazil and Kenya are presented.

Introduction

The sustainable management of the environment, and especially of soils, is obligatory with respect to the support and welfare of a growing world

 Soil Erosion at Multiple Scales
(eds F.W.T. Penning de Vries, F. Agus and J. Kerr)

population. Sustainability comprises socio-economic as well as biophysical aspects. Biophysically seen, the evaluation of sustainability is not easy. It can be realized on different levels of intensity, for instance by quantifying crop yields over years, soil analyses and fertilizer experiments carried out on plots. The results refer to the specific experimental plot and it is frequently difficult to extrapolate them to larger areas, especially in mountainous regions, where soil erosion is a major problem, causing serious on- and off-site damages. On-site effects concern, for instance, the chemical and physical fertility of soils by nutrient depletion and the reduction of water-holding capacity. Off-site effects include flooding of lowlands, damaging infrastructure, increasing health risks, siltation of dams and shortening the lifetime of reservoirs for drinking and irrigation water. For economic aspects of off-site effects see Craswell *et al.* (Chapter 11, this volume) and Enters (Chapter 1, this volume).

The analysis and understanding of the complexity of these on- and off-site effects is a prerequisite to improve land-use practices and their sustainability. For mountainous areas the most appropriate approach is based on the quantification of water and element fluxes in water catchments. In the following sections the potential of this catchment approach for the analysis of off-site impacts and nutrient dynamics of land-use systems will be discussed.

Prerequisites for Catchment Studies

Scientifically seen, catchment studies give reliable, ecosystem-relevant information on the effects of land-use changes only if the following prerequisites are taken into consideration:

1. To analyse the effects of well-defined land-use changes a minimum of two catchments must be selected; three to four catchments would be optimum.
2. At the beginning of the experiments, these catchments should be as similar as possible, with respect to geology, climate, relief, soils and vegetation/land-use.
3. Catchments should have well-defined boundaries and be compact underground to minimize water and element losses.
4. Effects of distinct land-use changes can be best quantified using microcatchments of about 5–10 ha. Using larger catchments makes it more difficult to interpret the effects of defined land-use changes.
5. Catchments must be accessible.
6. Due to the high costs for the installation (e.g. weirs, meteorological station), long-term management must be guaranteed by the land owners, including
  (i) protection against fire
  (ii) no uncontrolled land-use during the baseline measurement period (at least 2 years) and the introduction of new land-use techniques.

After this short description of major prerequisites for the successful realization of catchment studies, different levels of study intensities will be presented in the following sections, comprising:

- the quantification of the output only;
- the combined quantification of output and input; and
- the combination of output/input measurements with plot experiments.

Quantification of Output

Off-site effects of land-use practices can be characterized by quantifying water discharge and element output in solid and liquid state at weirs, installed at lower parts of the catchments. Weirs should be equipped with:

- an automatic water-level recorder for continuous reading of water discharge;
- an automatic water sampler for the continuous collection of bulk water samples, including dissolved, colloidal and particulate matter; and
- a sediment trap for the collection of bedload.

In Fig. 13.1 two similar model catchments are shown. Assuming that at the beginning of the experiments catchments 1 and 2 are used for shifting cultivation, the off-site effects of improved land-use techniques can be studied according to the following design.

1. Water discharge and element output are measured at the weirs of both catchments for 2 years. This approach gives baseline data.
2. In the third year the improved land-use techniques (e.g. agroforestry instead of slash and burn) will be introduced in catchment 1, whereas in catchment 2 the traditional land-use system (shifting cultivation) is continued. The water discharge and element output of catchment 2 provides reference

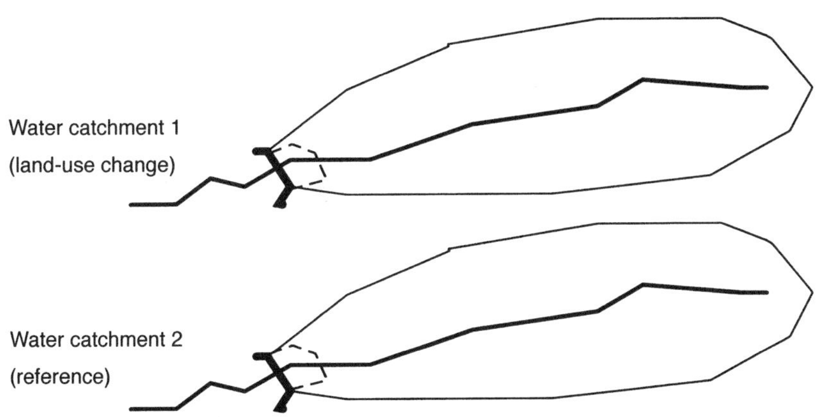

Fig. 13.1. Design of two model microcatchments for the quantification of off-site effects of land-use practices which are introduced in catchment 1 after a calibration time of at least 2 years. In catchment 2, land use is not changed; it serves as reference.

data to compare with catchment 1. The differences indicate off-site effects of the introduced, improved land-use practices.

Since off-site effects generally concern not only the mountainous areas but also the lowlands, the installation of additional weirs downstream is necessary if the whole range of the off-site effects from the microcatchments to the water reservoirs is to be examined (Fig. 13.2). In combination with remote sensing techniques, off-site effects (e.g. inundations, river bank dynamics) can be better evaluated.

Monitoring of water discharge (e.g. in $m^3\ h^{-1}$) and sampling (bulk water samples, bedload) can be done continuously and/or discontinuously using, for instance, automatic water-level recorders and automatic water samplers. To quantify rain storm effects, discontinuous sampling may be adequate. Bedload can be sampled weekly, monthly or yearly. Bulk water samples are analysed for:

1. pH, electric conductivity (EC)
2. dissolved fraction (i.e. 'truly' dissolved and colloidal matter < 0.45 μm)
   (i) total element concentrations:
      e.g. Ca^{2+}, Mg^{2+}, Fe^{3+}, Al^{3+}, K^+, Na^+, Mn^{2+}, with: atomic absorption spectrometry (AAS) or inductive coupled plasma (ICP)-mass spectrometry (MS) or atomic emission spectrometry (AES)
         e.g. S, P, with ICP-AES
         e.g. N, with N_{tot} analyser
         e.g. dissolved inorganic and organic carbon (DIC, DOC), with DOC analyser
   (ii) dissolved species concentrations:
      e.g. NH_4^+, Ca^{2+}, Mg^{2+} Fe^{3+}, Al^{3+} with: ion chromatography (IC)/ photometry
      e.g. NO_3^-, SO_4^{2-}, $H_2PO_4^-$, HCO_3^-, with IC/photometry
3. particulate matter (fraction) > 0.45 μm
   (i) total element concentrations:
      e.g. metals, P, S with wet digestion, AAS, ICP, photometry
      e.g. C, N with autoanalyser

For the analysis of bedload samples the same methods are used as for particulate matter.

Figures 13.3 and 13.4 give results from microcatchment studies carried out in the Bavarian Alps (Bäumler and Zech, 1997). The objective was to quantify the effects of forest clearing. In 1990, two similar, compact microcatchments (WC 1 and WC 2) of about 10 ha were selected and fully equipped (weirs, meteorological station, rain gauges, rainfall collectors, lysimeters, suction cups, tensiometers). After baseline measurement from January 1990 until the end of 1992, forests were cleared in WC 1. The pH of the stream water at the weirs was not affected by forest clearing (Fig. 13.3a), but the NH_4^+ concentration increased significantly (Fig. 13.3b).

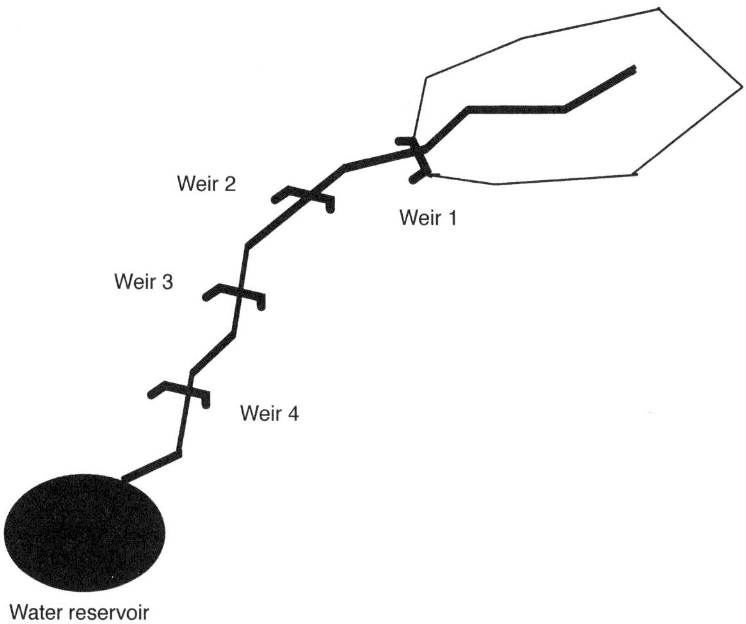

Fig. 13.2. Optimum conditions to evaluate off-site effects of land-use changes comprise additional equipment downstream in combination with remote sensing techniques, which provide information about the extent and duration of inundations, and river bank dynamics.

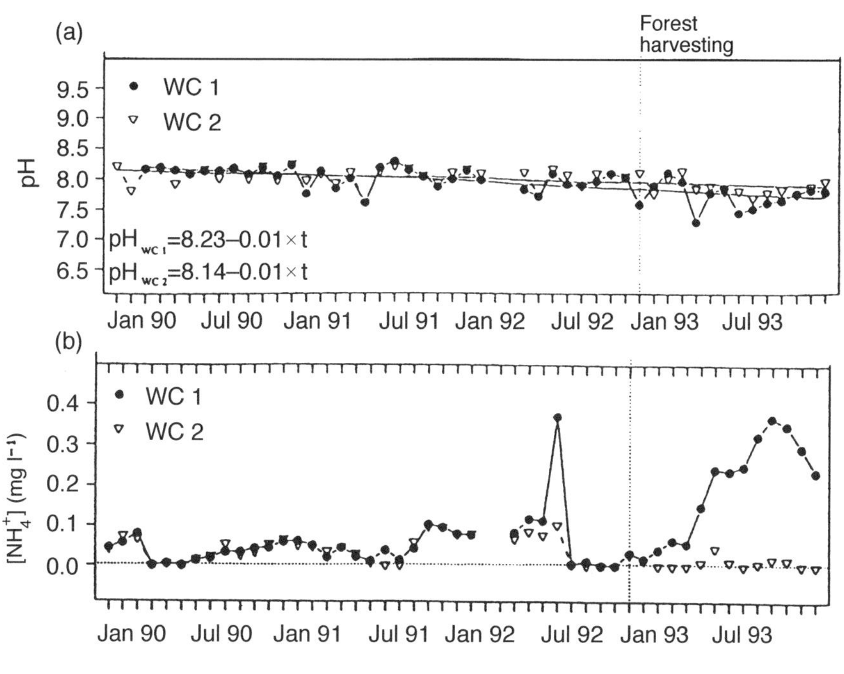

Fig. 13.3. Effects of selected forest harvesting in microcatchment studies carried out in the Bavarian Alps. Effects on pH of the stream water (a) and on its NH_4^+ concentration (b) (Bäumler *et al.*, 1995). WC 1, water catchment 1; WC 2, water catchment 2.

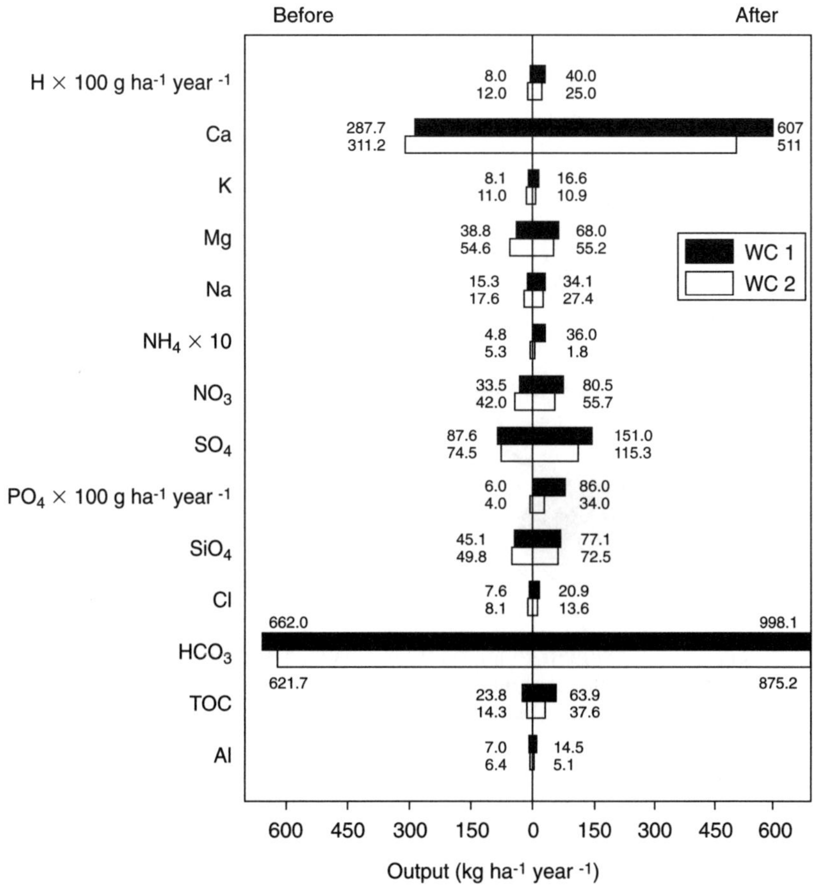

Fig. 13.4. Element output of a microcatchment (WC 1) before and after forest harvesting compared with an untouched reference (WC 2; Bäumler *et al.*, 1995). The values given indicate mean annual fluxes calculated from weekly measurements for 24 months before and 18 months after forest harvesting.

In Fig. 13.4 dissolved element output before and after forest harvesting in WC 1 is compared with the reference WC 2. The results indicate that forest harvesting increased the output of many solutes, especially of Ca, K, NH_4^+, $H_2PO_4^-$ and Al. Output of particulate matter and bedload is not considered in Fig. 13.4, but they contribute to the off-site effects of forest clearing.

Generally, output measurements as described here are combined with quantification of the inputs into catchments. This approach, characterized in more detail below, allows the calculation of water balances, element budgets and gives additional information about transformation processes within catchments. It thus contributes to the understanding of nutrient dynamics in catchment-oriented research (Likens *et al.*, 1969; Bruijnzeel, 1982; Feger *et al.*, 1990; Bäumler *et al.*, 1995; Likens and Bormann, 1995).

Comparison of Catchment Output with Catchment Input

Input data usually refer to atmospheric inputs, but other inputs such as fertilizers have to be taken into consideration. For the establishment of nutrient budgets the export via crops must be analysed. The quantification of gaseous inputs/outputs is generally difficult. Input studies need additional equipment:

- a meteorological station with rain gauges for the continuous measurement of precipitation and for the calculation of the water balance;
- rainfall collectors in several replicates to measure bulk precipitation (i.e. precipitation + dry deposition) and throughfall below the canopy (i.e. precipitation + dry and wet interception);
- litter collectors to record soil input via litter; and
- stem-flow collectors to record water and element inputs along tree stems.

Litter and stem-flow collectors are helpful in understanding redistribution processes within the catchment. They do not add information about input into the catchment.

In Fig. 13.5, the element inputs into forested alpine microcatchments are compared with outputs. It is obvious that N species and heavy metals are retained, whereas Ca, Mg, Na, SO_4^{2-}, SiO_4 and HCO_3^- are released. Comparing input/output data of control catchments with model catchments after the introduction of improved land-use practices supplies information about the effects of land-use changes on nutrient budgets.

Combination of Input/Output Studies with Plot Experiments

In input/output analyses, soils are considered as black boxes. For detailed information on nutrient dynamics within catchments the combination of input/output studies with plot experiments, established within the catchments, is necessary. This combination may also help to detect relationships between soil solution and stream water chemistry, that is, off-site effects.

Figure 13.6 shows a typical set-up of the instruments of a plot experiment in a forest. If surface runoff occurs, surface flow collectors must be installed additionally. With this equipment, the calculation of water and nutrient fluxes on plane sites is possible (Matzner and Ulrich, 1981); on sloping lands severe difficulties may arise, for example from interflow. If these difficulties cannot be solved by hydrologists or soil physicists, water and element fluxes within the soil compartments cannot be calculated. However, the installation of lysimeters, suction cups and tensiometers allows at least the quantification of the effects of land-use change on the nutrient concentrations in the soil solution at different soil depths. These data provide valuable information on sustainability (Meiwes *et al.*, 1984).

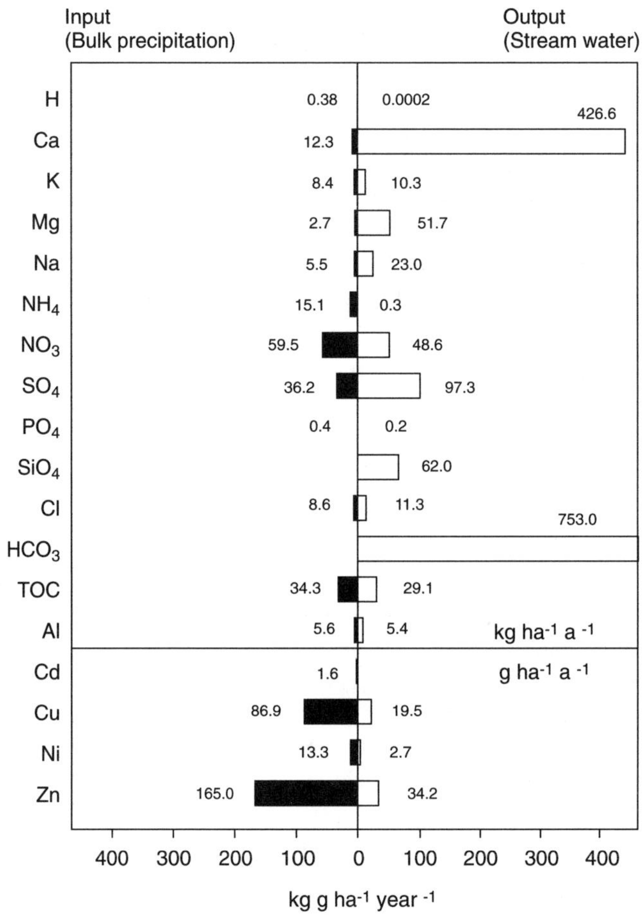

Fig. 13.5. Comparison of the element input and output of an undisturbed, forested, alpine microcatchment (Bäumler *et al.*, 1995). The values given indicate mean annual fluxes calculated from weekly measurement for 24 months before and 18 months after forest harvesting.

To fulfil the prerequisites for statistical evaluation, at least three plots within one microcatchment should be installed at as great a distance from each other as possible ('statistical independence').

Preliminary results of the Management of Acid Soils (MAS) project in the Brazilian Cerrados are shown in Fig. 13.7. The objective of this project is to find out how different land-use systems influence the water and nutrient fluxes in the Cerrado environment. Soils under pine plantations, and especially continuously cropped sites, reveal higher NO_3^- concentration (up to 14 mg NO_3^--N l^{-1} in the soil solution) than those covered with natural Cerrado vegetation and degraded pasture plants (< 1 mg NO_3^--N l^{-1}, Fig. 13.7). Soils of

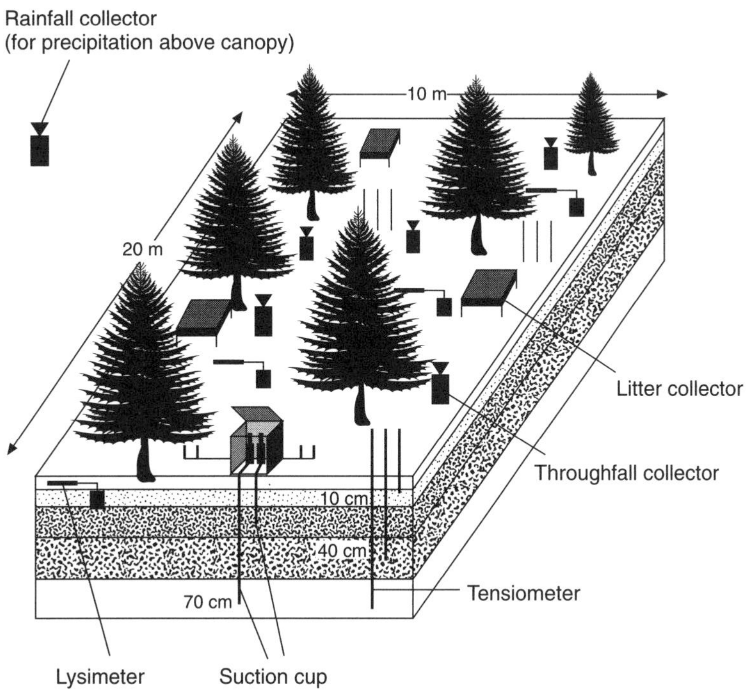

Fig. 13.6. Instrumental set-up of a plot experiment under forest cover.

the 'improved pasture' showed soil solution NO_3^- concentrations of about 2 mg N l^{-1} in 200 cm depth, indicating export of N out of the rooting zone.

Multiplying the nutrient concentrations in the soil solution with the water fluxes gives information on the nutrient fluxes, as presented in Figs 13.8 and 13.9 for an agroforestry system (*Acacia saligna* and *Sorghum*) in comparison with monocrops of *Acacia* and *Sorghum*, established on alluvial soils in northern Kenya (Lehmann, 1997). The results indicate that *Sorghum* mainly absorbs N from surface soil horizons, and *Acacia* from subsoil horizons, in agreement with the distribution of the fine root biomass (Lehmann, 1997). The combination of *Acacia* plus *Sorghum* drastically reduces the soil N concentrations and consequently N losses into the subsoil (Fig. 13.9).

For further results of plot experiments with similar equipment see Likens *et al.* (1977); Matzner and Ulrich (1981); Meiwes *et al.* (1984) and Likens and Bormann (1995).

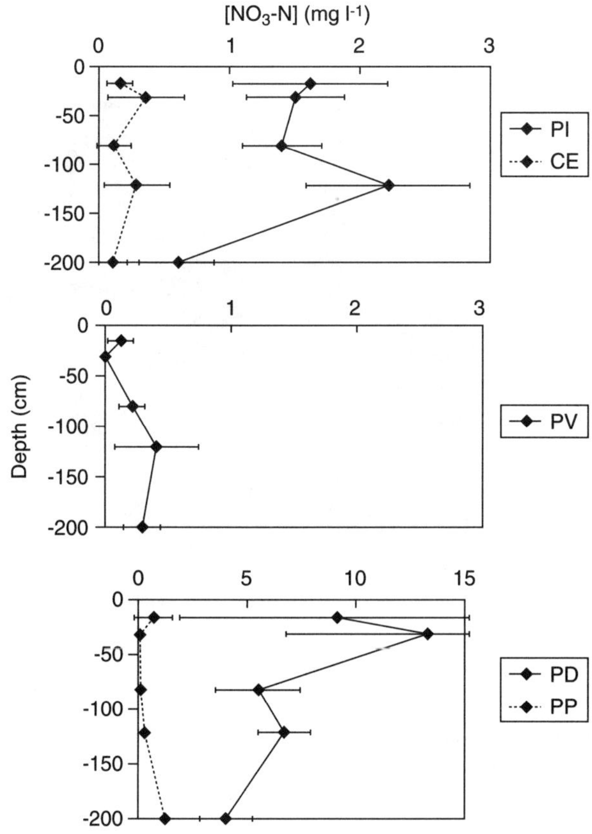

Fig. 13.7. Average NO_3^--N concentrations (and standard deviations) in the soil solution of differently used Cerrado oxisols (Brazil). Samples were taken weekly in April 1997. CE, natural Cerrado; PI, *Pinus caribaea* Morelet plantation; PV, degraded pasture; PP, improved pasture; PD, no-till continuous crop cultivation.

System Analysis, Prediction and Optimization of Management Systems

The design of sustainable, site-adapted and productive land-use systems requires not only the consideration of specific site and crop conditions but has also to take into account restrictions due to priorities of land users, the socio-economic background and the need to minimize off-site damage. Adapted simulation models provide useful tools for the optimization and risk assessment of different land-use systems. Furthermore, the application and transfer of research results to other sites and the regionalization of information is strongly supported by the use of simulation models (Schulz and Huwe, 1994; Huwe and Totsche, 1995; Schulz *et al.*, 1997, 1998a).

The scientific approach of our research is based on a conceptual model that integrates the main compartments of the systems under study, such as

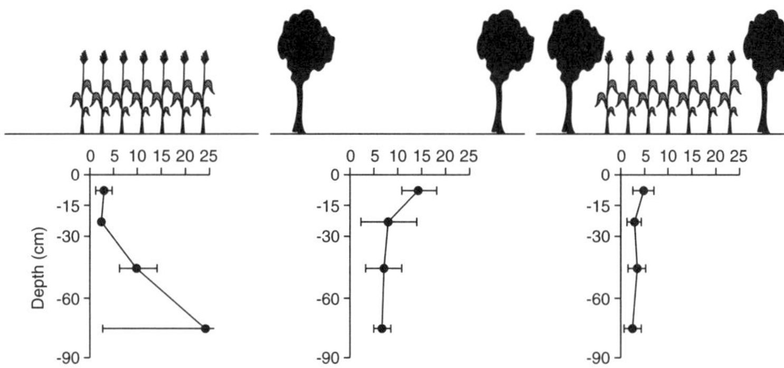

Fig. 13.8. Soil mineral N concentration in the soil solution (mg l^{-1}) of an agroforestry system composed of *Acacia saligna* and *Sorghum* (right), an *Acacia saligna* plantation (middle) and a *Sorghum* monocrop (left) (Lehmann, 1997).

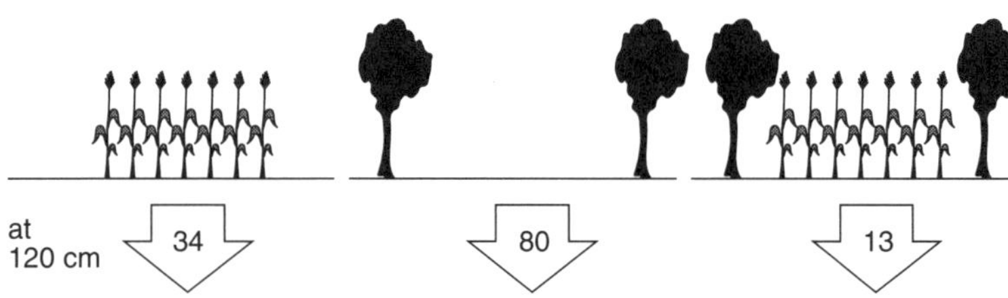

Fig. 13.9. N loss in the subsoil (kg ha^{-1} (100 day)$^{-1}$) of an agroforestry system composed of *Acacia saligna* and *Sorghum* (right)), an *Acacia saligna* plantation (middle), and a *Sorghum* monocrop (Lehmann, 1997).

atmosphere, vegetation, soil, lithosphere and hydrosphere, as well as the associated fluxes, interactions, phase transitions and turnover processes within the compartments. In the bulk soil, generally the solid, liquid and gaseous phases need to be distinguished. The corresponding integrative representation of processes in simulation models can be used for comprehensive system analyses, the valuation of single processes on different scales and the calculation of prognostic scenarios.

Simulation models yield quantitative estimations of water and matter fluxes as 'indicators of the stability of different land-use systems' and thus contribute considerably to system optimization and decision-making strategies. Regarding different scales (micro-, plot, field, catchment scale), models of different integration and complexity levels must be used. While on the plot level, deterministic and physically based process and system analysis models are generally preferable, grey and black box models or effective models, based on fuzzy set theory (Schulz and Huwe, 1994) may be more appropriate on the catchment scale.

Huwe and Totsche (1995) presented an application of an integrative approach to simulate field-scale nitrogen dynamics. They used a deterministic, process-based simulation to model water, heat and nitrogen dynamics simultaneously on the plot scale, where the assumption of parameter homogeneity holds true. On the field scale, heterogeneity, spatial variability and scale dependencies become more and more important. To cope with these characteristics, the model was combined with stochastic simulations based on Monte Carlo simulations to consider parameter variability. This was done by dividing the whole catchment into multiple, independent, spatial domains (subcatchments), each of a size for which the homogeneity assumption holds true. The deterministic model was then applied to each subcatchment, giving individual and independent simulation results for nitrogen dynamics and water flow. Using a geographical information system (ARC-Info), a synopsis of the individual results was achieved. Sensitivity analysis was used to quantify the impact of spatial variability on the model processes. This integrative approach resulted in good agreement between measured and modelled nitrogen deep seepage and groundwater recharge on both the plot scale and the overall catchment scale.

Alternatively, the plot to catchment problem might be solved by spatial extrapolation of point measurements. C. Würlen *et al.* and K. Schulz *et al.* (unpublished) presented approaches to extrapolate spatially point measurements of micrometeorological data for a small catchment site. As these parameters are subject to variations in space due to factors such as topography and vegetation, extrapolation schemes based on data from a central meteorological station and other easily measurable parameters (here the digital elevation model of the catchment) were first extracted. By comparing predicted (extrapolated) data with data from several additional stations, these extrapolation schemes were validated and thus allow for a data transfer (regionalization) from the plot to the catchment scale.

For the application of simulation models at different scales to analyse, predict and optimize management systems, measurements on the catchment scale as well as on the plot level must be designed in order to provide the necessary data sets for specification of initial and boundary conditions, parameter identification and model calibration. Special consideration has to be given to uncertainties in data caused by heterogeneity, measurement errors and the use of estimators as well as to situations where only 'soft' information (thematic maps, vaguely described information, expert knowledge) is available (this might be the rule on the larger scale). Here, fuzzy set theory may provide a useful tool for integrating uncertainties of different quality into model calculations on both catchment and plot scale. The fuzzy approach was successfully used to model water flow in the unsaturated zone (Schulz and Huwe, 1997) by integrating stochastically described measurement errors of boundary conditions and vaguely described soil hydrological properties into the calculation process. In a second application, the modelling of chemical equilibrium composition of aqueous systems (K. Schulz *et al.*, unpublished) under vaguely described thermodynamic parameters was presented and discussed in the light of risk analysis.

Conclusions

In this chapter an optimum design for the study of off-site effects and nutrient dynamics of different land-use systems using the catchment approach in combination with plot experiments is presented. The results of such an ecosystem-oriented study may help ecologists, agronomists and decision-makers to find appropriate solutions for the protection of natural resources (especially in mountainous areas with high erosion hazard) and for the welfare of farmers. Data will be used for the validation of existing models and/or for the establishment of new models which allow the prediction of the water and element fluxes, and thus may estimate the sustainability of a given land-use system and land-use changes under different site conditions.

References

Bäumler, R. and Zech, W. (1997) Atmospheric deposition and impact of forest thinning on the throughfall of mountain forest ecosystems in the Bavarian Alps. *Forest Ecology and Management* 95, 243–251.

Bäumler, R., Blessing, U. and Zech, W. (1995) Untersuchungen zur Stoffdynamik zweier bewaldeter Kleineinzugsgebiete im Flysch – Auswirkungen eines geregelten forstlichen Eingriffs. *Forstwissenschaftliches Centralblatt* 114, 261–271.

Bruijnzeel, L.A. (1982) Hydrological and biochemical aspects of man-made forests in south-central Java, Indonesia. *Final Report, Vol. 9, Nuffic Serayu Valley Project ITC/GUA/VU/1*, The Hague, The Netherlands, 250 pp.

Feger, K.H., Brahmer, G. and Zöttl, H.W. (1990) Element budgets of two contrasting catchments in the black forest (Federal Republic of Germany). *Journal of Hydrology* 116, 85–99.

Huwe, B. and Totsche, K.U. (1995) Deterministic and stochastic modelling of water, heat and nitrogen dynamics on different scales with WHNSIM. *Journal of Contaminant Hydrology* 20, 265–284.

Lehmann, J. (1997) Tree–crop interactions in a runoff agroforestry system in northern Kenya. *Bayreuther Bodenkundliche Bericht* 57, 151 pp.

Likens, G.E. and Bormann, F.H. (1995) *Biogeochemistry of a Forested Ecosystem.* New York, Springer-Verlag, 159 pp.

Likens, G.E., Bormann, F.H., Johnson, N.M., Fisher, D.W. and Pierce, R.S. (1969) Effects of forest cutting and herbicide treatment on nutrient budgets in the Hubbard Brook watershed-ecosystem. *Ecological Monographs* 40, 23–47.

Likens, G.E., Bormann, F.H., Pierce, R.S., Eaton, J.S. and Johnson N.M. (1977) *Biogeochemistry of a Forested Ecosystem.* Springer-Verlag, New York.

Matzner, E. and Ulrich, B. (1981) Bilanzierung jährlicher Elementflüsse in Waldökosystemen im Solling. *Zeitschrift für Pflanzenernährung und Bodenkunde.* 144, 660–681.

Meiwes, K.J., Hauhs, M., Gerke, H., Asche, N., Matzner, E. and Lamersdorf, N. (1984) Die Erfassung des Stoffkreislaufs in Waldökosystemen – Konzept und Methodik. *Berichte Forschungszentrum Waldökosysteme/Waldsterben*, Göttingen 7, 68–142.

Schulz, K. and Huwe, B. (1994) Regionalisierung von Transportvorgängen in der ungesättigten Bodenzone. *Eco-Informa-94, Band 7.* Publisher Umweltbundesamt, Bonn, Germany.

Schulz, K. and Huwe, B. (1997) Water flow simulation in the unsaturated zone with imprecise parameters using a fuzzy approach. *Journal of Hydrology* 20, 211–229.

Schulz, K., Würlen, C., Eiden, R. and Huwe, B. (1997) Regionalization of potential evapotranspiration prediction. In: Diekkrüger, B. and Richter, O. (eds) *Regionalization in Hydrology, Landschaftsökologie und Umweltforschung,* Heft 25. Institut für Geographie und Geökologie, TU Braunschweig, pp. 255–258.

Schulz, K., Würlen, C., Eiden, R. and Huwe, B. (1998) Windspeed regionalization and the influence on areal evapotranspiration prediction. *IAHS Publication* (in press).

14

On-site Nutrient Depletion: An Effect and a Cause of Soil Erosion

G.M. Hashim[1], K.J. Coughlan[2] and J.K. Syers[3]

[1]Malaysian Agricultural Research and Development Institute (MARDI), PO Box 12301, Kuala Lumpur 50774, Malaysia; [2]Australian Centre for International Agricultural Research (ACIAR), GPO Box 1571, Canberra, ACT 2601, Australia; [3]Department of Agricultural and Environmental Science, University of Newcastle, Newcastle upon Tyne NE1 7RU, UK

Nutrient depletion is an ongoing process in most parts of the developing world; a 'quiet crisis' which is threatening sustainable agriculture and food security.

Soil erosion is one of several processes contributing to on-site nutrient depletion. These processes are measurable, but are rarely all considered in the same experimental system, making assessment of their relative significance difficult.

Nutrient depletion can also be a contributing cause of soil erosion because, when nutrients are limiting, there is lower production of above- and below-ground biomass which protects the soil against erosion. This is less frequently recognized than the fact that erosion causes nutrient depletion.

The quantity of nutrient loss by soil erosion is the product of soil loss and the nutrient content of sediment, but may also be predicted from soil loss and topsoil nutrient content. However, erosion selectively removes topsoil. The nutrient-rich sediment is further enriched by the selective deposition of faster-settling particles during transport. Thus, prediction of nutrient depletion must take this enrichment process into account.

The concept of enrichment ratio (ER) is central to predicting nutrient loss from soil loss data, particularly with increasing scale. Unfortunately, ER is not constant, varying with soil type, erosion event and scale of measurement. Separating sediment into suspended load and bedload components is particularly useful in studying nutrient enrichment processes.

The problem of scaling from experimental plot to catchment level remains a major difficulty. As scale increases, erosion mechanisms change, producing sediment with ER values approaching 1.

 Soil Erosion at Multiple Scales (eds F.W.T. Penning de Vries, F. Agus and J. Kerr)

Nutrient balance studies, where the effect of added nutrients on soil loss can be quantified, help to identify those situations where nutrient depletion is a cause of soil erosion. In nutrient balances for low-input systems, nutrient losses by soil erosion can be very large. Such balances highlight which interventions should receive priority.

Introduction

Nutrient depletion is an ongoing process which is a characteristic feature of agriculture in most developing countries (Syers, 1997). This 'quiet crisis' (Smaling, 1991), is threatening sustainable agriculture and food security in many parts of the world, particularly in Asia and Africa.

On-site nutrient depletion does not only occur because of nutrient removal in harvested crop products and residues which are not returned. Several other processes can be involved and these include soil erosion, losses in runoff, leaching and gaseous losses (in the case of nitrogen, N). These processes can be measured with varying ease but rarely are they all considered adequately in the same experimental system. This makes assessment of their relative significance very difficult.

It is widely recognized that soil erosion is the most important type of soil degradation (Oldeman, 1994) and a major cause of on-site nutrient depletion, with direct implications for productivity. In this chapter, soil erosion is restricted to water erosion. Although wind erosion is widespread in arid and semi-arid areas, particularly on coarse-textured soils with a limited vegetative cover (Oldeman, 1994), water erosion occurs over more extensive areas and is a problem which affects many resource-poor farmers in the humid tropics. However, it must be emphasized that the reliability of information on current rates of erosion and their consequences for soil productivity is sometimes questioned (Crosson, 1995). In spite of this, it is clear that nutrients are depleted by soil erosion. What is less frequently commented on is the fact that nutrient depletion, particularly as a consequence of crop removal, can be a causative factor in soil erosion. The reasons for this are quite simple. Both phosphorus (P) and N play an important role in root growth and development (Barber, 1984). Roots are important for binding soil and promoting aggregation. Where P and N are deficient, root growth is reduced and the soil is more susceptible to erosion. Also, N, P and potassium (K) are required for above-ground biomass production, which is important in relation to canopy protection and for residue production and return. Soil nutrient depletion, to the extent that deficiencies arise, reduces below- and above-ground plant biomass with direct and often major implications for soil erosion. This is well illustrated by the fact that when nutrient depletion is reversed by the input of nutrients, soil erosion and nutrient loss can be reduced substantially. This is discussed below using an example from Indonesia (Santoso *et al.*, 1995)

There are two approaches to studying on-site nutrient depletion caused by soil erosion. These are:

- experimental monitoring, which involves the collection and chemical analysis of sediments transported in runoff; and
- prediction of nutrient loss from soil loss data (measured or predicted) and nutrient concentrations in topsoils, allowing for an enrichment factor.

This chapter briefly considers these two approaches to studying on-site nutrient depletion by soil erosion and also discusses nutrient loss by soil erosion as a component of the nutrient balance.

Sampling and Analysis of Sediment

Nutrient loss by erosion can be calculated from the following equation:

$$\text{nutrient loss} = (\text{soil loss}) \times (\text{nutrient concentration in sediment}) \quad (14.1)$$

This requires measurement of soil loss from experimental plots and the sampling of sediment and its chemical analysis. Plot size can vary enormously. For example in the studies reported by Hashim *et al.* (1997), this varied from 18 to 3500 m^2, according to the size of the production unit being studied for a range of sites in Malaysia, Thailand, the Philippines and Australia. Where runoff is also collected and analysed (after filtering out the sediment) it is possible to compare nutrient losses in runoff with those from soil erosion. Results for the farmers' practice plots at the IBSRAM sites at Vientiane and Luang Prabang, Lao PDR, for 1995 are shown in Table 14.1 (Phommasack *et al.*, 1996). At these two sites, a very high proportion of the total loss of N and P was accounted for by soil erosion; only in the case of K at the Luang Prabang site did loss in runoff exceed that in eroded soil.

At the Kemaman (Malaysia) experimental site, the loss of soluble K in runoff from treatment T1 (clean weeded with chemicals in a cocoa stand) for

Table 14.1. Soil loss, runoff, nutrient content of topsoil and nutrient loss in runoff and in eroded soil from the farmers' practice at Vientiane and Luang Prabang in 1995.

	Nutrient	Vientiane[a]	Luang Prabang[b]
Soil loss (t ha^{-1})		47.6	9.2
Runoff (m^3 ha^{-1})		3320	2120
Nutrients in topsoil	Total N(%)	0.14	0.25
	Extractable P (mg kg^{-1})	5.7	5.4
	Exchangeable K (mg kg^{-1})	292	305
Nutrient loss in runoff (kg ha^{-1})	N	1.02	2.4
	P	0.28	0.85
	K	14.3	26.1
Nutrient loss in eroded soil (kg ha^{-1})	N	104.2	53.9
	P	32.3	9.3
	K	358.7	24.0

[a] Experimental plots are 200 m^2; average annual rainfall is 2000 mm.
[b] Experimental plots are 240 m^2; average annual rainfall is 1300–1400 mm.

an event on 18 December, 1994 was 1.1 kg K ha^{-1}, whereas that in eroded soil was 0.16 kg K ha^{-1}, indicating that some 87% of K loss was as soluble K in runoff. Even though nutrient concentrations in runoff water may be low, loss in runoff can still be a significant, indeed the major, source of nutrient loss in environments such as Kemaman, where runoff is high. The estimated annual average percentage loss of K in the soluble form in runoff for the same treatment (T1) is 61%, a somewhat lower figure than that recorded for the event of 18 December, 1994, but still a substantial one. In contrast, results in the literature show considerable variation, but it is not uncommon for between 80 and 90% of the total P lost to be transported as particulate P, that is, associated with eroded sediment (Sharpley and Smith, 1990).

It is well known that sediment has a range of settling velocities and that material of different settling velocities has a different chemical composition. For this reason, it is quite common to separate sediment into bedload and suspended load. The methodology for measuring the two sediment fractions separately is described in Ciesiolka and Rose (Chapter 18, this volume). The importance of this separation is seen in the large differences in nutrient concentration between the two fractions. This becomes important in the calculation of the enrichment ratio (ER) of sediment for developing predictions of nutrient loss (discussed below). At Kemaman, suspended load was enriched in N with respect to the topsoil, but the bedload, in all cases was not (Table 14.2).

The question of frequency of sampling becomes important in relation to rainfall pattern and stage during the cropping cycle. In rainy, tropical areas, such as Kemaman, the number of erosion events can be large. For example, for an area of 1000 m^2 on an average slope of 18% without live vegetation cover, approximately 100 erosion events per year can be expected. It may be necessary to restrict samples to a manageable number and therefore a sampling strategy is required. The size of an event influences both the quantity and the composition of the transported sediment through differences in sediment settling velocity distribution. Thus event size is expected to influence ER and sampling should preferably include both large and small events. For small events, the problem of insufficient material may arise (particularly for suspended load samples) and it may be necessary to combine samples from a

Table 14.2. Nitrogen concentration (%) in bedload and suspended load from treatment T1 (1000 m^2 plot, where cocoa and *Gliricidia* are grown and clean-weeding with chemicals is practised) at Kemaman for selected events.

Date of event	Bedload	Suspended load	Topsoil
11 March, 1993	0.08	0.27	0.08
19 May, 1993	0.04	0.18	0.08
23 June, 1993	0.04	0.23	0.08
14 September, 1993	0.04	0.27	0.07
27 October, 1993	0.04	0.32	0.07
4 November, 1993	0.02	0.23	0.06

number of events. However, nutrient loss will be dominated by larger events, and accuracy of estimation of ER for these events is most important.

At different times of the year there may be differences in the rainfall pattern. This should be borne in mind when planning sampling events so that samples representing different periods are included. At Kemaman there is a heavy-rainfall period towards the end of the calendar year. During this period, 'wet spells' occur, when more than two consecutive rainy days are accompanied by almost continuous runoff. Wet spells are a major contributor to annual soil loss at Kemaman (Hashim, 1995). Daily sampling of sediments during wet spells (Hashim *et al.*, 1997) revealed a consistent pattern of N and organic C depletion (Fig. 14.1).

Changes in nutrient loss patterns may also be expected at different stages of the cropping cycle and with fertilizer addition. In the case of annual crops, these stages are more numerous and occur closer to one another. A sampling strategy for sediment should aim to account for all of the variations which are likely to occur in the farming system concerned.

Finally, in the context of calculating nutrient loss by sampling and analysis of sediments, it is important to be aware of the errors which can be introduced by scaling-up from small experimental plots. Scaling is briefly considered later in this chapter.

Predicting Nutrient Loss

Predictions of on-site nutrient loss can potentially be made using information on soil loss and the nutrient concentration of the surface soil layer transported by erosion processes (here termed the topsoil):

$$\text{nutrient loss} = (\text{soil loss}) \times (\text{nutrient concentration in topsoil}) \quad (14.2)$$

However, soil erosion is a selective process which enriches runoff sediment in finer-sized particles and lighter organic matter (Sharpley and Smith, 1990), both of which contain relatively higher levels of nutrients than mineral soil. Thus it is not possible to predict nutrient loss with Eqn 14.2. The selective transport of clay-sized particles in runoff has led to the concept of enrichment ratio (ER) where:

$$\text{ER} = \frac{\text{nutrient concentration in sediment}}{\text{nutrient concentration in topsoil}} \quad (14.3)$$

Thus Eqn 14.2 must be modified to include the ER as follows:

$$\text{nutrient loss} = (\text{soil loss}) \times (\text{nutrient concentration of topsoil}) \times (\text{ER}) \quad (14.4)$$

Sharpley (1985) has developed an equation which predicts the particulate P (PP) concentration in runoff from the P content of the surface soil and the ER for P (PER).

$$\text{PP} = (\text{surface soil P}) \times (\text{sediment concentration} \times \text{PER}) \quad (14.5)$$

(where the units of surface soil P are mg kg^{-1} and the units for sediment

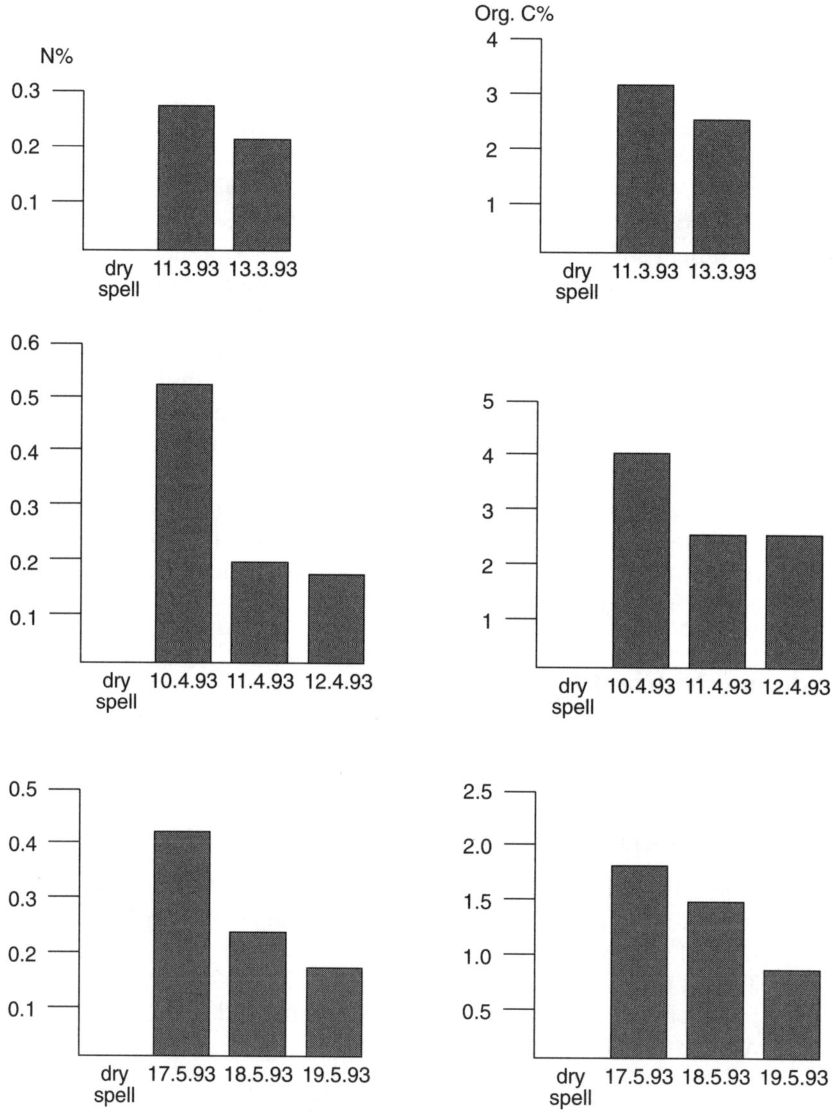

Fig. 14.1. Nitrogen and organic carbon contents of suspended load sediment sampled daily from T1 at Kemaman.

concentration in runoff are g l^{-1}). The enrichment ratios were calculated from soil loss (kg ha^{-1}) using the following equation:

$$\ln(\text{PER}) = 1.21 - 0.16 \ln(\text{soil loss}) \tag{14.6}$$

When measured and predicted annual PP concentrations, calculated from each runoff event, were compared for 22 watersheds in the Southern Plains of Oklahoma and Texas, values were not significantly different at the 1% level (Sharpley and Smith, 1990). This approach is promising, but ER is also a function of soil type and other factors. Because Eqn 14.6 was applicable to the 22 watersheds mentioned, it could be used to predict PER in other situations, provided soil and environmental factors are reasonably similar.

Values for ER for total N and exchangeable K on two plots at Kemaman are shown in Table 14.3. Values for total N were always higher from the T2 treatment (high surface cover) than the T1 treatment (low surface cover) and with one exception this was also the case for exchangeable K. The difference in ER between T1 and T2 is consistent with the relationship between ER and the accumulated mass of soil (M) reported by Knisel (1980):

$$\mathrm{ER} = AM^{-B} \tag{14.7}$$

A and B are empirically fitted parameters. However, there is no clear evidence for the influence of event size on ER within treatments T1 and T2. Instead, the ER values shown in Table 14.3 actually increase with an increase in the percentage of suspended load. Because of the surface cover in T2, runoff was less and litter from leaf fall (enriched in nutrients) would tend to accumulate on the soil surface. The higher ER values in T2 suggest that leaf litter may make a significant contribution to nutrient enrichment in T2. Palis *et al.* (1990) reported that as residue cover increased, the proportion of fine sediment lost by erosion also increased. In predicting nutrient loss for any particular situation, it is important to bear in mind the factors which influence ER.

Table 14.3. Values for enrichment ratio (ER) for total N and exchangeable K in two plots at the Kemaman site during 1993.

	Enrichment ratio (ER) for	
Date	Total N	Exchangeable K
Treatment T1 (low surface cover)		
11 March	2.84	1.00
13 March	2.03	0.80
14 September	2.35	2.57
28 October	2.39	2.14
31 October	2.39	1.57
1 November	2.27	1.29
2 November	1.70	1.14
4 November	2.16	1.29
9 November	2.27	1.43
Average	2.27	1.47
Treatment T2 (high surface cover)		
31 October	4.00	4.00
1 November	3.50	3.40
2 November	2.90	2.40
4 November	3.20	2.40
Average	3.40	3.05

Sampling Topsoils for Predicting Nutrient Loss

Depth of soil sampling is a key issue in calculating meaningful ER values. In a normal soil-sampling exercise for assessing soil nutrient status, it is common to sample the upper 15 cm and regard this as topsoil or the 'cultivated layer'. For soil erosion studies more shallow sampling depths are required. With perennial crops, such as cocoa (*Theobroma cacao*), as grown at the Kemaman site, the soil is not usually cultivated and litter from leaf fall continuously adds nutrients to the surface. Consequently, a gradient in nutrient concentration exists in the immediate surface layer. If the 0–15 cm layer is sampled in such situations, the denominator in Eqn 14.3 will be lower and ER will be overestimated. Because a large proportion of the feeding roots of cocoa proliferate in the top 2 cm of soil, large quantities of nutrients are thought to circulate between the soil and the tree in this layer. Thus at the Kemaman site, the top 2 cm was regarded as topsoil. Where annual crops are grown, the normal cultivated layer of 0–15 cm may be appropriate for sampling, although it is interesting that in the large-scale study reported by Sharpley and Smith (1990), a sampling depth of 0–5 cm was used in the 22 grassed and cropped watersheds studied.

Spatial variations in surface topography are likely when no cultivation is carried out and the area of measurement is relatively large. At the Kemaman site (1000 m^2 per plot), the surface geometry is characterized by preferred pathways of overland flow. As the intensities of various erosion processes in pathways are expected to differ from those in non-pathway areas, pathways were sampled separately from inter-pathway areas. Differences in nutrient concentrations are shown in Table 14.4. Pathways tended to have lower concentrations, but there are exceptions. There are also temporal differences in pathway nutrient contentrations which tended to be lower in January, 1993, a time immediately after high soil loss. In addition, the nutrient concentrations varied between two pathways or two inter-pathway areas within the same plot. For the calculation of ER at Kemaman, the nutrient concentration of topsoil was calculated on a proportional basis. Because pathways were estimated

Table 14.4. Topsoil (0–2 cm) concentration of total N (%) and exchangeable K (cmol(+)kg^{-1}) in inter-pathway areas (I) and pathways (P) at two locations (L1 and L2) in the same plot for treatments T1 and T2 at Kemaman.

		T1				T2			
Nutrient		July 1992		Jan. 1993		July 1992		Jan. 1993	
		L1	L2	L1	L2	L1	L2	L1	L2
Total N	I	0.05	0.10	0.07	0.05	0.18	0.12	0.10	0.10
	P	0.07	0.09	0.02	0.06	0.13	0.13	0.05	0.06
Exchangeable K	I	0.21	0.22	0.08	0.05	0.33	0.39	0.09	0.08
	P	0.11	0.14	0.04	0.08	0.23	0.42	0.21	0.06

to constitute 5% of the area, the nutrient concentration in pathways was given a 5% weighting.

The sampling at Kemaman was carried out every 2 months and ER was calculated based on the analytical result at the appropriate time. On each sampling occasion and for each location shown in Table 14.4, the soil was sampled from at least ten locations with a view to accommodating soil heterogeneity. For annual crops this temporal variation is expected to be more important and sampling should preferably be scheduled to coincide with major activities during the cropping cycle.

Factors which Influence ER

Unfortunately, ER is not constant for a given soil but is influenced by a number of factors apart from the size of the erosion event. These include soil type, erosion mechanism and scale of measurement, biological activity and surface cover. These are discussed below.

The influence of soil type stems from the fact that sediment with different settling velocity distribution has a different chemical composition. High clay content tends to reduce ER as clay aggregates and sand do not tend to segregate on wetting. As clay content increases, ER decreases, approaching 1 when clay content approaches 50%. In our studies at a range of sites in South-East Asia, this influence is generally shown by the lower ER obtained for Los Baños, where the soil is of clay texture, in comparison with that at Kemaman, which has only 20% clay (Hashim *et al.*, 1997). The ER values at Los Baños are only slightly greater than 1, while those at Kemaman are larger than 2. However, the large variations in ER for the very sandy soil at Khon Kaen with 80% sand and the sandy soil at Kemaman indicate that simple clay enrichment does not fully explain the variations in ER. Although the soil at Goomboorian is also sandy, ER is relatively low. The long slopes at Goomboorian suggest that entrainment processes are active and this may explain the low ER in spite of the texture being sandy. Thus the erosion mechanism also has a strong effect and may confound the effect of soil type. At Kemaman, there are two opposing influences. Although the soil is relatively sandy, and selective deposition of coarser material occurs on longer slopes, the long slopes and the presence of prominent preferred pathways of overland flow also suggest that entrainment processes are active. Long slopes tend to increase ER through selective deposition, whilst entrainment which becomes a more dominant mechanism at longer slope lengths, is associated with lower ER.

As water discharge in runoff increases, because of an increase in slope length, there is a probability of a change in erosion mechanism, with incision resulting in the formation of rills (ephemeral gullies) on large-scale gullies, and in stream-bank erosion. All of these erosion mechanisms remove sediment from deeper layers in the soil profile, which are lower in nutrients than the surface layer. Thus the ER of the transported material will decrease, making the prediction from Eqn 14.3 less reliable. In fact for large catchments, soil heterogeneity and uncertainty regarding erosion processes make it impossible to predict nutrient losses other than from empirical, location-specific relationships.

An erosion mechanism which tends to increase ER is 'raindrop stripping' (Ghadiri and Rose, 1991). In this process, raindrops remove the enriched outer layer of large, stable aggregates. Similarly at Kemaman, under cocoa and *Gliricidia* trees, there is abundant leaf litter and a high level of biological activity. The surface layer is therefore considerably enriched in nutrients. Raindrop stripping of the immediate surface soil could have occurred and contributed to a higher ER. Thus, the eroded surface material may be enriched in nutrients even compared with the 0–2 cm layer sampled as topsoil, and used for the calculation of Eqn 14.3.

Because of their burrowing activity, surface-casting earthworms can decrease runoff by increasing infiltration and the loss of soluble N and P in surface runoff (Sharpley *et al*., 1979). However, the soil and plant material ingested is excreted as surface casts which, when transported in overland flow because of their lower density, would increase the ER because of their higher nutrient content. The N concentration (%) in earthworm casts and surface soil at Kemaman is shown in Table 14.5. The N concentration was consistently higher in the casts than in the surface (0–2 cm) soil, higher in T2 (high surface cover) than in T1 (low surface cover), and higher in the inter-pathway than the pathway areas.

Preferential deposition of large soil particles plays a very important role in the nutrient enrichment of sediment. Contact cover, which tends to reduce flow velocity, plays a part in this process. With reduced flow velocity more of the coarse sediment is deposited. This effect is clearly demonstrated at Kemaman and Goomboorian. At Kemaman, the ER for N in T1 varied from 1.7 to 2.8. However, in T2, where live surface vegetation covers the soil, the ER was larger, ranging from 2.9 to 4. As suggested previously, there is a strong possibility of nutrient release from the larger amount of litter in T2. A similar effect is shown by the mulched plot at Goomboorian, where the ER for N was 1.5 to 7.5, in contrast to that for the bare plot, which was 0.5 to 1.8. (Hashim *et al*., 1997).

Table 14.5. Nitrogen concentration (%) in earthworm casts and surface soil (0–2 cm) at Kemaman in T1 and T2.

	Treatment	
Sampling time	T1	T2
Earthworm casts		
June, 1993	0.11	0.15
October, 1993	0.11	0.13
July, 1994	0.10	0.12
Surface soil sampled in November, 1993		
Pathway	0.04	0.09
Inter-pathway	0.09	0.10

Thus ER is influenced by several factors. Identification of the factors which are important for a specific site could be particularly useful in assisting with the prediction of nutrient loss.

The Issue of Scale

Scaling is arguably more relevant to model building for predicting soil erosion, runoff and nutrient loss than to actual measurements. However, it is essential to be aware of the limitations inherent in converting the data collected from small experimental plots, which can be as small as 1 m^2 in rainfall simulation studies, to the hectare and the catchment level. Many soil erosion studies employ plots which are of the order of 150 m^2 and the results obtained for soil and nutrient loss are often scaled-up without question. This is dangerous because the scale of experimental monitoring, that is, the area involved, can have an important effect on size sorting of sediment into classes with different chemical composition. As area increases, distance of transport increases and sediment tends to be finer because of the preferential deposition of faster-settling aggregates. This results in an increase in ER.

Erosion models have been developed to cover a range of spatial scales, but scale continues to pose a major problem in erosion research (Dickinson *et al.*, 1990). Information on erosion processes obtained from microscale field plots (1 m^2) using a rainfall simulator, from Wischmeier-type field plots and from catchments has allowed the use of knowledge obtained at the smaller scale to be used to refine models at the larger scale. However, the way in which results can be transferred is not always obvious. In the context of on-site nutrient depletion by soil erosion, the impact of up-scaling is potentially to reduce accuracy and increase heterogeneity. Thus the reliability of the data must be maintained or the model changed to accommodate a reduced data requirement. Integration of data from various scales remains a major challenge, particularly in developing nutrient balances.

Nutrient Loss by Soil Erosion as a Component of Nutrient Balance

Nutrient balances can be developed at a range of scales (Syers, 1996) including: (i) the experimental plot, (ii) field or farm, (iii) district, and (iv) country. Most of the nutrient balance studies reported in the literature relate to the experimental plot, where it is frequently less difficult to obtain reliable information than at larger scales. Quantification of the inputs and outputs required to establish the nutrient balance involves calculation (based on actual measurements), estimation, and even assumption, depending on the ease with which the component inputs and outputs can be assessed. Where experimental data are available, nutrient balance studies can give context to soil erosion as a causative factor in on-site nutrient depletion.

Santoso *et al.* (1995) have reported some interesting data for N balance in an alley-cropping experiment on sloping land at Kuamang Kuning in Indonesia, involving *Flemingia congesta* with crops of rice, peanut and mung bean (Table 14.6). Leaching, as in many similar studies, was not measured at

Table 14.6. Nitrogen balance in an alley-cropping experiment at Kuamang Kuning (Indonesia). Values are reported as kg N ha^{-1} over 5 years.

	No input	High input
Inputs		
Fertilizer	0	675
Crop residues		
rice straw	31	142
peanut vines	46	161
mung bean straw	3	38
Flemingia loppings	278	353
Total	358	694
Total	358	1369
Outputs		
Removal		
upland rice grain	10	74
peanut seed	17	108
mungbean seed	2	19
rice straw	31	142
Total	60	343
Leaching	–*	–*
Runoff	141	20
Soil erosion	175	7
Total	376	370
Balance	−18	999

* Not measured, estimated, or assumed.
Data from Santoso *et al.* (1995).

the site. The loss of N by soil erosion from the no input plot over 5 years was 175 kg N ha^{-1}, which accounted for 47% of the output of total N. In this and similar experiments, the extent to which the N in crop residues should be regarded as an input is open to question, as part of the N is likely to be soil derived, adding to that contributed by biological fixation. Certainly, if crop residues are removed, this represents a loss of nutrients from the system.

The loss of N by soil erosion at the Kemaman site (Table 14.7) was larger than that contributed by fertilizer N input when clean weeding with chemicals was practised (T1). Improved cover management (T2) at this site reduced the loss of N by soil erosion from 194 to 93 kg N ha^{-1} year^{-1} in 1994. Unfortunately, loss of N in runoff and leaching was not estimated, making it difficult to assess the contribution of soil erosion to the output of N.

In a rather more detailed study of the K balance in the Kisii District in Kenya, where little fertilizer K is added, Smaling (1993) was able to demonstrate a substantial negative balance for K (−70 kg K ha^{-1} year^{-1}), pointing to the non-sustainability of the cropping system involving maize and beans. The results point to the importance of erosion as a major pathway of K loss (36 kg K ha^{-1} year^{-1}), being second only to removal in the harvested crop

Table 14.7. Nitrogen balance at Kemaman in 1994. All values given as kg N ha^{-1} $year^{-1}$.

	T1	T2
Inputs		
Fertilizer	180	184
Cocoa litter	51	79
Gliricidia litter	21	26
Rainfall	12	12
Rainwash from leaves	12	12
Total	276	313
Outputs		
Harvest of banana	2	6
Soil erosion	194	93
Total	196	99
Balance	80	214

Data from Hashim *et al.* (1997).

(43 kg K ha^{-1} $year^{-1}$), about which nothing can be done given the comparatively low inputs of K. These results led to recommendations for interventions involving soil conservation practices and agroforestry to reduce nutrient depletion, which is probably one of the most useful outcomes of nutrient balance studies.

The N balance study reported by Santoso *et al.* (1995) provides confirmation of the earlier suggestion that nutrient input, in this case of N, can reduce soil erosion. The high input treatment (90 kg N ha^{-1} $year^{-1}$ + 40 kg P ha^{-1} $year^{-1}$ + 20 kg K ha^{-1} $year^{-1}$ for upland rice, 40 kg N ha^{-1} $year^{-1}$ + 25 kg K ha^{-1} $year^{-1}$ + 2 t lime ha^{-1} $year^{-1}$ for peanut, and 0 kg N ha^{-1} $year^{-1}$ for mung bean) reduced the soil loss from 46 t ha^{-1} $year^{-1}$ to 13 t ha^{-1} $year^{-1}$ between 1990 and 1992 (Paningbatan, 1995). Over a 5-year period (1990–1994), the loss of N by soil erosion was reduced from 175 kg N ha^{-1} on the no input plot to 7 kg N ha^{-1} on the high input, pointing to the substantial reduction in soil erosion resulting from fertilizer applications, by reversing nutrient depletion. This is perhaps not surprising if N, and possibly other nutrients, are limiting both above- and below-ground plant production. These results raise some important questions regarding the potential for using nutrient input to reduce soil loss by erosion: a little studied and still poorly understood topic.

Conclusions

With knowledge of soil loss and the nutrient concentration of sediment, on-site nutrient depletion in an erosion event can be measured. However, to predict it from knowledge of soil loss and the nutrient concentration in topsoil, one needs to consider the enrichment factor. Nutrient enrichment of sediment is a central concept in nutrient depletion by soil erosion. It is influenced by a number of factors which are now reasonably well known. Because of the enrichment process and the fact that as scale increases erosion processes change, scaling remains an issue for assessment and prediction.

Information on nutrient depletion by soil erosion is an important component in nutrient balances, which serve as a means of assessing the relative importance of soil erosion as a mechanism for nutrient loss from any agricultural system. A good nutrient balance, containing most of the major input and output components, can be used as a guide towards making decisions as to what priority to give to soil erosion management so as to maintain a positive balance.

References

Barber, S.A. (1984) *Soil Nutrient Bioavailability.* John Wiley, 398 pp.

Crosson, P. (1995) Future supplies of land and water for world agriculture. In: Islam, N. (ed.) *Population and Food in the Early Twenty-first Century.* International Food Policy Research Institute, Washington, DC, pp. 143–159.

Dickinson, W.T., Wall, G.J. and Rudra, R.P. (1990) Model building for predicting and managing soil erosion and transport. In: Boardman, J. *et al.* (eds) *Soil Erosion on Agricultural Land.* John Wiley, pp. 415–428.

Ghadiri, H. and Rose, C. (1991) Sorbed chemical transport in overland flow. I. A nutrient and pesticide enrichment mechanism. *Journal of Environmental Quality* 20, 628–633.

Hashim, G.M. (1995) Soil erosion under different cover management practices during an extreme storm event. In: *Proceedings of the Soil Science Society Conference, Malaysia,* Malaysian Society of Soil Science, Kuala Lumpur, pp. 145–154.

Hashim, G.M., Ciesiolka, C.A.A., Rose, C.W., Coughlan, K.J. and Fentie, B. (1997) Loss of chemical nutrients by soil erosion. In: Coughlan, K.J. and Rose, C.W. (eds) *A New Soil Conservation Methodology and Application to Cropping Systems in Tropical Steeplands.* ACIAR Technical Report 40, Canberra, pp. 79–100.

Knisel, W.G. (1980) *CREAMS: A field-scale model for chemicals, runoff and erosion from agricultural management systems.* United States Department of Agriculture, Conservation Research Report 26.

Oldeman, L.R. (1994) The global extent of soil degradation. In: Greenland, D.J. and Szabolcs, I. (eds) *Soil Resilience and Sustainable Land Use.* CAB International, Wallingford, UK, pp. 99–118.

Palis, R.G., Okwach, G., Rose, C.W. and Saffigna, P.G. (1990) Soil erosion processes and nutrient loss. II. The effect of surface contact cover and erosion processes on enrichment ratio and nitrogen loss in eroded sediment. *Australian Journal of Soil Research* 28, 641–658.

Paningbatan, E. (1995) The *ASIALAND* management of sloping lands network: soil erosion under different soil-conservation practices. In: *International Workshop on Conservation Farming for Sloping Uplands in Southeast Asia: Challenges, Opportunities, and Prospects. IBSRAM Proceedings No. 14.* Bangkok, Thailand, pp. 109–126.

Phommasack, T., Sengtaheunghung, O. and Phanthaboun, K. (1996) *Management of sloping lands for sustainable agriculture in Laos. IBSRAM Network Document No. 20.* Bangkok, Thailand.

Santoso, D., Wigena, I.P., Eusof, Z. and Chen, X.H. (1995) The *ASIALAND* management of sloping lands network: nutrient balance study on sloping lands.

International Workshop on Conservation Farming for Sloping Uplands in Southeast Asia: Challenges, Opportunities, and Prospects. IBSRAM Proceedings No. 14. Bangkok, Thailand, pp. 93–108.

Sharpley, A.N. (1985) The selective erosion of plant nutrients in runoff. *Soil Science Society of America Journal* 49, 1527–1534.

Sharpley, A.N. and Smith, S.J. (1990) Phosphorus transport in agricultural runoff: the role of soil erosion. In: Boardman, J. *et al.* (eds) *Soil Erosion on Agricultural Land.* John Wiley, pp. 351–366.

Sharpley, A.N., Syers, J.K. and Springett, J.A. (1979) Effect of surface-casting earthworms on the transport of phosphorus and nitrogen in surface runoff from pasture. *Soil Biology and Biochemistry* 11, 459–462.

Smaling, E. (1991) Africa's soils are being mined. *ILEIA Newsletter*, May, 9–11.

Smaling, E. (1993) Soil nutrient depletion in sub-Saharan Africa. In: von Reuter, H. and Pring, W.H.P. (eds) *The Role of Plant Nutrients for Sustainable Food Crop Production in Sub-Saharan Africa.* UKP, the Netherlands, pp. 53–68.

Syers, J.K. (1996) Nutrient budgets: uses and abuses. In: *Soil Data for Sustainable Land Use: A Training Workshop for Asia. IBSRAM Proceedings No. 15.* Bangkok, Thailand, pp. 163–168.

Syers, J.K. (1997) Managing soils for long-term productivity. *Philosophical Transactions of the Royal Society of London B* 352, 1011–1021.

15

Erosion and Sedimentation as Multiscale, Fractal Processes: Implications for Models, Experiments and the Real World

M. van Noordwijk[1], M. van Roode[2], E.L. McCallie[3] and B. Lusiana[1]

[1]*International Centre for Research in Agroforestry (ICRAF), PO Box 161, Bogor, Indonesia;* [2]*ICRAF, Nairobi, Kenya;* [3]*Cornell University, Ithaca, USA*

Past research and existing models allow a fair prediction of the soil and water balance at plot level under different land-use types, but do not apply to the landscape mosaics of the real world. We propose a classification scheme for landscapes on the basis of the presence of 'filter' strips in between cropped fields and explore what research is needed to 'scale-up' from homogeneous experimental plots to heterogeneous landscapes. Both slope length and slope width can affect the net soil loss per unit area from a mosaic element.

Total water and sediment yields from landscapes with a partial tree cover (agroforestry) may respond differently to interventions. For sediment load of rivers the location of trees may be more important than the area covered by trees, as riparian strips can intercept substantial sediment movement from upslope. By contrast, total water yield and base flow of rivers may be determined by the percentage forest cover, virtually independent of tree location.

Farm productivity effects of vegetative soil conservation measures are a net effect of the labour and other costs involved and the yield. Harvested yield per unit area depends on the area occupied by the soil conservation strips, the effects on crop yields of soil redistribution within the slope, effects of heterogeneous water infiltration, shading and competition for water and nutrients between vegetation and crop, and effects on the soil organic matter balance. The WaNuLCAS model for spatially zoned agroforestry systems can predict where the net physical yield effect can be positive.

Erosion and sedimentation processes primarily affect the heterogeneity of soil fertility within a landscape. The main issues thus are whether or not fertility can be exploited on the place where it ends up, and whether or not this outweighs the yield opportunities lost at the sites of net soil loss.

 Soil Erosion at Multiple Scales (eds F.W.T. Penning de Vries, F. Agus and J. Kerr)

Introduction

As we witness a transition in paradigm from 'soil conservation' (keep everything in place) to 'managing soil erosion' (reduce negative impacts, exploit new opportunities), many established concepts and calculation procedures are up for review. In this chapter we explore concepts of scale in the relations between models, experiments and the real world. Erosion and its counterpart process of sedimentation cannot be simply 'scaled-up' on an area basis from observation plots to farms and watersheds; we will discuss whether or not dimensions other than 2.0 (used in area) can be used instead. Erosion concerns also differ widely between human interest groups, and that makes 'scaling-up' erosion control experiences a far from trivial exercise.

In articulating the policy context of erosion we are faced with the following paradox: Erosion would not be a problem worth the attention of international research bodies if it were a problem to all – there is no lack of technical solutions and these would be implemented if they would benefit all members of society. The problem is that perceptions differ on how serious a problem erosion is.

Concerns on soil conservation on agriculturally used lands are usually based on a combination of:

A. on-site loss of land productivity,
B. off-site concerns on water quantity:
  B1. annual water yield,
  B2. peak (storm) flow,
  B3. dry season base flow,
C. off-site concerns on water quality as it leads to sedimentation on lowlands and siltation of reservoirs.

Concerns B and C are mainly valued by lowland interest groups who will perceive changes when a natural (forest) vegetation is converted into agriculture, whereas aspect A is mainly an uplander issue. The combination of these concerns has led to a widespread concept of erosion as the major contributor to loss of productivity of uplands as well as the cause of lowland problems with water quantity and quality. Government interventions in supporting soil conservation in the uplands would address both poverty alleviation in the uplands and protection of lowland interests, and have a direct appeal of a win–win model at all scales in which analysis of the details may not be necessary. Countless 'participatory' integrated watershed management projects have been built on this paradigm.

Although no one will deny that there is a core of truth in this soil conservation paradigm, the case has probably been oversimplified, and a number of studies have indicated that the supposed link between lowland and upland concerns is not necessarily as clear cut. For example, Diemont *et al.* (1991) argued for Java that the loss of productivity of uplands was mainly due to nutrient depletion and low fertilizer use, while dam siltation was largely due to sediments from roadsides and 'morpho-erosion' in the landscape, rather than coming directly from upland soils (see also Enters, Chapter 1, this volume). Zinke *et al.* (1978) described serious field-level erosion in shifting cultivation systems in Northern Thailand, in a landscape where most (if not all) of the topsoil accumulated in a 'soil deposition fan' at the bottom of the slope – in a potentially useful location. Many hydrological studies have shown that widespread beliefs on 'forest functions' with respect to the regional water balance (concerns B) are not supported by facts (Hufschmidt, 1986; Bruijnzeel, 1990): forests use more water than other vegetation and thus total water yield will generally increase after deforestation (B1), with a small increase or little change in dry-season base flow (B2) and increase in storm flow (B3). The last effect may be as much due to other landscape changes accompanying agriculture as it is to what happens in the fields *per se*.

Erosion is normally quantified as an amount of soil lost from a given area over a specified period of time and expressed in standardized units, mostly in tonnes per hectare per year, as if net soil loss is independent of the size of the measured area. Erosion is based on the movement of soil particles, from a place of origin to a place of sedimentation. This raises issues of how many soil particles are involved and how far they travel during a unit time – an integral of mass × distance might be a more meaningful way of quantifying the process, leading to a dimension of mass × length per unit area, or kg m^{-1} $year^{-1}$. The amount of soil depends on the balance between topsoil cohesiveness and the displacement forces from splash (or wind) impacts and entrainment, the distance travelled on surface conditions along the transport pathway and the rate of flow of the carrier (water or wind). Unless all particles which start movement immediately disappear beyond the boundaries of consideration, there is likely to be a form of double counting when considering larger areas as a simple summation of their constituting small sample areas. The whole may be much less (or more) than the sum of its parts suggests and indeed surface area may be a very ambiguous basis for expressing results of erosion measurements.

Human influence on erosion is based on influencing both soil cohesiveness and the transport pathway, the latter often in complex ways:

> The new creek bed is ditched straight as a ruler: it has been 'uncurled' by the country engineer to hurry the runoff. On the hill in the background are contoured strip crops; they have been 'curled' by the erosion engineer to retard the runoff. The water must be confused by so much advice.
>
> Leopold, 1949

Soil conservation measures will not address all concerns (A, B and C) to the same degree and we have to differentiate between them in evaluating environmental impacts of land-use change and in considering options for maintenance of 'forest functions' in agricultural landscapes. Our analysis may take the following steps:

1. How do soil and water movement at patch level combine to overall effects at landscape scale, and to what degree can 'filter' strips or zones make up for incomplete soil and water conservation upstream?
2. Which aspects of soil and water conservation at patch and landscape scale depend on land-use decisions?

3. How do interventions aimed at off-site concerns B and C influence on-site productivity A?
4. How can desirable land-use decisions be influenced by a conducive policy environment?

In this chapter we will first discuss the potential role in erosion management of agroforestry in a landscape context, followed by a further discussion on scaling rules and units for measuring erosion (mainly concerns B and C, across scales). We will present model calculations on the effects of slope length and width and soil properties on the scaling rules for surface runoff and soil loss (concerns B and C at plot scale). We will proceed with a modelling frame for predicting soil productivity aspects (function A at plot scale) and review literature on how farmers respond to heterogeneity in soil fertility. Finally we will consider how the effects of 'filters' in the landscape can be further incorporated into models, experiments and real world decisions on soil and water conservation.

Erosion and Sedimentation in Landscapes with Some Trees

Agroforestry research is gradually evolving from a focus on (field-level) technologies and domesticated trees, to a more complete consideration of landscape-level processes and constraints to a farmer-led process of technology development. Just as we need a nomenclature and taxonomy for trees, soils and climates before we can make meaningful comparisons, a classification of landscape types or stages in typical dynamic processes is a first step in making inferences at landscape level.

In a schematic form most agriculturally used landscapes have undergone a gradual process of intensification (Fig. 15.1) with a gradual loss of the 'forest functions' in soil and water balance, leading to concerns A, B and C. As a first approximation we will use the five stages in a typical dynamic process identified in Fig. 15.1 as a landscape typology. If we want to operationalize this scheme, we need to know not only what happens on the various fields, but also what happens in between these fields and the point of outflow of the river.

Much research on soil erosion by surface runoff is primarily concerned with a comparison of runoff and sediment loss from plots with different land-cover types ('treatments') with a bare fallow or farmer practice 'control'. Plots usually have standard dimensions and are normally protected from surface runon from upslope. Although the results of such experiments are useful in testing hypotheses on the effects of treatments *per se*, difficulties arise when it comes to scaling-up these results to what should be the target areas: farmers' fields, hill sides or catchments. In 'paired catchment' research there has been a tendency to select for homogeneous land uses within each of the (micro-) catchments – yet there may be far more challenging questions in how to predict the result of a mosaic of differently managed fields in a matrix of field edges, paths with hedgerows, etc. It is unlikely that a mosaic will behave as a simple addition of its component parts. For a better understanding of erosion

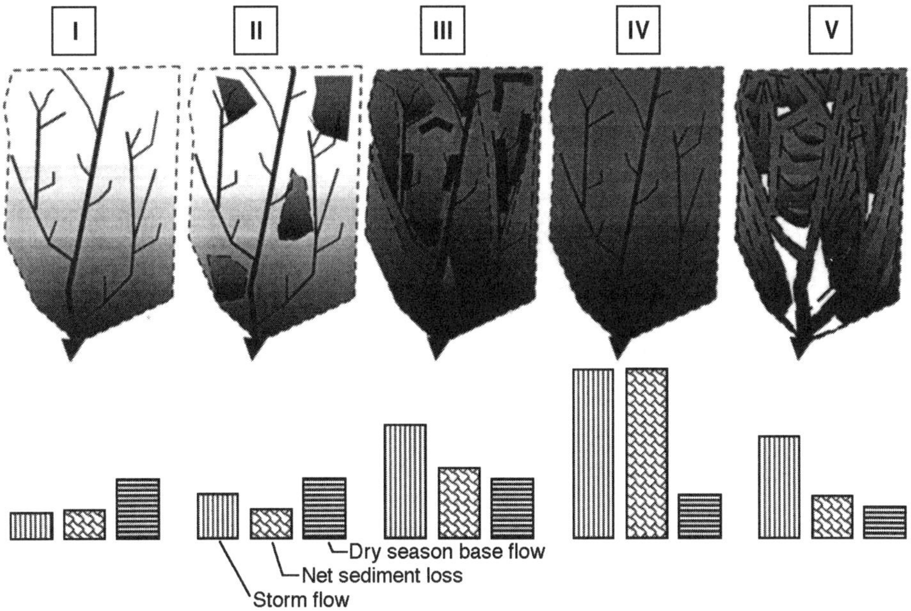

Fig. 15.1. Schematic development of the landscape in a subwatershed and its effects on storm flow, net sediment loss and dry-season base flow: I. original forest cover, II. patches of forest opened for shifting cultivation, III. intensification of land use has brought most land into cultivation, except for riverain borders and hedges along paths, IV. reclamation of all 'wastelands' has removed all filter strips causing a disproportional rise in net sediment loss, V. restored agroforestry landscape with permanently vegetated contour strips and riparian woodlands.

as a multiscale process we may also have to analyse processes within the standard measurement plots of conventional research.

Agroforestry technologies are widely believed to contribute to (cost-) effective solutions to soil conservation problems. Young (1989) defined soil conservation as a combination of controlling erosion and maintaining soil fertility; agroforestry techniques can have effect on both sides of the coin, whereas physical barriers only contribute to erosion control. The SCUAF (Soil Changes Under Agro-Forestry) model was developed as a first approach to the effects of agroforestry practices on soil changes and combines a 'universal soil loss equation' (USLE) approach and a soil organic matter balance model. The SCUAF model does not account for negative effects of trees via competition and the specific mechanisms involved in contour hedgerow systems. Vermeulen *et al.* (1993) noted deficiencies in the soil C balance description of the SCUAF model, but the model continues to be used (Menz and Grist, Chapter 3, this volume). In the decade since the SCUAF model was formulated, substantial progress has been made both in the understanding of how agroforestry systems can affect the movement of soil and water across landscapes, and in predicting the effects of soil conservation practices on crop productivity by a combination of positive and negative interactions (competition and complementarity).

Garrity (1996) discussed the effects of soil movement within contour hedgerow systems on crop production (concern A), emphasizing the scouring effect on the upper terrace while topsoil accumulates on the lower terrace. Contour hedgerow systems in this sense increase the spatial heterogeneity of soil fertility and create new problems as well as new opportunities. Below we will explore a modelling framework for assessing the net balance of these effects on productivity.

Kiepe and Rao (1994) came to the conclusion that the erosion control effect of contour hedgerows in the semi-arid/subhumid environment of Machakos (Kenya) was based both on effects of surface litter ('curling' the pathway of water on the scale of a few centimetres) and on barrier and filter effects of the contour hedgerow (on the scale of several metres). The latter is based on three effects: (i) an increased ability of the soil to store water due to its previous depletion, (ii) an increased infiltration due to macropores formed by both decaying tree roots (van Noordwijk *et al.*, 1991) and earthworms living off the leaf mulch, and (iii) a physical barrier effect by stems of the hedgerow trees. An increased water storage capacity is especially likely during heavy storms in the early part of the rainy season due to dry-season water use by trees. A high surface infiltration rate without storage capacity, however, will still lead to surface ponding and runoff as soon as all macropores are filled and outflow from them becomes the bottleneck. The potential filter role of contour hedgerows will thus become less if we move to higher rainfall zones.

In the context of agroforestry options, a major question is how landscapes with trees are positioned in the continuum of pure crop land to pure forest. Is the proportion of space occupied by trees a reasonable predictor of environmental service functions of the landscape or is the spatial pattern of additional (or even overriding) importance? A convenient way of presentation is based on a replacement series concept with the width of the envelope of possible solutions indicating the importance of spatial patterns at any given tree density (Fig. 15.2).

To answer such questions we need to understand why and how trees interact with surface and subsurface transfers of water and sediment. We will explore a modelling framework which incorporates these mechanisms and can be used to explore the effect of tree locations in real landscapes.

Scaling Rules and Units for Measuring Erosion

The term *scaling* can be used in many different contexts and apparent meanings. Here we use it to describe changes in average unit value if the extent (scale) of a phenomenon changes. An observed change in unit value can indicate a real world response to contraction or expansion of the phenomenon itself, or be a consequence of changed (larger or smaller) sample size. There are many biological as well as economic reasons why the average unit

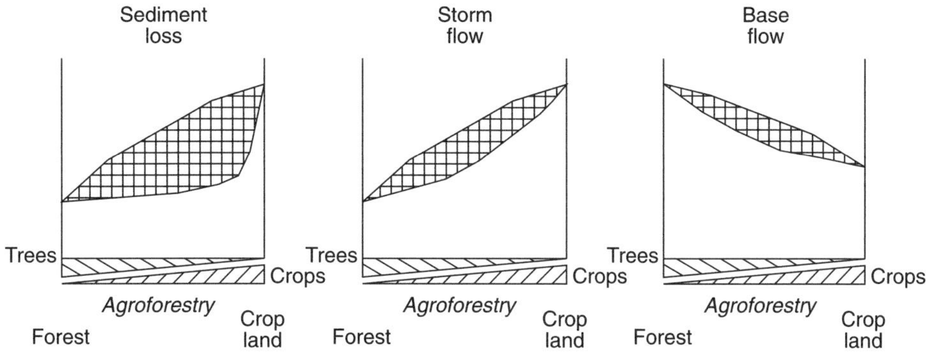

Fig. 15.2. Tentative soil and water conservation functions of landscapes consisting of both crops and trees (replacement series, Yin–Yang design); the width of the envelope indicates different spatial patterns at equal total tree density at a given scale of measurement.

value of farm components (e.g. crops or trees) depends on the extent at which these components occur in the landscape: the epidemiology of pests and diseases, thresholds for mechanization or marketing, price elasticity, saturation of the environmental service function provided, etc. Such real changes in average unit value will normally involve a time aspect and may show hysteresis. To measure the effects, we first have to consider scaling rules for samples and observations.

In many areas of research the implicit assumptions are often made that spatial phenomena can be expressed per unit area, length or volume, that the average value per unit area (length, volume) does not change systematically if sample size is increased and that the total value over the whole domain can be obtained by multiplying an average value per unit area (length, volume) and total area (length, volume). Use of the SI (or similar) system of units in expressing results of agronomic or ecological research often suggests that the value of a response variate is independent of the exact sample size which was measured. Statistical sampling theory is normally built on an assumption of 'independent' members of a population of observable objects. For spatial phenomena a 'minimum representative area' (mra), 'minimum representative length' (mrl) or 'minimum representative volume' (mrv) is normally postulated to exist and sampling theory is used as if there were a population of independent mra's, mrl's or mrv's. Increasing sample size will normally improve precision of an estimate, unless different members of the population of mra's (mrl's, mrv's) have unequal chances of being observed in 'biased' sampling schemes, but not change the average unit value. Stratified sampling or geostatistics can make sampling schemes more efficient, but still involve a multiplication of average unit value per stratum and total stratum weight, or an integral of a value per unit area over a certain domain.

The recent interest in *scaling rules* is based on the suspicion that there may be deficiencies in such an approach and that the average unit values can in fact depend both on the real extent of a phenomenon and the scale at which it is observed. The main problem is probably introduced by the assumption of independence of mrv's (mrl's, mra's). Although it appears to be convenient to distinguish such supposedly representative elements, or even to physically separate sample areas from their surroundings (as in conventional runoff and erosion plots, lysimeters or in well-trenched agroforestry experiments), the efforts to exclude *lateral interactions* and consider a superposition of elements with a sphere of influence, where the influence decreases with distance necessitates a procedure to bring them back into consideration when observations are used to make real-world inferences (Fig. 15.3; van Noordwijk and Ong, 1996). It may be desirable to avoid any discretization and consider a superposition of elements with a sphere of influence, where the influence decreases with distance.

It has been widely accepted that there is not necessarily a linear relationship between the amounts of runoff and soil loss and the length of the slope, or the size of the area. Although total soil loss will normally increase with plot size, soil loss per unit area may increase, remain constant or even

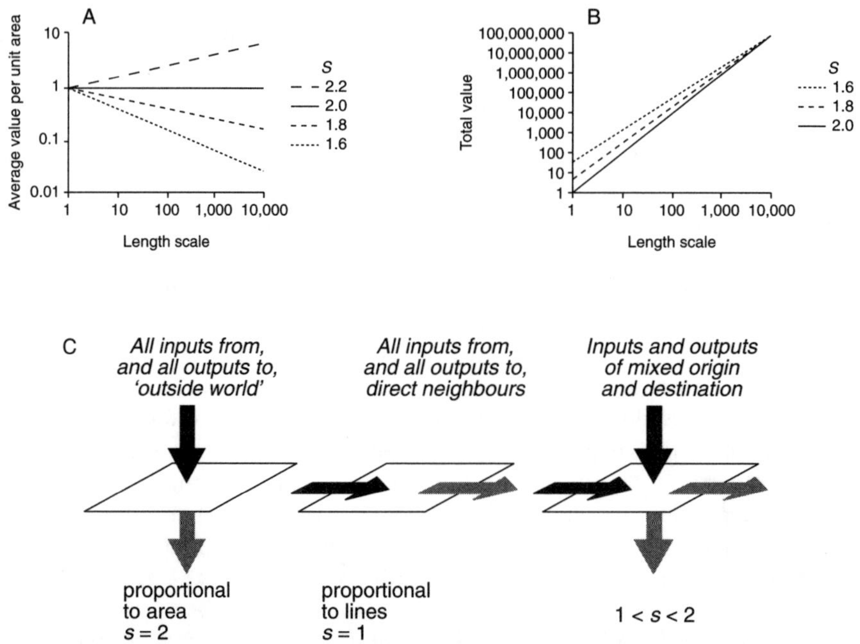

Fig. 15.3. A, B. Scaling rules potentially lead to changes in average value per unit area. C. If all inputs come from, and outputs go to the outside world, an area-based approach with dimensionality of $s = 2$ may apply (left); if all inputs and outputs are laterally transmitted, a linear approach ($s = 1$) is appropriate (middle); if inputs and outputs are of mixed origin and destination, a dimension between 1 and 2 may apply (right).

decrease when larger areas are considered. Area can be seen as a product of slope length and slope width. On a long slope, there is a bigger build-up of the amount of surface runoff and its velocity, leading to more scouring of soil, as the amount of sediment potentially carried in surface runoff increases more than proportionally to the volume of water flow. This effect can be counteracted by effects of surface roughness with locally convex areas and patches of high infiltration capacity where sedimentation can occur. Spatial redistribution of runoff and soil during the process of transport downhill can lead to a consistent decrease in net sediment loss per unit area if larger areas are likely to contain disproportionately more effective 'filter' zones. The latter is possible if some of the filters are not saturated from the direct neighbourhood but can become more effective in a larger context. Movement of water and soil particles from one place to another within the measuring area is not regarded as soil loss, simply because it is not measured at the point of outflow. As slope length increases, the time of concentration (longest time needed for overland flow to reach the outlet of the area) increases. If the runoff is a direct result of the rainfall capacity exceeding the soil's infiltration capacity, the destiny of the runoff water largely depends on the duration of the peak rainfall rates in relation to the travel time to the outlet. The shorter the rainfall peaks and the longer the time to the outlet, the higher the chances that the surface water will infiltrate on its way (spatial redistribution) and not reach the outlet. The same applies to soil particles carried by the runoff water; infiltration of water leads to increased chances for sedimentation of soil particles. So, on long plots, runoff and soil losses (per unit area), measured at the outlet of the area, can be less than on short plots.

The overall slope length effect depends on the interaction between the two processes described, acceleration and spatial redistribution. It is not possible to make a general statement on which of the two processes is stronger, as this depends on the environment in which the erosion takes place. On a rough surface with many patches of vegetation, more water and sediment can be trapped than on a smooth bare slope. In the first, redistribution may be dominant, whereas the second is more likely to be dominated by the acceleration effect.

Zingg (1940) found soil loss per unit area (E) on measurement plots of different slope length to be proportional to $L^{0.6}$, while in the 'universal soil loss equation' (USLE) Wischmeier and Smith (1962) used a proportionality with $L^{0.5}$ (square root of slope length).

$$E = a\,L^{0.5}$$

where E is annual net soil loss per unit area and the parameter a incorporates all other terms of the USLE,

$$E = E \times \text{area}$$

E is the total annual net soil loss for a given area,

$$\text{area} = L \times W$$

where L is slope length and W slope width, and

$W = L/(l/w)$

where the length/width ratio (l/w) is a shape factor.

These equations can be combined into:

$$E = \frac{a\,L^{2.5}}{l/w} \tag{15.1}$$

According to the USLE, soil loss would thus be proportional to length to the power 2.5, and inversely proportional to a field length/width ratio. By contrast, McLaughlin *et al.* (1992) found an empirical proportionality of net sediment loss and length to the power 1.5–1.8 for Australian watersheds on a landscape scale (van Noordwijk *et al.*, 1997). It thus seems that within a field, net sediment loss per unit area will increase with increasing areas of consideration (the acceleration argument overrides the results), while at landscape scale, net losses decrease as the redistribution argument predominates.

In the next section, we will look at runoff and soil loss in relation to size and shape of the measuring area and examine whether simple scaling rules can be used for predicting erosion losses from plots of different size. In most existing erosion models, the width of the field does not appear in any of the equations, as it is generally assumed that width does not affect the runoff and soil losses. This would be true if erosion occurred on perfect straight slopes, without any lateral interactions. In the real world, such slopes do not exist and runoff water meanders around local obstructions. The lateral interaction increases as the water moves down, until eventually a drainage pattern – sometimes with rills and gullies – develops. A faster transport network is then created, leading to an overall increase in the velocity of the runoff.

The type of soil strongly affects the behaviour of runoff and soil movement, and may influence the scaling properties (and fractal dimension). If the soil is vulnerable to sealing, runoff water can move much further down the slopes than on soils where infiltration rates are not drastically reduced. Soil type is also a primary determinant for soil movement, influencing the soil detachment and transport processes. It is more difficult to detach clay particles than silt particles, but it is much easier to transport the clay than the silt. The relation between soil loss and plot size is therefore different for detachment-limited and transport-limited soils. Another factor to be considered is the extent of rilling; due to the concentration of flow in rills, the scouring effect is stronger in rill erosion than in inter-rill (or sheet) erosion.

The erosion/sedimentation literature has generally acknowledged that there is potentially a wide gap between plot level predictions from the USLE or one of its many modifications, and sediment measurements in streams (see Agus *et al.*, Chapter 16, this volume). The concept of a 'sediment delivery ratio' has been introduced for the change in value between two scales of consideration. This ratio is, however, not a fixed value but depends on the scales considered. 'It is well known that measured soil erosion rates vary according to the scale of measurement, the sediment delivery ratio decreasing as the plot size increases' (Imeson and Cammeraat, 1998).

The observation that the net sediment loss due to erosion changes with scale, but apparently in a predictable way, may be based on a fractal (self-similar across scale) process (Burrough, 1993). In self-similar processes the relation between a response variate Y and scale can be expressed as

$$Y = Y_1 L^s = Y_1 \text{ Area}^{s/2} = Y_1 \text{ Volume}^{s/3} \tag{15.2}$$

where Y_1 gives the Y value for a unit of length and s gives the fractal dimension. The sediment delivery ratio between scales 1 and 2 is thus $Y_2/Y_1 = (L_2/L_1)^s$. The 'fractal' dimension can be obtained from a log–log relationship of Y and L (or area or volume), as

$$\log Y = s \log L + \log Y_1$$

A fractal dimension of 1.5–1.8, as found for Australian watersheds, can be used to redefine a 'specific sediment yield' per length scale to the power, say 1.6, and this value can be used for extrapolation to other scales than the one used for measurements. If specific sediment loss were expressed as Mg $\text{km}^{-1.6}$ year^{-1}, results would be approximately independent of the length scale in the data sets of McLaughlin *et al.* (1992). For a fractal dimension of 1.6, net erosion losses per unit area are halved when the area increases by a factor of 32, reduced by 75% for a thousand-fold increase in area and reduced by 94% for a million-fold increase in area (a thousand-fold increase in length scale). The fractal dimension thus sets a rule for estimating a sediment delivery ratio for any scale transition.

Many erosion control measures reduce the amount of material flowing downhill, by providing local catchments. Consequently, a bigger fraction of erosion losses from a small unit area will be deposited in a neighbouring area, changing the fractal dimension of the erosion and sedimentation patterns. If an intervention is made at one scale only, the scale relations of erosion will show break-points. In fact, fractal properties will only apply over specific scale ranges and log–log plots may indicate a change of slope at certain points.

Drainage networks and rivers have strong self-similarity across scale, as at any point in the branched system the (peak) flow capacity should be adjusted to the (peak) flow rate, similar to the situation in trees (Mandelbrot, 1982; Phillips, 1993). Gray (1961) found an empirical relationship between the length of the principal stream L_p and the area of a watershed area; data from 47 watersheds over five orders of magnitude could be described by:

$$L_p = 131\ A^{0.568} \tag{15.3}$$

with L_p in km and A in km^2 (converted from: $L_p = 140\ A^{0.568}$, with L_p in miles and A in square miles and 1 mile is 1.609 km).

The relation between water and sediment flow in streams and rivers has been studied as well; generally there is a more than proportional increase of sediment load with increasing stream flow:

$$Q_s = a\ Q_w^b \tag{15.4}$$

where Q_s = sediment flow, Q_w is water flow and a and b are fitted parameters.

Morgan (1986) quoted *b* values of 1.86–2.46 depending on channel stability. There are various physical reasons for this non-linear scaling rule, related to thresholds for the transition from laminar to turbulent flow. A phase diagram of Q_s versus Q_w may give an efficient summary of a stream system (Fig. 15.4)

As was already mentioned, the environment in which the erosion occurs plays a big role in the acceleration and spatial redistribution of water and soil, thus determining the fractal behaviour of erosion. One can therefore not compare two catchments with different soil, vegetation and/or topographic properties in order to investigate the fractal properties of erosion. Similarly, soil erosion rates from a plot on a straight slope cannot be compared with the erosion from a catchment with a complex topography. This is not because of the different scale of measurement, but because of a wrong representation of the catchment. The conditions on a plot are generally controlled by the researcher: the slope is straight, there are no local depressions, no unwanted vegetation barriers (unless this is a treatment), no ditches etc. However, on larger sites, these 'disturbances' are there and they create opportunities for flow delay, infiltration and deposition of soil. The enhanced redistribution of soil increases with increasing area size, causing a decrease of soil loss per unit area (Table 15.1). In the extreme case of an area as big as the whole world, there is no net soil loss; all the soil is simply redistributed from the mountains to the seas!

Sampling and extrapolation scheme for landscape-level studies

Naïve methods of 'scaling-up', with or without a geographical information system, may be based on a 'stratified' sampling approach, estimating a total

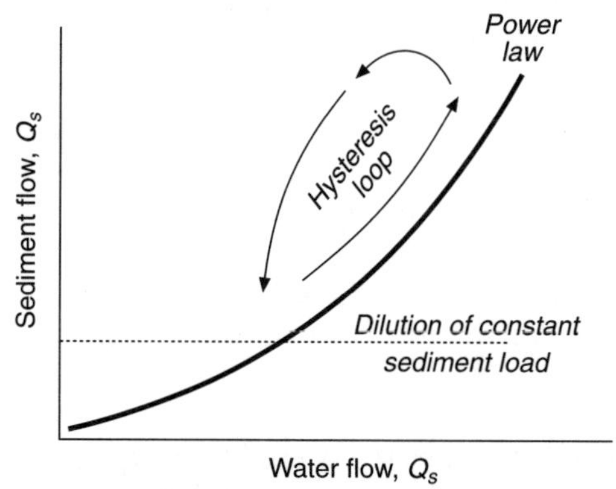

Fig. 15.4. Phase diagram of solute transport Q_s versus water flow Q_w of a stream system; the dotted line indicates a theoretical river where the streambed does not provide a buffer and where more water would simply dilute the sediment load; the curved line indicates a river where higher flow rates cause a more-than-proportional increase in sediment flow; real-world rivers may show hysteresis loops, as the sediment flow may respond differently to an increase than to a decrease in water flow.

Table 15.1. Examples of lateral interactions between landscape elements at different scale, influencing scaling rules for landscape functions in soil and water conservation.

	Scale					
	Patch	Field		Farm	Village	District
Function	0.001	0.01	0.1	1	10	100 km
A. On-site productivity	Topsoil loss, transfer, capture and utilization					
B. Water					Low spatial correlation of seasonal rainfall	
B1. Annual water yield						
B2. Dry-season base flow			Subsoil recharge via infiltration 'hot spots' from surface runon			
		Interception of subsurface flow by deep-rooted trees				
B3. Storm flow (floods)	Infiltration of surface runon			Low spatial correlation of hourly–daily rainfall		
C. Water quality: off-site sedimentation	Acceleration of surface runoff		Field boundary and riparian filter strips; capture and utilization		River bed sediment buffer; capture and stabilization	

sediment yield Q_t from the area fractions (A_i) and sediment yields per unit area (Q_i) for each of n classes of land use types:

$$Q_t = \sum_{i=1}^{n} A_i Q_i \tag{15.5}$$

The preceding discussion leads to the following adjustments to this scheme:

$$Q_t = \sum_{j=1}^{m} A_j^{0.5s} Q_j - \sum_{k=1}^{p} \Phi_k \tag{15.6}$$

adding a fractal scaling factor s to be applied to all m separate area units j (instead of total areas per land use type), and a cumulative effect of all p filters Φ_k.

In summary, for an estimate of a property like sediment flow from a landscape mosaic considered at different scales, we will have to get data on:

- a meaningful typology of land use, which can be used for all strata in a sampling scheme,
- estimates of the total area for each of the strata identified, as well as the size distribution of individual units (fields) A_j,
- estimate of net sediment loss Q_j for a given unit scale for all strata,
- estimates of the fractal scaling rule s for the landscape under study,
- the number or total length of filters in the landscape and the landscape units they separate, and
- the net reduction in sediment flow due to the filter effects for each class of filters.

As there is considerable literature on the Q_j values and remote sensing and GIS systems can be used for the A_j values, the values for s and Φ_k are the weakest part of the chain and should receive priority attention in future work.

Fractal Dimension *s* for Sediment Loss and Runoff in a Spatially Distributed Model

To explore effects of length, width and soil type on the scaling of runoff and soil losses, a spatially distributed erosion model was used. This model (van Roode, in preparation) is developed in PC-Raster, a GIS-based modelling tool, for studying the spatial behaviour of erosion in relatively small areas. The model has been calibrated for runoff with field data from erosion plots in an experiment in Machakos, Kenya. At this stage, the model has not yet been calibrated for soil loss. A model of similar scope is described by Collins *et al.* (Chapter 19, this volume).

In the model (Figs 15.5 and 15.6), runoff occurs when rainfall and incoming runoff from neighbouring upper cells exceed the infiltration capacity of the soil. The velocity of the runoff is calculated with the Manning formula, with the mean water depth in a cell as the hydraulic radius. The routing of runoff is determined by the steepest slope gradient as calculated from the elevation data within the plot. The elevation map is therefore one of the major inputs in the model, determining the flow pattern of the runoff. Soil can be detached by raindrops and by overland flow (Meyer and Wischmeier, 1969) and is thereafter transported using the Yalin transport capacity equation (Yalin, 1963). When the soil load exceeds the capacity, the excess of soil is deposited.

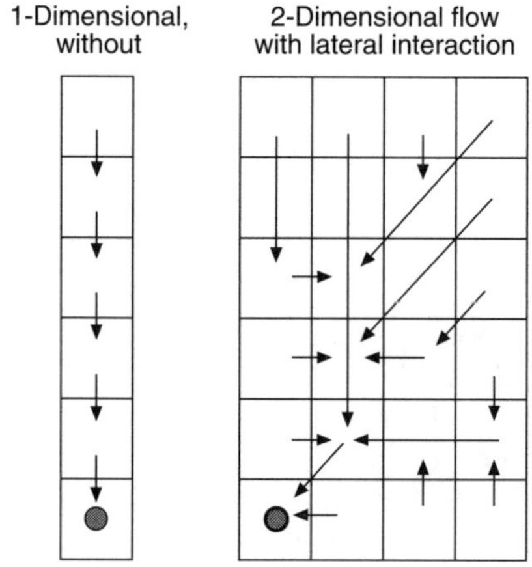

Fig. 15.5. Lateral interactions in a 1-dimensional runoff model of a series of grid cells and a 2-dimensional frame where each outflow from grid cell goes to the lowest of its eight neighbouring cells.

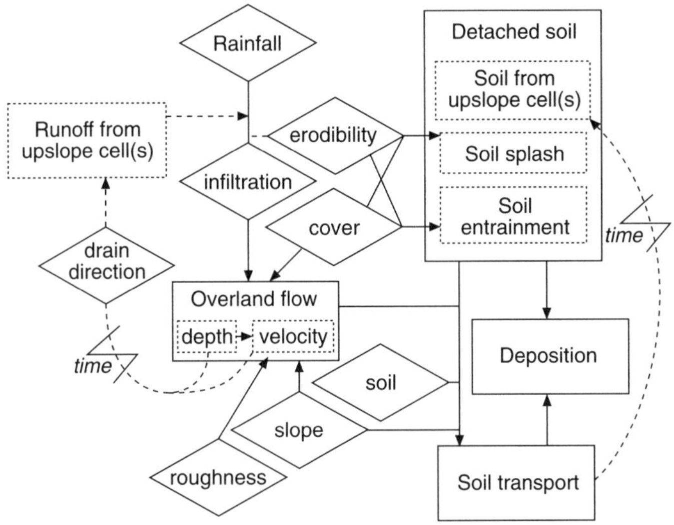

Fig. 15.6. Elements of a spatially distributed runoff and erosion model developed in PC-Raster; the parallelograms refer to maps used as parameters while the rectangles refer to functions using these parameters; arrows indicate influences within each time step used in the calculation, the arrows with a 'time' mark are implemented with a delay of one time step.

For the purpose of investigating the single effects of plot length, width and soil type on the fractal properties of erosion, other landscape variables were set as simple as possible. Only straight slopes were considered, without vegetation or other 'disturbing' patches. However, the elevation data of the slope had to be from an existing field, so that a natural flow pattern could be created.

Twenty different plot sizes were used, varying from 2.5 m long and 0.5 m wide to 28 m long and 5 m wide. For all plots, 0.25 m^2 was chosen as the standard cell size. The elevation data, measured on a 0.5 × 0.5 m grid, are from an erosion plot in Machakos. For the rain input, real data from a storm on 18 February 1996 in Machakos was used. In this storm, total rainfall was 50 mm within 4 h, divided over four high rain intensity peaks. The time steps in the simulations were 1 s.

By choosing different soil parameters, two soil-type scenarios were run: a transport-limited situation (soil is easily detached, but not easily transported) and a detachment-limited situation (soil is not easily detached, but once detached it is easily transported). Results are expressed as fractal dimension s_l and s_w, obtained by linear regression of log Y (with either runoff or sediment loss as the response parameter) versus log L or log W, respectively.

Runoff per unit area *decreased* slightly with **plot length**; the decrease is strongest on narrow plots with minimal lateral interaction. The fractal dimension of the length effect on a one-cell-wide plot, s_l was 0.93. Apparently, the effect of spatial redistribution of water was slightly stronger than the runoff acceleration effect for the model formulation and parameters used. On wider plots the s_l approached 1.

Runoff per unit area *increased* with **plot width**; the increase was strongest on long plots because of more available water and a stronger developed flow network. On extremely short plots, there was no width effect and therefore s_w is 1.00; on a 28-m-long plot D_{width} is 1.05. When length and width were increased simultaneously with a length:width ratio of 5:1, showing a size effect rather than length or width effect, s_l equaled 0.99 and runoff seemed not affected by plot size.

Under **transport-limited** conditions, soil loss per unit area strongly *decreased* with increasing plot length, like runoff. The effect was much stronger than for runoff, as s_l equalled 0.81 on a one-cell-wide plot. Contrary to the runoff results, the decrease was even stronger on wider plots. On a 5-m-wide plot, s_l had already decreased to 0.64.

Soil loss was also strongly affected by width on long plots: s_w was 0.92 on a 10-m-long plot and it was 0.85 on a 28-m-long plot. This shows that soil losses strongly *decreased* with width under transport-limited conditions. On very short plots, the width effect was negligible as no lateral interaction could occur. The combined effect of length and width (ratio 5:1) had a fractal dimension s_l of 0.84, which means that an increase in plot size led to a decrease in soil loss per unit area from the total plot.

Under **detachment-limited** conditions, the soil loss per unit area strongly *increased* with plot length, in contrast to runoff, the transporting agent. The effect was stronger if the plot was wider; s_l was 1.07 for a one-cell-wide plot and it was 1.13 for a 10-cell-wide plot. The build-up of runoff on long plots and the existence of a flow network in larger plots allowed for the increase in soil detachment by overland flow. Once the soil was detached, it could be transported out of the system and little redistribution of soil within the plot occurred, causing the strong increase in soil losses with plot size.

The plot width effect on soil loss under detachment-limited conditions was negligible on short plots, but it was stronger on longer plots, up to an s_l value of 1.06 on a 28-m-long plot. The combined effect of length and width (ratio 5:1) resulted in a slight increase in soil loss per unit area, with an s_l of 1.06.

Our main aim in presenting these data is not that they represent values of the fractal dimension s_l for universal use. Rather, we want to demonstrate that models such as the one we used can generate quantitative hypotheses on how the scaling rules will depend on soil, rainfall pattern and landscape properties. These hypotheses can be tested by taking measurements in a nested design, measuring the same process at at least three different scales (preferably differing by at least one order of magnitude – factor 10).

Filter Effects

For the watershed-level predictions of erosion effects we also need an assessment of the number and effectiveness of filters (Φ in Eqn 15.6). Filter efficiency depends on lateral transfers between sources and sinks. Total filter efficiency of a landscape depends on the radial or directional transport velocity, the total density of emitters and filters, their spatial pattern (correlation) and the saturation response of filters and their regeneration as a function of time. Filters in this general sense can include a range of landscape elements: depressions, cut-off drains, ditches (without or with plant residues as 'vertical mulch'), embankments, vegetated strips, hedgerows and riparian vegetation.

The importance of spatial positioning of the filters is reflected in the width of the envelopes in representations such as used in Fig. 15.2, where the overall frequency of filter units is gradually increased. For a circular geometry and radially symmetric transport situations, well-established methods of nearest neighbour analysis in spatial point patterns (Diggle, 1983; van Noordwijk *et al.*, 1993) can be used. If filter effectiveness can be defined as some function of radial distance, the overall effect at any given filter density can be compared with that of the same number of filters in a regular spacing (which forms the most efficient distribution in this case). For the types of filter effects used in Fig. 15.7, truly random distribution patterns are only slightly less efficient than regular spacings, while for (extremely) clustered distributions, overall effectivity is reduced to 40% (Fig. 15.7B). This type of filter effect would thus lead to fairly close linkage of total filter effect and the density of filter points (and thus to fairly narrow envelopes, in the sense of Fig. 15.2).

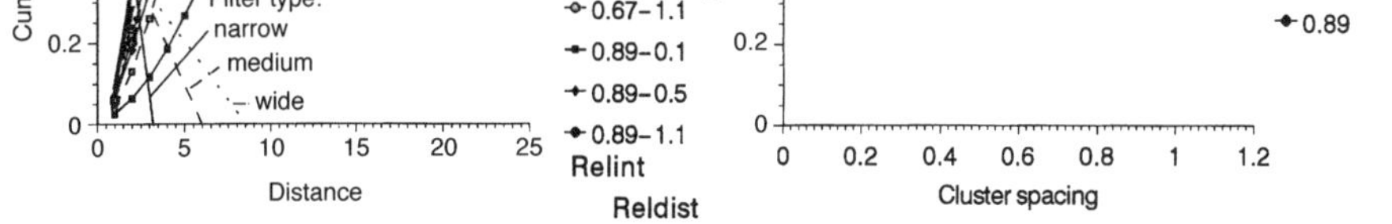

Fig. 15.7. Effects of spatial distribution of filters with a radially symmetric effect on their neighbourhood; A. the frequency distribution of nearest neighbour distances can provide an efficient summary statistic of the spatial distribution for a regular distribution, a pure random distribution and a family of 'clustered Poisson' processes, characterized by the Relint parameter (the number of 'daughter' points per cluster) and the Reldist (the spatial distribution within a cluster) (van Noordwijk *et al.*, 1993); three types of spatial filter effects are marked as 'narrow', 'medium' and 'wide'; B. filter efficiency for the various point patterns as a function of the Reldist parameter, relative to that of a regular distribution of the same average point density.

The theoretical formulation of Fig. 15.7 cannot be directly extended to a directional transport process, such as soil and water transport along slopes. It seems likely that under this type of transport a closed line of filters, along well-spaced contour lines, may be the most efficient configuration. The spatially distributed runoff/erosion model introduced in the preceding section was used for a number of model runs after adding 'tree' patches in the grid, to explore the envelopes of Fig. 15.2. From many simulation runs with different tree distributions over 25, 50 and 75% of all grid cells, we only show the results for the 'best' and 'worst' tree pattern.

The model yielded 'envelopes' for storm flow and sediment loss when applied to agroforestry (mixed tree–crop) landscapes which confirm the expectations of Fig. 15.2. As Fig. 15.8 demonstrates, in the model a tree cover of 50% with the most favourable spacing had the same effect as a full forest, while a tree cover of 25% could reduce the negative impacts of crops to 20% for sediment loss and 30% for storm flow. The least efficient tree positions lead to combined effects slightly less than proportional to the percentage tree cover. The model can be used in the future to explore how the widths of these envelopes vary with climate, soil type and topography. The strongest reductions in net sediment loss were obtained when trees were placed at the bottom of the slope (to intercept sediment) and in well-spaced contours (to prevent gully formation and breakthroughs in the riparian strips and other filters). Again, the model results should not be taken as absolute truths, but rather as a demonstration that for a given soil, landscape and climate specific hypotheses can be formed.

The spatially distributed model thus allows us to explore how the parameters for patch-size scaling (s) and filter effectiveness (Φ) can be related to known processes of soil and water transport on the surface of soils. The specific results will depend on such factors as soil type, surface roughness and rainfall pattern, but a quantitatively coherent framework is possible.

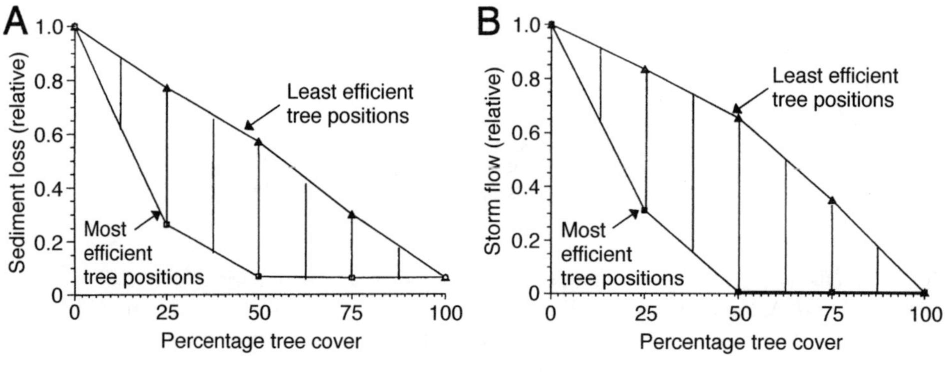

Fig. 15.8. Model calculations (compare with Fig. 15.2) of the effect of tree cover on sediment loss (A) and storm flow (B) in the spatially distributed model of Fig. 15.6 and a set of parameters for Machakos experimental station (Kenya).

When linked with a GIS and remote sensing facility for the more conventional parameters A_i and Q_i per land-use type, we may expect to make progress in scaling-up for the concerns B and C, introduced above. For farm productivity, however, we need a different level of complexity of model to predict the outcome of positive and negative interactions on crop yield.

Modelling Crop Production on Fields with Permanently Vegetated Contour Strips

Well-spaced trees or other permanent vegetation along contours can be effective in reducing erosion. But what about their effects of crop production (and labour demand)? The overall effect is likely to be a balance of short- and long-term changes, and will include negative effects through competition between the hedgerow and crop, as well as positive effects via maintenance of soil fertility. Tree–soil–crop interactions in hedgerow intercropping on flat land already are a complex of positive and negative interactions (Van Noordwijk *et al.*, 1998). At ICRAF a scheme is used for separating the overall interaction effects (*I*) on crop yield into positive terms related to maintenance of soil fertility (*F*) and negative terms related to competition effects (*C*). Putting hedgerows on sloping land adds a further complication. Whereas hedgerow intercropping on flat land has not generally lived up to its expectations, contour hedgerows on sloping land are supposed to have positive effects on crop yield (Sanchez, 1995). The yield increase may still not be worth all the labour invested in pruning the trees (Garrity, 1996), but physical yields are expected to increase as the *F* term will be positive (or at least not deteriorate as fast as in the pure-crop control).

$$Y_{hic} = (1 - \alpha_t)(Y_c + F - C + R_w + R_s + M) \tag{15.7}$$

where: Y_{hic} = crop yield in contour hedgerow intercropping, α_t = area fraction used for hedgerows, Y_c = crop yield in pure-crop control (on unprotected slope), F = fertility effect (short + long term), C = competition effect (above + below ground), R_s = effect of uneven water infiltration, R_w = effect of topsoil redistribution during terrace formation, M = microclimate effect (including effects via pests, diseases and weeds).

The net effect will be positive if:

$$\frac{F - C + R_w + R_s + M}{Y_c} > \frac{\alpha_t}{1 - \alpha_t} \tag{15.8}$$

The sign of the R_w and R_s term is uncertain. Uneven water infiltration due to contour hedgerows may reduce water availability in the upper alleys, but may in high rainfall zones also lead to reduced leaching of N. Redistribution of topsoil will benefit the lower alley at the expense of the upper alley zone. We may expect that the overall effect of contour hedgerows depends on even more factors than that of flat-land alley cropping and strong generalizations should be avoided.

The WaNuLCAS model for predicting water, nutrient and light interactions in agroforestry systems (van Noordwijk and Lusiana, 1998) can be used to explore some of the interactions at process level. The model is set up for spatially zoned tree–crop systems and can be easily adapted to sloping lands. The model can deal with four zones (one for the hedgerow, three for the crop), of different topsoil depth, with heterogeneous water infiltration, shading and below-ground competition between regularly pruned hedgerows, where the prunings are returned to the alley and will gradually mineralize N. Figure 15.9B gives initial results, accumulated over four crops. If we follow the lines from left to right, we see that the effect of not growing crops on the space reserved for hedgerows is negative, but that the uneven water infiltration (R_w) can make up for the yield loss in the humid series (it reduces N leaching from the crop zone). Considering a regularly pruned hedgerow on the contour instead of a bare strip has a moderate positive effect on crop

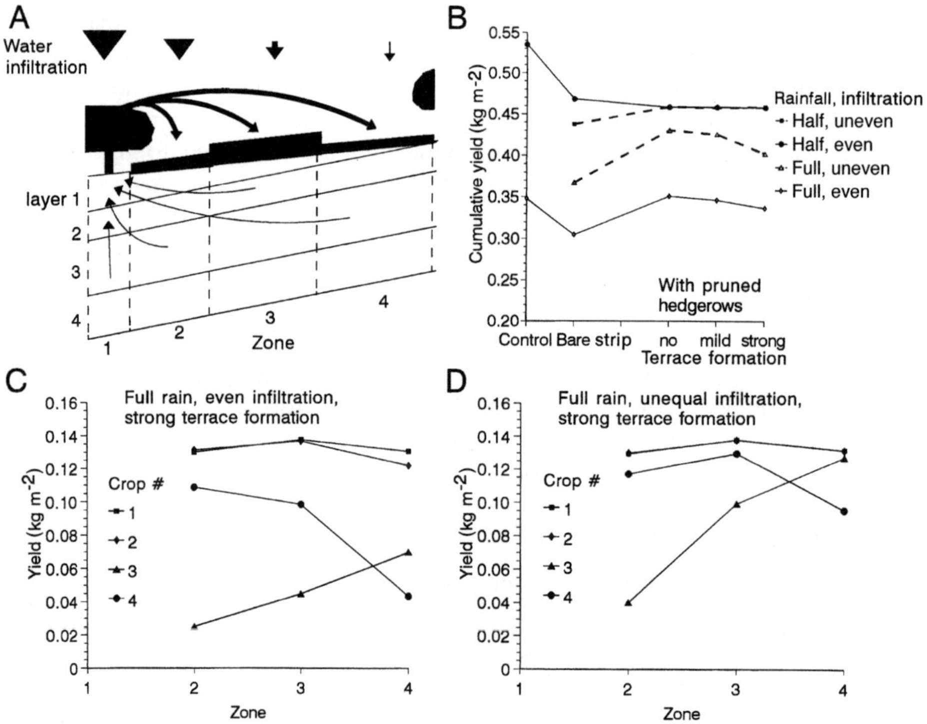

Fig. 15.9. Calculations with the WaNuLCAS model (van Noordwijk and Lusiana, 1998) of crop yield in a contour hedgerow system on sloping land; A. Model scheme for applications on sloping land; B. Cumulative yield over four crops (2 years) for a humid (3000 mm year^{-1}) and subhumid (1500 mm year^{-1}) climate, with and without uneven infiltration of rainfall over the respective zones; C, D. results per crop and zone (tree, crop and soil parameters are taken from the P series in the standard setting of van Noordwijk and Lusiana, 1998).

yields, but terrace formation (R_s) has a negative effect on yields. For the subhumid series all effects are weak, and no treatment combination can make up for the space lost to make the contour strip. The results per crop zone (Figs 15.9C and D) contain some surprises, as they show a range of patterns between crops: for some crops the middle of the alleys gives the highest yield, for others the lower alley, or even the upper alley. Although all types of patterns can be observed in real-world experiments, it is surprising that the balance of positive and negative interactions can, apparently, change so easily in the complexity of the WaNuLCAS model.

We conclude that the WaNuLCAS model can be used to explore the balance of positive and negative interactions for contour hedgerow intercropping and that the net effect may well be small, as strongly positive and strongly negative effects on crop yield strive for prominence. Further model validation is necessary before any soil, climate, tree and crop specific model predictions should be seen as more than 'interesting hypotheses'.

Farmer Recognition and Exploitation of Within-field Variability of Soil Fertility

Assessments of the impact of erosion on crop productivity have often been made by planting a uniform crop across a toposequence, to measure both the possible benefits of accumulated topsoil at deposition sites and loss of productivity uphill (Lal, Chapter 10, this volume). Although this is an improvement over past approaches where only negative effects were considered, it may still overestimate the negative effects, as the response of crop growth to topsoil depth will generally have a 'diminishing returns' shape, rather than an S-shape, and the losses uphill cannot be fully compensated downhill. In practice, however, farmers could utilize the enriched zone for different crops and different crop management than the impoverished zones. The end result may then be positive – as is the case where the enriched zone can be used for paddy rice.

Erosion and sedimentation processes at field and landscape scale primarily affect the heterogeneity of soil fertility within these land units. The main issues thus are whether or not fertility can be exploited on the place where it ends up and whether or not this outweighs the yield opportunities lost at the sites of net soil loss. Redistribution of fertile topsoil at certain spatial and temporal scales can be very beneficial from a productivity perspective. For example, without erosion in the hills, there would not have been the good yields in river flood plains and in the paddy rice fields of the world, which allowed permanent crop production, before chemical fertilizers were used. In this section we will review literature on farmer strategies to recognize and exploit microscale variability of soil fertility.

It can safely be said that farmers around the world do recognize and exploit elevational, lateral and temporal variations in soils. Recent literature documents indigenous classifications of soils and farmers' rationale for cropping patterns in relation to soil differences (Alcorn, 1989; Grobben, 1992; Diallo and Keita, 1995; Moran, 1995; Drechsel, 1996; Joshi *et al.*, 1996).

For example, *elevational* gradients in high altitude areas of the tropics are exploited with 'barley (*Hordeum vulgare* L.) and other cold-tolerant crops grown on relatively level, frost-prone fluvial surfaces, whereas maize is grown on steeper slopes, down which cold air drains' (Sandor and Furbee, 1996). In terms of *laterally* discontinuous soil properties, outside of Kupang, West Timor, Indonesia, farmers have three major classes of soils: red, black and white; each is recognized to have different physical, chemical and biological properties (McCallie, personal observation). Red soils are shallow, rocky soils with low fertility. They are appropriate for supporting cattle grazing and/or maize production if they are managed well, which includes regular fallowing. Black soils are reserved for cash crops, primarily vegetable and rice production, as the soils are fine textured, deeper and more fertile than the red soils. White soils, on the other hand, occupy much less area and are not specifically cropped. *Temporal* differences in soils are illustrated in upland, shifting cultivation systems where soil fertility is recognized as highest during the first cropping season following a burn, and diminishing with each subsequent crop. Upland rice is first planted, then maize and beans, and finally cassava, the heartiest of the crops.

On a 2–50 m scale, microvariability in soil properties has also been documented (Brouwer *et al.*, 1993). In developed countries of the temperate zone, farming systems are exploring and implementing precision agriculture, also known as variable rate technology (VRT), in which locations within individual fields are mapped according to physical and chemical properties and then fertilizer management is adjusted according to the microsite requirements in order to produce maximum yields (Anderson and Bullock, 1996). Precision agriculture can lead to higher production per unit land, higher nutrient-use efficiency and higher economic returns per unit land area (Tweeten, 1996). Such within-field soil microvariability has also been documented in developing countries of the tropics (Eswaran *et al.*, 1992; Brouwer *et al.*, 1993; Lamers and Feil 1995) and is generally attributed to microscale erosion and sedimentation, manure application (intentional and by animals) and ash deposition from burning. Do farmers in the tropics recognize and exploit soil variations occurring on a microscale, just as they do with the larger scale differences as discussed above?

In addressing this question, we first look at the generalizable context of the small farmer in developing countries in the tropics. Though soils in the tropics vary widely in terms of their physical, chemical and biological properties (Cassel and Lal, 1992; Eswaran *et al.*, 1992; Sanchez and Logan, 1992), the small farmer, typically resource-poor with little to no capital or access to credit, typically works on marginal lands, uses few external inputs, relies on rainfed systems, and produces low yields per unit area (Sivakumar *et al.*, 1992). Given this context, subsistence and marginal farmers are more concerned with assuring sufficient yields in poor years than in good averages over all years (van Noordwijk and van Andel, 1988; Brouwer *et al.,* 1993; van Noordwijk *et al.*, 1994). Therefore, recognizing and acting on microvariability in soil properties may be an asset to the small farmer, as is crop diversification, staggering of planting time, or having several fields dispersed within a large area (Brouwer *et al.*, 1993). Do small farmers recognize these fertility differences and, if so, how are they exploited?

Two broad strategies can be identified. Strategy 1, similar to that of precision agriculture, is to recognize locales of lesser soil quality and enhance its productivity by spot management, often through organic soil amendments. While precision agriculture is applied to maximize yields through achieving optimal fertility throughout the field, small farmers, on the other hand, with limited access to inputs, may rely on enhancing certain locales within a field. This is exemplified by a study of land-use planning of Mazihwa in southern Zimbabwe in which a local farmer described the traditional practice of managing problem soil microsites within fields by adding organic material to build up the area physically, protecting it from further decline by erosional forces (Wilson, 1995). Thus, while high-input agricultural systems look to optimize production all over the field, low-input systems may look toward stop-gap measures to prevent further decline, and thus variability, within a field.

Contour hedgerow systems, especially in combination with ploughing, can lead to a rapid terrace formation, on the basis of topsoil material from the upper alley zone. This process creates a fertile lower and 'scoured', unfertile upper alley (Garrity, 1996). Farmers recognize this differential fertility and respond by differentiating strategies for crop planting density and fertilizer applications in the two zones (Garrity *et al.*, Chapter 5, this volume).

In the second strategy, farmers take advantage of more productive microsites within a field. Several methods of strategy 2 are apparent:

A. planting the entire field so that some yield can be expected no matter what the environmental conditions of the particular season;
B. modifying planting densities according to soil properties; and
C. planting crop species or varieties in accordance to specific soil microsites within a field.

As an example of method A, Brouwer *et al.* (1993) assert that subsistence farmers in the semi-arid tropics (Sahel) of Africa recognize and exploit soil microvariability by planting the entire field. With uncertain rainfall and rainfall patterns (Sivakumar *et al.*, 1992), the combination of seed and soil variability should result in some yield, though never bumper crops. Sandor and Furbee (1996) and Eash and Sandor (1995) describe method B by which Lari farmers in the Andes of southern Peru exploit what they recognize as thicker, more fertile soil near the front of terraces by planting crops more densely moving from the back to the front of agricultural terraces. Method C is nearly the reverse of strategy 1 and precision agriculture as it strives to take greatest advantage of the most fertile sites. Classic examples of method C are home gardens and other agroforests in which a combination of soil and climate microsites are exploited through diverse species plantings (Nair, 1993). Microsite exploitation is also done by small farmers in annual crop systems. Vegetables such as aubergines and red peppers are planted in small ash piles which result from the previous field burning in Rantau Pandan, Sumatra, Indonesia (van Noordwijk, personal observation). Literature on erosion, sedimentation and microsites is primarily from semi-arid sandy soils of Africa. Farmers in the Sahel of Niger, Africa recognize microtopography within fields, where they attribute enhanced crop or vegetation production to thick fertile sand, compared with low-lying, thin-sand and infertile microsites (Lamers and Feil, 1995). Locally more fertile sites in this study were caused by organic materials or residues capturing wind-blown particles. Taylor-Powell (1991) documents within-field micro-lows and micro-highs caused by wind and by sheet erosion. Geiger and Manu (1993) followed up on the farmers' claim of enhanced crop production on designated microtopographic sites, finding millet yields to be 120% higher than the average. Brouwer *et al.* (1993) corroborate similar assertions with soil property analyses in such microsites.

By applying local knowledge and exploitation of within-field soil variability to research and development programmes, low-input cropping systems may be enhanced through increased production and/or cultural acceptability, while irrigation systems could be improved. Technologies such as planting in rows to make optimal use of area limitations may actually compromise maximal yields under input- or seed-limited conditions. In such cases, exploiting soil microvariability may allow the resource-poor farmer to improve production levels. In less marginal conditions, exploiting microvariation, except in the instances of preventing further degradation, may not effectively increase production levels. For example, in years of particularly low rainfall, poorly drained microsites may actually lead to higher yields than well-drained microsites. The reverse is also true. In years of high rainfall, extremely well-drained sites may be the high producing sites. To reduce risk of crop failure, a variety of microsites must be planted. If nutrient levels limit a crop's production, recognizing and exploiting within-field microsite variation may allow the resource-poor farmer to optimize and/or diversify yields.

An Agenda for Landscape Ecological Research

The three soil and water conservation concerns (A, B and C, see above) probably have different 'scaling properties' and this leads to a different array of management options at different points of intervention (Table 15.1). Siltation of reservoirs can be avoided or at least reduced by having trap dams just before the reservoir, whereas disturbances in seasonal patterns of water flow after forest conversion are much more difficult to correct, without installing reservoirs. For sedimentation, the specific position of trees and effective barriers in the landscape is of major concern, whereas for the total water balance, the percentage tree cover is more important. In developing a quantitative landscape ecology we thus need to separate the various concerns and opportunities for intervention.

In landscapes with an abundance of vegetated field-border strips and hedgerows along paths (II, III and V in Fig. 15.1), the link between on-site and off-site effects can be easily overstated. Even with the most erosive field practices, contributions to sediment load in rivers probably depend on 'incidents' (footpaths, breakthroughs, land slides of poorly stabilized river banks) rather than the amounts of sediment captured in conventional 'soil conservation studies'. Indeed, there appears to be a major gap in the literature between plot/patch-level studies and studies based on stream flow. The whole matrix of field borders in which the plots are set appears to have been ignored in most previous work. If we see this as a priority area we should aim at:

- developing a method for 'reading the landscape', an effective descriptive system (language) for the hot spots of sediment loss and the possible points of intervention;
- developing spatially distributed models which have 'cells' for plots, as well as explicit rules for sediment flow across plot boundaries (resistances); and
- analysing stream flow/sediment load data for characteristic patterns of breakthroughs as an overall descriptor of landscape quality.

Rigorous quantitative analysis can and should go hand in hand with developing easily applied methods for recognizing field phenomena. A related point is an analysis of where and how sediment capture can provide a resource:

- on site, where specific components benefit more than proportionally from increased topsoil fertility (for example, fruit trees may not meet their minimum required topsoil depth on the field as a whole, but may do so where topsoil has accumulated in front of a 'naturally vegetated strip', NVS);
- on farm, where enrichment of lowlands creates opportunities for rice paddies;
- off farm, where river floodplains are prime farmland; and

(• off farm if sediment is used for brick making, etc.)

In water flow *per se*, there are less opportunities for making up downstream for changes in hydrology upstream. At the same time, landscape-level hydrology is more dependent on the subsoil properties and underlying geomorphology, on which land-use change is superimposed. Whereas removing forests and trees generally increases total water yield from a catchment (by reducing interception losses and transpiration), it will do so mainly by increasing storm flow and it may reduce base flow if subsoil infiltration becomes less effective. The main entry point for field studies and modelling is the effective rate of infiltration which separates runoff from soil recharge. Runoff will lead to runon and neighbourhood interactions until 'channelling' occurs. Mapping the dendritic (fractal branching) properties of channels up to their finest extension is probably a first requirement to understand how they function.

The models discussed above can help in quantifying scale relationships and theoretical options for agroforestry systems to contribute to on-site and off-site interests. Model development should go hand in hand with a more applied approach to improve our ability to 'read' the landscape and recognize the 'hot spots' of problems as well as the 'cool zones' of filters.

Recognizing hot spots of sediment loss and gully formation should not look at cropped fields alone, but include road sides, footpaths and morpho-erosion (Diemont *et al.*, 1991). For example, a local informant participating in a study of land-use planning of Mazihwa in southern Zimbabwe indicated that some of the worst erosional processes in the area are related to footpaths; where once footpaths were zigzagged up slopes and well cared for, including abandoning of usage in order to decrease erosion, changes in the local socio-political system now allow for steep footpaths that are not managed to prevent erosion (Wilson, 1995). Local regulations on public access to land may thus have an important effect.

We need a better way of recognizing and testing the effectiveness of filter strips in the landscape. The accumulated soil wealth in riparian filters can be exploited:

- via biomass transfer, as with *Tithonia diversifolia* growing in roadside filter strips in western Kenya, (ICRAF, 1997); or
- by planting of nutrient-demanding crops, such as fruit trees.

In evaluating the longer-term effectiveness of these strips, we need to know how they are stabilized, for example by tree roots as well as surface vegetation and effects of soil fauna. Under what conditions will breakthroughs and landslides occur? Filter strips may accumulate sediments rich in soil organic matter. The rate of decomposition in this environment is poorly known, and thus it is hard to judge whether erosion/sedimentation processes contribute to C sequestration at a landscape scale or to net C loss (Paustian *et al.*, 1997; van Noordwijk *et al.*, 1997).

In developing this new tool box for research, we have to acknowledge the different intensities at which landscapes are used, as recognized in the stages of Fig. 15.1. Ongoing research in the multi-institutional 'Alternatives-to-Slash-and-Burn' programme is aimed at both 'forest margin' and 'degraded lands' zones. The subwatersheds selected for research reflect this typology (Table 15.2).

Most ongoing 'soil conservation' projects appear to aim for a transition of stage IV to V. Avoiding the transition from III to IV, and thus avoiding the worst stage of the soil degradation process, may be at least as urgent a problem (and a more efficient avenue for government intervention), but public support for solving evident problems is always easier to generate than for problem prevention.

There is less literature on stage II, apart from an analysis of logging impacts and the large reductions possible by 'low-impact' logging techniques which maintain an effective riparian filter strip (Bruijnzeel, 1990, 1997). Effects of plot-level opening of fields by smallholders in slash-and-burn systems deserve further attention. Where slash-and-burn agricultural systems have evolved from a crop–fallow rotation into agroforests, as is common in Sumatra, the 'rejuvenation' stage of the agroforest is the main stage in which erosion will occur. In the ASB research area in Jambi, a stage II landscape on slopes is studied, with 'jungle rubber' as the dominant land-use system.

Most jungle rubber systems in Sumatra depend on plot-level slash-and-burn when trees are to be replanted. Erosion rates can be high directly after the slash-and-burn event and much of the charcoal and partly burnt slash will be redistributed in the field. Usually trees are felled downslope, but if

Table 15.2. (Sub-)Watersheds studied within the Alternatives-to-Slash-and-Burn programme in three countries of South-East Asia; the landscape typology refers to Fig. 15.1.

Landscape type	Indonesia	Thailand	Philippines
II. Distributed patches of open (cropped) land in an (agro)forest context	Bungo tributary to Batang Hari river (Jambi, Sumatra)	–	–
III/IV. Agricultural landscapes in various stages of intensification and removal of vegetative filter strips	Tulang Bawang river (Lampung, Sumatra)	Mae Chaem river	Manupali river in Pulangui river basin (Mindanao)
IV/V. Recovery in degraded landscapes by stimulating permanently vegetated contour strips and protection of riparian strips	–	–	Claveria (Mindanao)

occasionally they fall along a contour line, they can provide a fairly effective filter, accumulating topsoil. Heterogeneity of topsoil fertility in this system is a resultant of heterogeneity before opening the field, differential inputs via the (sl)ash, differential mineralization if nutrients (P) from organic pools are released by the heat during the burn, and a redistribution by erosion/sedimentation. In this chaotic pattern, rubber trees and other crops are planted, while the spontaneous regenerating secondary forest vegetation is regulated – some trees are allowed to grow, others not. Does this system allow farmers to make use of fertility patches and thus give little incentive to reduce erosion *per se*? We have initiated research to explore these issues.

Conclusions

The degree to which soil productivity and regional water balance are uncoupled is directly relevant to the policy debate, as it determines whether or not on-site (upland) and off-site (downstream) groups will have a parallel interest in finding solutions to their respective problems. Scaling rules for water and sediment transport in landscapes can be rather different, and this may lead to scale-dependent value judgements.

The total water and sediment yields from landscapes with a partial tree cover (agroforestry) may be partly unrelated. For sediment load of rivers, the location of trees may be far more important than the area covered, as riparian strips can make up for substantial sediment movement from upslope. For total water yield and base flow of rivers, the percentage forest cover may be more important than the location of trees within the watershed. A 'replacement series' design can be used to illustrate this effect. The proportion of 'lateral interactions' in the process under study appears to be directly linked to the sensitivity of the overall outcome to spatial distribution of landscape elements.

Farm productivity effects of vegetative soil conservation measures are a net effect of the labour and other costs involved and the yield. Harvested yield per unit area depends on: the area occupied by the soil conservation strips; the effects on crop yields of soil redistribution within the slope; effects of heterogeneous water infiltration, shading and competition for water and nutrients between vegetation and crop; and effects on soil organic matter balance. The WaNuLCAS model for spatially zoned agroforestry systems can now be used to explore these interactions and define conditions where the net physical yield effect can be positive, as a starting point for financial evaluations.

Erosion and sedimentation processes at field and landscape scale primarily affect the heterogeneity of soil fertility within a field and a landscape, in combination with a certain net loss of sediment. The main issues thus are whether or not fertility can be exploited at the place where it ends up, and whether or not this outweighs the yield opportunities lost at the sites of net soil loss. Future landscape ecological models should give a fuller account of the life expectancy of filter strips and of farmer opportunities for capitalizing on the soil accumulated in specific zones in the landscape.

Acknowledgements

Thanks are due to Fred Stolle and Subekti Rahayu for help with the calculations for Fig. 15.7.

References

Alcorn, J. (1989) Indigenous agroforestry systems in the Latin American tropics. In: Altieri, M.A. and Hecht, S.B. (eds) *Agroecology and Small Farm Development.* pp. 203–218.

Anderson, L. and Bullock, D. (1996) Use caution in interpreting clusters of similar values in soil fertility maps. *Better Crops* 80(3), 26–29.

Brouwer, J., Fussell, L.K. and Herrmann, L. (1993) Soil and crop growth micro-variability in the West African semi-arid tropics: a possible risk-reducing factor for subsistence farmers. *Agriculture, Ecosystems, and Environment* 45, 229–238.

Bruijnzeel, L.A. (1990) *Hydrology of Moist Tropical Forests and Effects of Conversion: a State of Knowledge Review.* HTP, UNESCO, Paris.

Bruijnzeel, L.A. (1997) Hydrology of forest plantations in the tropics. In: Nambiar, E.K.S. and Brown, A.G. (eds) *Management of Soil, Nutrients and Water in Tropical Forest Plantations.* ACIAR, Canberra, Australia, pp. 125–167.

Burrough, P.A. (1993) Fractals and geostatistical methods in landscape studies. In: Lam, N.S. and de Cola, L. (eds) *Fractals in Geography.* Prentice-Hall, Englewood Cliffs, New Jersey, USA, pp. 87–121.

Cassel, D.K. and Lal, R. (1992) Soil physical properties of the tropics: common beliefs and management restraints. In: Lal, R. and Sanchez, P.A. (eds) *Myths and Science of Soils of the Tropics: Proceedings SSSA, Las Vegas, Nevada, 17 October 1989. SSSA Special Publ. 29,* pp. 61–89.

Diallo, D. and Keita, D. (1995) An indigenous soil classification system in the agroecological zone of Djitoumou, Mali. *Cahiers Agricultures* 4(5), 371–375.

Diemont, W.H., Smiet, A.C. and Nurdin (1991) Re-thinking erosion on Java. *Netherlands Journal of Agricultural Science* 39, 213–224.

Diggle, P.J. (1983) *Statistical Analysis of Spatial Point Patterns.* Academic Press, London.

Drechsel, P. (1996) AFRICALAND: Traditional knowledge for participatory approaches to soil management. *IBSRAM Newsletter* 41, 8–9.

Eash, N.S. and Sandor, J.A. (1995) Soil chronosequences and geomorphology in a semiarid valley in the Andes of southern Peru. *Geoderma* 65, 59–79.

Eswaran, H., Kimble, J., Cook, T. and Beinroth, F.H. (1992) Soil diversity in the tropics: implications for agricultural development. In: Lal, R. and Sanchez, P.A. (eds) *Myths and Science of Soils of the Tropics. Proceedings SSSA, Las Vegas, Nevada, 17 October 1989. SSSA Special Publ. 29,* pp. 1–16.

Garrity, D.P. (1996) Tree–soil–crop interactions on slopes. In: Ong, C.K. and Huxley, P.A. (eds) *Tree–Crop Interactions – a Physiological Approach.* CAB International, Wallingford, UK, pp. 299–318.

Geiger, S.C. and Manu, A. (1993) Soil surface characteristics and variability in the growth of millet in the plateau and valley region of Western Niger. *Agriculture, Ecosystems and Environment* 45, 203–211.

Gray, D.M. (1961) Interrelationships of watershed characteristics. *Journal of Geophysical Research* 66, 1215–1223

Grobben, P. (1992) Using farmers' knowledge about soil classification. A case-study of farming systems research (FSR) project in Java. *AT Source* 20(1), 6–9.

Hufschmidt, M.H (1986) A conceptual framework for watershed management. In: Easter, K.W., Dixon, J.A. and Hufschmidt, M.H. (eds) *Watershed Resource Management, an Integrated Framework with Studies from Asia and the Pacific.* Westview Press, Boulder, pp. 17–31.

ICRAF (1997) *International Centre for Research in Agroforestry, Annual Report 1996.* ICRAF, Nairobi, Kenya.

Imeson, A.C. and Cammeraat, L.H. (1998) Upscaling erosion processes from the plot to the catchment scale. *Proceedings of the GCTE Soil Erosion Network Conference,* Utrecht, the Netherlands, 14–18 April 1997.

Joshi, K.D., Tuladhar, J.K. and Sthapit, B.R. (1996) Indigenous soil classification systems and their practical utility: a review. In: *Formulating a Strategy for Soil Fertility Research in the Hills of Nepal. Proceedings of a workshop held in Lumle Agricultural Research Centre, Pokhara, Nepal, 17–18 August 1995,* pp. 36–42.

Kiepe, P. and Rao, M.R. (1994) Management of agroforestry for the conservation and utilization of land and water resources. *Outlook on Agriculture* 23(1), 17–25.

Lamers, J.P.A. and Feil, P.R. (1995) Farmers' knowledge and management of spatial soil and crop growth variability in Niger, West Africa. *Netherlands Journal of Agricultural Science* 43, 375–389.

Leopold, A. (1949) *A Sand County Almanac.* Oxford University Press, Oxford.

Mandelbrot, B.B. (1982) *The Fractal Geometry of Nature.* W.H. Freeman & Co., San Francisco.

McLaughlin, M.J., Fillery, I.R. and Till, A.R. (1992) Operation of the phosphorus, sulphur and nitrogen cycles. In: Gifford, R.M. and Barson, M.M. (eds) *Australia's Renewable Resources Sustainability and Global Change. Bureau of Rural Resources Proceedings No. 14. AGPS,* Canberra, pp. 67–116.

Meyer, L.D. and Wischmeier, W.H. (1969) Mathematical simulation of the process of soil erosion by water. *Transactions of the American Society of Agricultural Engineers* 12, 754–762.

Moran, E.R. (1995) Socio-economic aspects of acid soil management. In: *Plant–Soil*

Interactions at Low pH: Principles and Management. Proceedings of the Third International Symposium, Brisbane, Queensland, Australia, 12–16 September 1993, pp. 663–669.

Morgan, R.P.C. (1986) *Soil Erosion and Conservation.* Longman, UK, 298 pp.

Nair, P.K.R. (1993) *An Introduction to Agroforestry.* Kluwer Academic Publishers, Boston, 499 pp.

Paustian, K., Andrén, O., Janzen, H.H., Lal, R., Smith, P., Tian, G., Tiessen, H., van Noordwijk, M. and Woomer, P.L. (1997) Agricultural soils as a sink to mitigate CO_2 emissions. *Soil Use and Management* 13, 230–244.

Phillips, J.D. (1993) Interpreting the fractal dimension of river networks. In: Lam, N.S. and de Cola, L. (eds) *Fractals in Geography.* Prentice-Hall, Englewood Cliffs, New Jersey, USA, pp. 142–157.

Sanchez, P. (1995) Science in agroforestry. *Agroforestry Systems* 30, 5–55.

Sanchez, P.A. and Logan, T.J. (1992) Myths and science about the chemistry and fertility of soils in the tropics. In: Lal, R. and Sanchez, P.A. (eds) *Myths and Science of Soils of the Tropics: Proceedings SSSA, Las Vegas, Nevada, 17 October 1989.* SSSA *Special Publ. 29*, pp. 35–46.

Sandor, J.A. and Furbee, L. (1996) Indigenous knowledge and classification of soils in the Andes of southern Peru. *Soil Science Society of America Journal* 60, 1502–1512.

Sivakumar, M.V.K., Manu, A., Virmani, S.M. and Kanesmash, E.T. (1992) Relation between climate and soil productivity in the tropics. In: Lal, R. and Sanchez, P.A. (eds) *Myths and Science of Soils of the Tropics: Proceedings SSSA, Las Vegas, Nevada, 17 October 1989. SSSA Special Publ.* 29, pp. 91–119.

Taylor-Powell, E. (1991) Integrated management of agricultural watershed: land tenure and indigenous knowledge of soil and crop management. *TropSoils Bulletin* 91–04. Department of Soils and Crop Sciences, Texas A&M University College Station, 30 pp.

Tweeten, L. (1996) Is precision farming good for society? *Better Crops* 80(3), 3–5.

van Noordwijk, M. and Lusiana, B. (1998) *WaNuLCAS, a Model of Water, Nutrient and Light Capture in Agroforestry Systems.* International Centre for Research in Agroforestry (ICRAF), Bogor, mimeograph.

van Noordwijk, M. and van Andel, J. (1988) Reduction of risk by diversity: a theoretical basis for age-old farming systems. *ILEIA Newsletter* 4(4), 8–9.

van Noordwijk, M. and Ong, C.K. (1996) Lateral resource flow and capture – the key to scaling up agroforestry results. *Agroforestry Forum* 7(3), 29–31.

van Noordwijk, M., Widianto, Heinen, M. and Hairiah, K. (1991) Old tree root channels in acid soils in the humid tropics: important for crop root penetration, water infiltration and nitrogen management. *Plant and Soil* 134, 37–44.

van Noordwijk, M., Brouwer, G., Zandt, P., Meijboom, F.W. and Burgers, S. (1993) *Root Patterns in Space and Time: Procedures and Programs for Quantification. IB-Nota 268.* IB-DLO, Haren, the Netherlands.

van Noordwijk, M., Dijksterhuis, G. and van Keulen, H. (1994) Risk management in crop production and fertilizer use with uncertain rainfall: how many eggs in which baskets. *Netherlands Journal of Agricultural Science* 42, 249–269

van Noordwijk, M., Woomer, P., Cerri, C., Bernoux, M. and Nugroho, K. (1997) Soil carbon in the humid tropical forest zone. *Geoderma* 79, 187–225

van Noordwijk, M., Hairiah, K., Lusiana, B. and Cadisch, G. (1998) Tree–soil–crop interactions in sequential and simultaneous agroforestry systems. In: Bergström, L. and Kirschner, H. (eds) *Carbon and Nutrient Dynamics in Natural and*

Agricultural Tropical Ecosystems. CAB International, Wallingford, UK, pp. 173–190.

Vermeulen, S.J., Woomer, P.L., Campbell, B.M., Kamukondiwa, W., Swift, M.J., Frost, P.G.H., Chivaura, C., Murwira, H.K., Mutambanengwe, F. and Nyathi, P. (1993) Use of SCUAF model to simulate natural miombo woodland and maize monocultural ecosystems in Zimbabwe. *Agroforestry Systems* 22, 259–271.

Wilson, K.B (1995) 'Water used to be scattered in the landscape': local understanding of soil erosion and land use planning in Southern Zimbabwe. *Environment and History* 1(3), 281–296.

Wischmeier, W.H. and Smith, D.D. (1962) Soil loss estimation as a tool in soil and water management planning. *Int. Assoc. Scient. Hydrol. Publ.* 59, 148–159.

Yalin (1963) An expression for bed-load transportation. *Proceedings of the American Society of Civil Engineers. Journal of Hydraulics Division* 89, 221–248.

Young, A (1989) *Agroforestry for Soil Conservation.* CAB International, Wallingford, UK, 276 pp.

Zingg, A.W. (1940) Degree and length of slope as it affects soil loss in runoff. *Agricultural Engineering* 21, 59–64

Zinke, P.J., Sabhasri, S. and Kunstadter, P. (1978) Soil fertility aspects of the Lua' forest fallow system of shifting cultivation. In: Kunstadter, P., Chapman, E.C. and Sabhasri, S. (eds) *Farmers in the Forest: Economic Development and Marginal Agriculture in Northern Thailand.* East-West Center, University of Hawaii, Honolulu, pp. 134–159.

16

Soil Erosion Research in Indonesia: Past Experience and Future Direction

F. Agus[1], A.N. Gintings[2], U. Kurnia[1], A. Abdurachman[1] and P. van der Poel[3]

[1]*Centre for Soil and Agroclimate Research, Jln. Juanda 98, Bogor 16123, Indonesia;* [2]*Center for Forestry and Natural Conservation Research and Development, Jln. Gunung Batu No. 5, Bogor, Indonesia;* [3]*National Watershed Management and Conservation Project, Jln. Taman Makam Pahlawan Kalibata No. 10, Jakarta, Indonesia*

Soil conservation research is conducted to (i) test alternative techniques to control soil erosion; (ii) understand erosion processes – their causes and consequences – in order to design better conservation techniques or to increase theoretical knowledge; (iii) monitor changes in watershed runoff and sediment load as caused by changes in land use; and (iv) find more effective approaches to introduce soil conservation techniques. In Indonesia, research has been concentrated on the first objective, while the second has also received some attention. It is conducted mostly on plot scale. Some research has been conducted at watershed scale to address the third objective. Recently, emphasis has been given to address the fourth objective in combination with the first one, but most of this research is still researcher managed. Assessing on-site causes and impacts of soil erosion at plot scale remains important, but demands for applicable research results are made for upstream as well as downstream areas. Soil conservation planners and implementers in the upstream areas often request conservation packages for specific locations/ecological conditions. This requires field-scale research covering agronomic as well as socio-economic factors. Socio-economic factors influencing implementation of conservation technology are best assessed through farmer-managed trials. Downstream water users may want to know the predicted lifespan of dams and this requires the use of watershed-scale models. However, most of the use of computer models is still aimed at model validation. It appears that the emphasis should be on designing and testing, together with farmers, appropriate field- and household-scale conservation measures and on improving existing watershed scale models. Both require a long-term strategy and long-term commitments from researchers and farmers.

 Soil Erosion at Multiple Scales (eds F.W.T. Penning de Vries, F. Agus and J. Kerr)

Introduction

Soil erosion research in Indonesia has taken place for decades. Despite the extensive body of literature on the subject, published by many institutions under the Ministry of Agriculture, Ministry of Forestry, Ministry of Public Works and universities, erosion remains one of the major problems threatening the sustainability of agriculture and the quality of the environment. Erosion not only causes a decrease in soil fertility, but also threatens the lifespan of dams and reduces the capacity of rivers. Despite intensive efforts to conserve and rehabilitate soil, the impact of such efforts seems limited.

Erosion might have taken place at a much higher rate than the government and/or farmers can handle. In addition, several conservation packages might not be effective, but we have not conducted enough research to provide scientific evidence that they are merely myths (Lal, 1991). If a conservation measure does not work at the farmers' level, one may not be able to explain the failure if scientific evidence of the success under experimental conditions has not been obtained. Thus, applied research should provide clear scientific explanations.

This chapter discusses the status of erosion research in Indonesia and the gaps in existing knowledge, and suggests directions for future research and methodologies. The main sources of literature are the publications of the Agency for Agricultural Research and Development (AARD) of the Ministry of Agriculture (MoA), the Forestry Research and Development Agency (FORDA) of the Ministry of Forestry (MoF), and a few universities. The discussions are structured according to the following objectives of soil conservation research:

- find or test more effective alternative conservation techniques to control soil erosion;
- understand erosion processes;
- monitor and, in turn, predict changes in watershed runoff and sediment yield as caused by changes in land use;
- find more effective approaches to introduce soil conservation techniques.

Status of Soil Conservation Research in Indonesia

Alternative conservation techniques

Past and present research concerns individual measures or the introduction of packages of conservation measures. Within this context, much attention has been given to bench terraces and more recently to vegetative soil conservation measures. Measures such as mulching and conservation tillage have received less attention.

Literature on vegetative soil conservation measures is voluminous. Most of these studies are short term (i.e. less than 3 years) and do not give an understanding of the sustainability of the systems. Long-term (6 years) research by Haryati *et al.* (1995) compared the effects on soil erosion and crop yields of level bench terracing, outward sloping bench terracing, ridge terracing and alley cropping with double hedgerows, spaced 50-cm apart, of

Flemingia congesta. In the three terracing treatments, the terrace lips were planted to *Setaria sphacelata* and the terrace raisers to *Brachiaria ruziziensis*. The experiment was conducted on a typical eutropept with slopes between 15 and 40% in Central Java. They found that with a rainfall of 2515 mm in the first year (1988/1989), level bench terracing had a surface runoff coefficient of 0.20, while for outward sloping bench terracing ridge terracing, and alley cropping it was 0.39, 0.34 and 0.38, respectively. In the same treatment order, the amounts of soil loss were 26, 45, 51 and 110 t ha^{-1}, respectively. The runoff and soil loss decreased with time. Soil losses below the tolerable limit of 13.5 t ha^{-1} $year^{-1}$ were reached in the second year for all but the alley cropping treatment. The alley cropping treatment reached the tolerable limit in the fourth year. No difference was observed in the crop yields during the 6-year experiment.

Erosion reduction by grass strips was discussed by Abujamin *et al.* (1983), Garrity (1991), Talaohu *et al.* (1993) and Sukmana and Suwardjo (1991). Little information is available on the effects of perennial trees on soil erosion, although many people assume that tree stands are among the most effective means of soil conservation.

Effects of bench terracing vary from one study to the next, and from one locality another. A review by Doolette and Smyle (1990) explained that there was no change in surface runoff due to bench terracing, especially on soils with low water-holding capacity. Other studies indicated that the wetting front after rainfall events was much deeper for terraced compared with unterraced land. Erosion reduction due to bench terracing has been demonstrated (Kurnia and Suwardjo, 1985; Haryati *et al.*, 1995).

Application of 9 t ha^{-1} of mulch delayed drought stress in soybean crops for 23 days compared with no mulch (Suwardjo *et al.*, 1988). Furthermore, no-till systems combined with mulch application reduced surface runoff by as much as 43 to 92% compared with traditional cultivation on slopes between 4–14% (Suwardjo and Abujamin, 1983). Mulch protects the soil surface from raindrop impact, reduces evaporation from the soil surface and slows down runoff, giving more time for water to infiltrate, and thus increases soil water storage available for plants.

In steep vegetable-growing areas, where farmers cultivate up and down the slope, soil loss could be as high as 200 t ha^{-1} (Suwardjo and Abudjamin, 1983). Sinukaban *et al.* (1994) explained that orienting the ridges along the contour and planting vegetable crops with a massive canopy, such as potato, could reduce soil loss to the acceptable limit (25 t ha^{-1} $year^{-1}$). Suganda *et al.* (1994) reported similar results but they included the alley cropping treatment in vegetable-growing areas. Both experiments showed that there was no difference in crop yields between control and treatments and this makes the alternative techniques hard to promote. The perception of farmers is that drainage gets worse due to reorientation of the ridges, which promotes the incidence of diseases. Thus, conservation research for steep vegetable-growing areas is not yet conclusive in the sense that adoptable techniques have not yet been identified.

Agus and Dariah (1997) reviewed publications on conservation/minimum tillage techniques. Conservation tillage is a labour-efficient practice and may conserve soil. Conservation tillage could increase the proportion of macropores and, in turn, infiltration capacity and water recharge of the subsoil. Thus, it reduces water stress during the dry season. For non-compacting soils, minimum tillage could give comparable or sometimes even higher crop yields than with conventional tillage. Weed encroachment is the greatest problem with minimum tillage practices. Mulching was suggested as a technique to control weeds, yet insufficient amounts of mulch could stimulate weed growth rather than control it. The review indicated the need for more research on:

- verification of minimum tillage practices on different soil types;
- environmental impact from and economic consequences of the use of herbicide; and
- non-chemical approaches to weed control in minimum tillage.

Causes and consequences of erosion processes

Research on erosion processes is mostly linked to the use of the universal soil loss equation (USLE) (Wischmeier and Smith, 1978; Collins *et al.*, Chapter 19, this volume). At present, CSAR is calculating the rainfall erosivity, *R*, factor for various islands in Indonesia for use in the revised universal soil loss equation (RUSLE) (Abdurachman *et al.*, 1996). However, the decision has not been taken whether the use of the RUSLE should be promoted. Abdurachman *et al.* (1989) evaluated various rainfall erosivity indices for Indonesia. They concluded that the total storm energy multiplied by the maximum 30-min rainfall intensity, EI_{30}, was most strongly correlated with erosion. EI_{30} can be estimated from monthly rainfall data (Bols, 1978) as long as we limit its use to areas with long-term rainfall records.

In an effort to optimize the applicability of the USLE, Abdurachman *et al.* (1985) found that relay cropping combined with bench terracing and mulch application gave a *CP* (cover factor times conservation practice factor) value of 0.01 compared with a *C* value of 0.33 for relay cropping only. This means that the combined treatment could reduce soil loss from 424 t ha^{-1} in a bare plot to 5 t ha^{-1}. Similar experiments were conducted by Abdurachman *et al.* (1984) to find *C* and *P* values.

Values of *C*, *P*, and *CP* have been issued by several sources. van der Poel (personal communication) found that *CP* values used in the RTL manual (*Rencana Teknik Lapangan* – Field Technical Plan, Dephut, 1994) do not correspond with the product of the individual *C* and *P* values presented in the same manual (Table 16.1). For a few conservation practices the apparent *P* values have been deducted, leading in some cases to unrealistic *P* values (e.g. > 1.0 indicates that conservation measures increase erosion) or to quite variable values for one practice. These discrepancies may have contributed to unrealistic estimates of soil loss and sedimentation as described in the next section.

Table 16.1. *C*, *P* and *CP* values according to the Field Technical Plan tables.

Crop	*C* table	Conservation practice	*P* table	*CP* calculated	*CP* table
Upland rice	0.53	Mulched rice straw	0.43	0.23	0.096
Groundnut	0.20	Mulched rice straw	0.43	0.09	0.049
Upland rice	0.53	Grass strips	0.53	1.59	0.841
Groundnut	0.20	Mulched groundnut	1.30	1.30	0.259

Numbers in column four (*P*) are obtained by dividing column six by column two. Numbers in column five (*CP*) are obtained by multiplying column two by column four. (Source: van der Poel, personal communication.)

Erosion assessment at the watershed scale

Murtilaksono and Nugroho (1996) used the USLE to predict soil loss and recommended soil conservation measures in the 130,000 ha Upper Solo Subwatershed. The USLE was also used by Rachman *et al.* (1991) and Kurnia *et al.* (1991) to calculate sediment yields. These latter investigators attributed the difference between the USLE predicted soil loss and measured sediment yield at the outlet to sediment deposition and suggested the use of a sediment delivery ratio.

Sukresno (1993) calculated the effect of several soil conservation measures on sedimentation reduction in the multipurpose (flood control, hydroelectricity, irrigation, fishery and tourism) Gajah Mungkur Dam. The dam was built in 1981 and its lifespan was estimated to be 100 years. In 1985, the Ministry of Forestry found that sedimentation in the reservoir amounted to 4.9×10^6 t year^{-1}, which would reduce the dam's lifespan to only 27 years (see Enters, Chapter 1, this volume). A proposed package of land rehabilitation and soil conservation measures would extend the dam's lifespan to more than 60 years. The proposal was approved in 1988 and the project was implemented until 1995. Sukresno (1993) evaluated the effect of the project on sedimentation reduction. Sediment yield, *Y*, (t year^{-1}) from the 130,000 ha watershed was calculated by:

$$Y = A \times \mathrm{SDR} + Gl + Sb + Rs + L1 \qquad (16.1)$$

Where *A* is annual rill and inter-rill soil loss (t year^{-1}) as predicted using the USLE; SDR is the sediment delivery ratio (0.124 for the Subwatershed); and *Gl*, *Sb*, *Rs* and *L1* are sediment yield due to gully, stream bank, road side, and other forms of erosion, respectively. PPIU (1992) issued several tables to describe the effectiveness of gully control, check dams, terracing and cropping pattern, and perennial tree planting in controlling erosion. With these tables, the change in sedimentation due to several sources could be estimated. Sukresno (1993) found that the total sediment yield in 1992 was 4.9×10^5 t year^{-1}, which is one order of magnitude smaller than the calculated value in 1985. Unfortunately, there are no observations of the actual sediment yield.

Analysis of river flow using actual measurements was done by Murtiono (1993) in the Tapan Subwatershed (part of the Upper Solo Watershed in Central Java). His research was in agreement with others in that several conservation measures delayed the occurrence of the peak flow and that the peak flow decreased as a result of conservation efforts. However, the use of harmonic analysis (no further explanation of the procedure was given) underestimated sediment yields during the peak flows.

Jaya (1994) conducted research on on- and off-site effects of erosion in a microcatchment of 10.5 ha. He found that overland flow and soil loss increased logarithmically with rainfall and with the proportion of bare land. Nutrient concentration in the runoff water decreased logarithmically, while sediment concentration was proportional to the nutrient concentration in the top 5 cm of the soil.

Several other models have been tested to predict soil conservation impact. Naik Sinukaban (personal communication) used the Agricultural Nonpoint Source Pollution Model (AGNPS) to predict overland flow, soil loss, nitrogen and phosphorus losses. He concluded that the predicted and observed soil loss, runoff, and nutrient losses from a 10.4-ha catchment were not statistically different and thus the model could be used to aid in watershed management planning. Ngaloken Gintings and Arief Ilyas (personal communication) simulated the effects of different scenarios of soil management using the Areal Nonpoint Source Watershed Environment Response Simulation (SWERS) and believed that the model needs further validation. Other models, such as TOPOG and TOPMODEL, were discussed by Pawitan and Murdiyarso (1996); they suggested the necessity to validate the models further. The Water Erosion Prediction Project (WEPP) was reviewed by Naik Sinukaban (personal communication) and Agus *et al.* (1997a), and both suggested testing of this model. Each model for prediction of erosion may be best at performing a particular task and no single model meets all needs.

One of the problems encountered when using these models was that they were developed in countries where agricultural management, such as terracing, strip cropping and crop rotations, is totally different from that in Indonesia. For example, in Indonesia, the horizontal interval of bench terraces could be as low as 3 m, while in the countries where the models were developed, these very narrow terraces are never encountered.

In Indonesia, the use of the USLE has gone beyond field-level prediction of soil loss. Due to the use of maps that are mostly less detailed than 1:25,000, it has been used incorrectly for slopes much longer than 300 m and much steeper than 30%. Its absolute values for estimated soil loss from watersheds have been used for important decision-making. Thus, validation and/or development of appropriate watershed-scale models is very urgent. R. Lal (personal communication) stated that the best model for Indonesia is a model that is developed in this country based on the local conditions. However, with the limited research resources, a locally developed and/or calibrated model calls for collaborative efforts from both scientists and donors in this region.

The limited availability and reliability of data is an obstacle for validation, and as a consequence, for applying the models. With the increasing demand to evaluate environmental quality at a watershed scale, research should be conducted in a few 'representative' watersheds.

Approaches to introduce soil conservation techniques

Today's field performance in Java shows extensive adoption of bench terraces (with or without fodder grasses or legumes for terrace stabilization) and tree planting, especially at field borders and in home gardens. For the establishment of level bench terraces, ridge terraces and hedgerows, the labour required was 662, 227 and 116 man-days ha^{-1}, respectively (Haryati *et al.*, 1995). Farmers who own ruminants quite readily adopt the use of fodder species for terrace stabilization (Agus *et al.*, 1997b; Cairns, 1997). On the outer islands there is little adoption of terracing and hedgerow systems, but there is more adoption of tree planting in home gardens (Colfer *et al.*, 1988), in the rubber forests (Penot, 1997) and at field borders, and of fallow rotation with *Leucaena* (Agus, 1997).

The general adoption of bench terraces in Java has influenced much of the conservation extension efforts in the outer islands, even for conditions in which they are unsuitable. For example, on some shallow entisols in eastern Nusa Tenggara, bench terraces were demonstrated as a conservation measure (Agus *et al.*, 1997b). This is mostly caused by lack of understanding of soil erosion processes and of the socio-economic and biophysical requirements of soil conservation measures.

On an andisol with slopes between 15 and 25% in West Java, farmers made ridges parallel to slope direction. This traditional technique did not give any lower yields than alternative techniques such as rearrangement of ridge direction, mulch application, and hedge systems with *Vetiveria zizanioides*, Guatemalan grass or *Calliandra calothyrsus* (Suganda *et al.*, 1994). Of great concern is the high soil loss under the farmers' practice, but this aspect was not reported by these authors. One could speculate that the alternative techniques could reduce soil loss, but again, this higher-cost technology will face a hard sale because farmers cannot expect benefits from their investment. Andisols in the research area are very deep and their fertility hardly decreases with depth. As such, a fairly high amount of sheet erosion may not affect yields.

The issue of non-adoption in general is always related to whether the alternative techniques promise direct and in many cases short-term benefits to farmers. In addition, farmers' socio-economic and cultural backgrounds influence their ability to adopt techniques. Therefore, in conducting verification trials, matching the alternatives with the conditions of farmers is a prerequisite. Field-level implementation requires very detailed data to reflect socio-economic and cultural conditions, which could be gathered through participatory rural appraisals.

Future Direction

Despite a large body of literature on soil erosion control technology, implementation problems of conservation measures, erosion processes and modelling, there are still gaps in our knowledge. These gaps, along with some brief research proposal are presented in the following sections.

Plot-scale study of soil physical and chemical processes

Studies on better alternative conservation measures have concentrated on crop yields and crop growth. Little attention has been given to the soil physical and chemical processes, especially to the water balance and nutrient dynamics. This is partly due to the interest of the owners in obtaining practical recommendations and thus, little room is given to the study of processes. Furthermore, process-oriented research is more costly, time-consuming and sometimes tedious. Thus, several measures have been promoted without having enough scientific basis (Lal, 1991).

Erosion is a dynamic process, and to understand it well we need intensive monitoring of several important variables. Soil water content, soil water potential, and solute transport in the soil matrix are among the soil conditions which undergo rapid spatial and temporal fluctuations. Due to the need for intensive measurements, non-destructive techniques should be employed. The use of time domain reflectometry (TDR) for soil water content determination (Topp *et al.*, 1980), tensiometry for soil water potential, and a ceramic cup suction lysimeter for soil water sampling (Klute, 1986), and/or free drainage lysimeter, are proposed. Alternative techniques for such measurements and advantages and disadvantages of each technique are explained in Klute (1986) and Topp *et al.* (1992). These proposed measurements are complementary to the typically used measurements such as crop yields, and soil chemical and physical properties.

Verification of conservation measures through farmer-managed participatory research (plot and household scales)

In the studies to verify soil conservation measures for a certain locality, the farmers' socio-economic and cultural backgrounds must be taken into consideration. The available standard data are not detailed enough. In true participatory rural appraisals (PRA), these variables would be properly considered because the farmers select measures that are fitting for their situation. The PRA should be facilitated by a sociologist, an agronomist and a soil conservationist. The role of the last individual is to assist in the discussion on proposed techniques and to explain new alternatives in a very practical fashion. The PRA facilitators should be able to encourage farmers to express their concerns, limitations, likes and dislikes about the present and proposed techniques. Feedback on proposed and introduced technologies should be gathered by the sociologist. The result of the PRA would be a list of conservation technologies which the farmers want to try and planning of these activities.

Farmer-managed trials could then be designed with farmers' fields as replicates. The number of replicates could range from ten to any manageable number. Incentives are seldom included, unless they will be available for long-term use. The agronomic and economic variables from T0 (farmers' practice/control) plots are usually compared with one alternative technique (T1) selected by the farmers. Researchers and extension workers monitor the test variables and farmers' activities. Monitoring may include:

- labour requirements and other inputs;
- crop yields and soil properties on plots T0 versus T1;
- modifications made for T1 by the farmers;
- problems encountered in implementation;
- expansion of the introduced treatment to other fields/farms; and
- farmers' opinions of the new technology.

The advantage of this design is that we can verify both agronomic as well as socio-economic and cultural responses to the alternative technique and that the participating farmers (preferably key farmers) can evaluate the alternative treatments more closely. If superior, they could influence other farmers to adopt the technique. A disadvantage is that the replicates are scattered over several farmers' fields, and thus are more time-consuming to monitor than simple plot studies. Also, some participating farmers may modify their treatments to such a degree that they could no longer be used for statistical analysis.

Watershed-scale research and modelling

Watershed-scale research is intended to study temporal variation in discharge, and sediment and chemical transport, temporal and spatial variation in overland flow, erosion and soil physical and chemical characteristics and their causes and effects.

The larger the watershed, the more difficult it becomes to monitor the effects of the on-site factors, and treatments may only cover a small percentage of the area. Thus, microcatchments (< 10 ha) are generally used if the on-site and off-site impacts of soil erosion are equally important. The on-site causes and impacts could be gathered using the techniques described in the two previous sections, but possible using less intensive observation per unit area. The study of the off-site impacts might involve monitoring of sediment and water discharge. Standard techniques to determine sediment concentration are given by Walling (1988) Dickinson and Collins (Chapter 20, this volume).

One problem in conducting watershed-scale research is that each watershed has unique characteristics. The area of watersheds varies from a few hectares to thousands of square kilometres. Extrapolation of research results from a limited area to a large area could lead to erroneous results (Lal, 1988; van Noordwijk *et al.*, Chapter 15, this volume). Standardization in the research methodology, although very difficult, is necessary. It is easier to standardize research in a network and almost impossible in isolated research programmes (Craswell *et al.*, Chapter 11, this volume).

As explained in the section on erosion assessment at the watershed scale, demands for predicting off-site impacts of soil erosion are increasing. Experiences in Indonesia confirm that the USLE should not be used for predicting soil loss at scales beyond the field level. Validation and/or development of watershed-scale computer models must be intensified and be supported by watershed-scale research. Coordination of these efforts is required in order to limit testing to those models most likely to be applicable under varying Indonesian conditions and not requiring excessive amounts of data and to prevent the waste of money and effort on unnecessary repetitive research.

Conclusions

1. The study of soil physical and chemical processes related to erosion in Indonesia has been very minimal and merits more attention. Lack of scientific explanations of the processes involved leads to immature promotion of conservation measures.
2. Agronomic and economic soundness of alternative technology under experimental conditions do not warrant field-level implementation. Participatory appraisal should follow to test its adoption at farmer level.
3. Watershed-scale research should be intensified in a network to understand processes associated with causes and impacts of soil erosion on-site and off-site and to validate and/or develop watershed-scale computer models.

References

Abdurachman, A., Barus, A. and Sudirman (1985) Peranan pola tanam dalam usaha pencehahan erosi pada lahan pertanian tanaman semusim. *Pemberitaan Penelitian Tanah dan Pupuk* 4, 41–46.

Abdurachman, A., Abujamin, S. and Kurnia, U. (1984) Pengelolaan tanah dan tanaman untuk usaha konservasi tanah. *Pemberitaaan Penelitian Tanah dan Pupuk* 3, 7–12.

Abdurachman, A., Sutono, Dariah, A., Haryati, U., Vadari, T., Nurida, N.L., Haryono, Juarsah, I. and Pramudia, A. (1996) *Penentuan indeks erosivitas beberapa propinsi di Indonesia: P. Jawa dan Madura.* Pusat Penelitian Tanah dan Agroklimat, Bogor.

Abdurachman, A., Kurnia, U. and Sukmana, S. (1989) Evaluasi berbagai index erosivitas hujan untuk penerapan di Indonesia. *Pemberitaan Penelitian Tanah dan Pupuk* 8, 25–31.

Abujamin, S., Abdurachman, A. and Kurnia, U. (1983) Strip rumput permanen sebagai salah satu cara konservasi tanah. *Pemberitaan Penelitian Tanah dan Pupuk* 1, 16–20.

Agus, F. (1997) Intensification of indigenous fallow rotation using *Leucaena leucocephala. Paper presented at the workshop 'Indigenous Strategies for Intensification of Shifting Cultivation in South East Asia Fallow Management, Bogor, Indonesia, June 23–27.*

Agus, F. and Dariah, A. (1997) Prospek pengembangan teknologi olah tanah konservasi di lahan kering. In: *Proceedings Simposium Nasional dan Kongres VI PERAGI, Jakarta, 25–27 June.*

Agus, F., Subagyono, K., and van der Poel, P. (1997a) Menuju model pendugaan erosi distributif: kendala dan peluangnya. In: *Proceedings Pertemuan Pembahasan dan Komunikasi Hasil Penelitian Tanah dan Agroklimat, Bogor, Indonesia 4–6 March*, pp. 69–87.

Agus, F., Watung, R. and Abbas, A.I. (1997b) Implementation problems of soil conservation measures in Indonesia. *Paper presented at 9th Annual Meeting of Asialand Sloping Lands, Bogor.* IBSRAM, Thailand (in press).

Bols, P. (1978) *The Iso-erodent Map of Java and Madura. Report Belgian Technical Assistance Project ATA 105.* Soil Research Institute, Bogor, Indonesia, 39 pp.

Cairns, M. (1997) Fallow, fodder and fences: the critical elements of integrating livestock into swidden systems. *Paper presented at the workshop Upland Farming Systems in the Lao PDR: Problems and Opportunities for Livestock*, Vientiane, Lao PDR, 19–23 May.

Colfer, C.J.P., Gill, D.W. and Agus, F. (1988) An indigenous agricultural model from West Sumatra: a source of scientific insight. *Agricultural Systems* 26, 191–209.

Dephut (Departemen Kehutanan) (1994) *Pedoman penyusunan rencana teknik lapang rehabilitasi lahan dan konservasi tanah daerah aliran sungai.* Departemen Kehutanan, Jakarta, Indonesia

Doolette, J.B., and Smyle, J.W. (1990) Soil and moisture conservation technologies: review of literature. In: Doolette, J.B. and Magrath, B.W. (eds) *Watershed Development in Asia: Strategies and Technologies. World Bank Technical Paper 127.* World Bank, Washington, DC, pp. 35–70.

Garrity, D.P. (1991) Sustainable land use systems for the sloping uplands of Southeast Asia. In: *Technologies for Sustainable Agriculture in the Tropics.* American Society of Agronomy, Madison, Wisconsin, USA.

Haryati, U., Haryono, and Abdurachman, A. (1995) Pengendalian erosi dan aliran permukaan serta produksi tanaman pangan dengan berbagai teknik konservasi pada tanah Typic Eutropepts di Ungaran, Jawa Tengah. *Pemberitaan Penelitian Tanah dan Pupuk* 13, 40–50.

Jaya, A. (1994) Dinamika Aliran Permukaan, Erosi serta Kehilangan Hara dalam Aliran Permukaan pada Daerah Tangkapan Citere, Pangalengan. MS. Thesis, Bogor Agricultural University, Indonesia.

Klute, A. (ed.) (1986) *Methods of Soil Analysis.* Part 1, 2nd edn. *Agronomy* 9, 563–596. American Society of Agronomy, Madison, Wisconsin, USA.

Kurnia, U. and Suwardjo, H. (1985) Pengaruh beberapa cara konservasi mekanik terhadap erosi pada tanah Tropudalf dan Troporthent di Yogyakarta. *Pemberitaan Penelitian Tanah dan Pupuk* 4, 46–50.

Kurnia, U., Sudharto, T., and Vadari, T. (1991) Penelitian pengkajian metode prediksi erosi pada tanah mergel di Sub DAS Ciseel. In: Sudharto, H. (ed.) *Prosiding Pertemuan Teknis Penelitian Tanah Bidang Konservasi Tanah dan Air, Bogor, 22–24 August, 1989.* Pusat Penelitian Tanah dan Agroklimat, Bogor, Indonesia, pp. 107–122.

Lal, R. (1988) Soil erosion by wind and water: problems and prospects. In: Lal, R. (ed.) *Soil Erosion Research Methods.* Soil and Water Conservation Society, Ankeny, Iowa, USA, pp. 1–6.

Lal, R. (1991) Myths and scientific realities of agroforestry as a strategy for sustainable management for soils in the Tropics. *Advances in Soil Science* 15, 91–137.

Murtilaksono, K. and Nugroho, C.S.P. (1996) A simulation model of optimal land use (biophysical criteria) for land rehabilitation in Upper Solo Subwatershed. *Watershed Management Journal* II(4), 11–23.

Murtiono, O.H. (1993) Kajian karakteristik aliran sungai Sub DAS Tapan. In: *Prosiding*

Seminar Hasil Kajian Balai Teknologi Pengelolaan DAS Surakarta, 19–20 Maret, pp. 56–66.

Pawitan, H. and Murdiyarso, D. (1996) Monitoring dan evaluasi komponen biofisik DAS. In: *Prosiding Lokakarya Pembahasan Hasil Penelitian dan Analisis Pengelolaan Daerah Aliran Sungai, Garut, 20–24 November 1995.* Pusat Penelitian Tanah dan Agroklimat, Bogor, pp. 47–60.

Penot, E. (1997) From shifting cultivation to sustainable jungle rubber in Indonesia: a history of innovations integration for smallholders in the peneplains of Sumatra and Kalimantan since the turn of the century. *Paper presented at the workshop Indigenous Strategies for Intensification of Shifting Cultivation in South East Asia Fallow Management*, Bogor, Indonesia, 23–27 June.

PPIU (1992) *Institutional Strengthening for Evaluation of Effectiveness of Upper Solo (Wonogiri) Watershed Project. Mid Term Review.* Ditjen RRL, Ministry of Forestry, Jakarta.

Rachman, A., Suwardjo, H. and Kurnia, U. (1991) Studi perbandingan metode prediksi erosi menurut USLE dan metode pengukuran erosi di lapangan di Sub Das Cimuntur, DAS Citanduy. In: Suhardjo, H. (ed.) *Prosiding Pertemuan Teknis Penelitian Tanah Bidang Konservasi Tanah dan Air, Bogor, 22–24 Agustus, 1989.* Pusat Penelitian Tanah dan Agroklimat, Bogor, pp. 123–131.

Sinukaban, N., Pawitan, H., Arsyad, S., Armstrong, J.L and Nethry, M.G. (1994) Effect of soil conservation practices and slope lengths on runoff, soil loss, and yield of vegetables in West Java. *Australian Journal of Soil and Water Conservation* 7(3), 25–29.

Suganda, H., Abujamin, S., Dariah, A. and Sukmana, S. (1994) Pengkajian teknik konservasi tanah dalam usahatani tanaman sayuran pada Andisols Batulawang, Pacet. *Pemberitaan Penelitian Tanah dan Pupuk* 12, 47–57.

Sukmana, S. and Suwardjo, H. (1991) Prospect of vegetative soil conservation measure for sustainable upland agriculture. *Indonesian Agricultural Research and Development Journal* 13 (1 and 2).

Sukresno (1993) Pengaruh konservasi tanah terhadap pengurangan sedimen. In: *Prosiding Seminar Hasil Kajian Balai Teknologi Pengelolaan DAS Surakarta, 19–20 Maret*, pp. 67–79.

Suwardjo and Abujamin, S. (1983) Crop residue mulch for conserving soil in upland of Indonesia. pp. 607–614. In: Swaify, S.A., Moldenhauer, W.C. and Lo, A. (eds) *Soil Erosion and Conservation.* Soil Conservation Society of America, Ankeny, Iowa, pp. 607–614.

Suwardjo, Suhardjo, H., and Talaohyu, S.H. (1988) Pengaruh panjang lereng dan cara pengelolaan lahan terhadap erosi dan pertumbuhan tanaman kedelai. In: *Prosiding Pertemuan Teknis Penelitian Tanah, Cipayung, 18–20 Maret, 1986.* Pusat Penelitian Tanah, Bogor.

Talaohu, S.H., Abdurachman, A., and Suwardjo (1993) Pengaruh teras bangku, teras gulud dan slot mulsa *Flemingia* dan strip rumput terhadap erosi, hasil tanaman dan ketahana tanahTropudult di Sitiung. In: *Proceeding Pertemuan Teknis Penelitian Tanah, Bogor, 22–24 Agustus, 1989*, Pusat Penelitian Tanah dan Agroklimat, Bogor.

Topp, G.C., Davis, J.L. and Annan, A.P. (1980) Electromagnetic determination of soil water content measurement in coaxial transmission lines. *Water Resource Research* 16, 574–582.

Topp, G.C., Reynolds, W.D. and Green, R.E. (eds) (1992) *Advances in Measurement of Soil Physical Properties: Bringing Theory into Practice. SSSA Special Publication No. 30.* SSSA, Inc. Madison, Wisconsin, USA.

Walling, D.E. (1988) Measuring sediment yield from river basins. In: Lal, R. (ed.) *Soil Erosion Research Methods.* Soil and Water Conservation Society, Ankeny, Iowa, USA, pp. 39–73

Wischmeier, W.H. and Smith, D.D. (1978) *Predicting Rainfall Erosion Losses – a Guide to Conservation Planning. Agricultural Handbook 537*, USDA, Washington, DC, 58 pp.

17

Dynamic Process Modelling of Hydrology and Soil Erosion

C.W. Rose and B. Yu
Faculty of Environmental Sciences, Griffith University, Nathan, Queensland 4111, Australia

This chapter outlines a new soil conservation methodology based on current understanding of the processes involved in water erosion. The methodology has been applied to a series of projects supported by the Australian Centre for International Agricultural Research, concerned with seeking sustainable cropping systems for tropical steeplands. Research outcomes are reviewed for collaborative programmes in Malaysia, the Philippines, Thailand and Australia. In each context, feasible and effective, agronomically based, soil-conserving systems have been achieved (Coughlan and Rose, 1997). The issues involved in the prediction of runoff and soil loss at the plot scale with a series of levels of data availability are also addressed and progress reviewed.

Introduction

One of the major changes in approach which has been taking place in soil conservation research is the move from suitably summarizing a sufficiently large and long-term experimental database, typified by the universal soil loss equation approach, towards using knowledge of erosion process in individual erosion events. Examples of more process-oriented, event-based erosion models are given by Foster (1982), Govers and Poesen (1988), Nearing *et al.* (1989), Hairsine and Rose (1991, 1992a), and Morgan *et al.* (1992). These descriptions are all deterministic in nature, though Lisle *et al.* (1998) have given a stochastic derivation of the type of theory used in the projects which are the main topic of this review.

(eds F.W.T. Penning de Vries, F. Agus and J. Kerr)

In the growing use of mathematical models in soil conservation, there is a discernable catena of activity, from model development and testing, evaluation of the physical meaningfulness of experimentally evaluated parameters in the model, through to use of the model in a predictive sense, perhaps in design or prediction in time.

This chapter outlines an example of this change in research methodology and provides a comparative summary of the results of application of this new approach in a multicountry, cooperative, soil conservation research programme involving the Philippines, Malaysia, Thailand and Australia. The objectives of this research programme were to:

- quantify field hydrological and sediment loss processes with a view to matching soil conservation technologies to dominant processes;
- test and interpret data on the effectiveness of locally applicable technologies to reduce soil loss to acceptable rates; and
- develop predictive methodologies for runoff, soil and nutrient loss.

This chapter both reviews outcomes of this research programme, and indicates how the dynamic understanding of hydrological and soil erosion processes illustrated in and assisted by this project can be applied more generally in sustaining tropical steepland cropping systems endangered by water erosion. The issues involved in the prediction of runoff and soil loss at the plot scale of cultivation are also addressed and progress reviewed.

More details of the methodology used and results obtained are given in the ACIAR Technical Report (Coughlan and Rose, 1997).

Soil Erosion Research Methodology Employed

The scale of the cultivation unit in the project was 72–1000 m^2, consistent with farmer practice in the area. Because of the modest size of the cultivation unit, flow can be regarded as essentially one-dimensional, but cultivation patterns or erosional features such as rills that affect overland flow need to be recognized. The steepness of the plots or the rate of overland flow is typically such that erosion driven by overland flow dominates the contribution from rainfall impact. This dominance of flow-driven erosion processes justifies use of a simpler option in the program GUEST (Griffith University Erosion System Template) (Misra and Rose, 1990, 1996) to analyse soil loss and runoff rate data from bare soil plots to yield a measure of soil erodibility, termed β.

The program GUEST uses a theoretically derived expression for the maximum possible sediment concentration that can be sustained due to flow-driven erosion processes which has been shown to agree well with experimental measurement. This upper limit to sediment concentration, c_t, has been called the 'transport limit' or 'transport capacity' by Foster (1982). This limiting value is reached when the rate of removal of weak sediment, such as occurs in the re-entrainment of sediment previously deposited over the entire eroding surface, is equal to its rate of deposition. Theory developed by Rose and Hairsine (1988) and Hairsine and Rose (1992a, b) shows that c_t is given to a good approximation by:

$$c_t = \frac{F\sigma SV}{\phi(\sigma/\rho \bullet 1)} \tag{17.1}$$

where F is the fraction of the stream power of overland flow used in erosion (≈ 0.1); σ and ρ are the sediment and water density, respectively; S is the land slope; V is the velocity of flow; and ϕ is the 'depositability' of sediment (the mean settling velocity of sediments), measured as described by Lisle *et al.* (1996). The velocity V can be inferred from measured runoff rate per unit area, Q, and observed features of rills, should they occur. Neglecting rills for simplicity, then the volumetric flux per unit width at exit from a plot of downslope length L, $QL = DV$, where D is the depth of flow. It then follows, using Manning's equation to relate D and V, that:

$$V = k_m Q^{0.4} \tag{17.2}$$

where k_m is calculated from L, S, and Manning's n. Substituting for V from Eqn 17.2 into Eqn 17.1 gives:

$$c_t = kQ^{0.4} \tag{17.3}$$

where k is approximately constant in any given context of slope, slope length and rill configuration, and is essentially independent of the runoff rate (Yu *et al.*, 1997d).

If sediment concentration remains at the transport limit, the total sediment flux for the erosion event is $kQ^{0.4}Q$, so that, by definition, the flow-weighted time-integrated mean sediment concentration, $\bar{c}_t$, is given by:

$$\bar{c}_t = \frac{\Sigma kQ^{1.4}}{\Sigma Q} \tag{17.4}$$

where the summation is for the event duration.

Comparing Eqns 17.4 and 17.2, and taking k as a constant, $\bar{c}_t$ can be written as:

$$\bar{c}_t = kQ_e^{0.4}, \tag{17.5}$$

where Q_e is an effective runoff rate given by:

$$Q_e = \left(\frac{\Sigma Q^{1.4}}{\Sigma Q}\right)^{2.5} \tag{17.6}$$

The rate Q_e can be interpreted as the steady-state runoff rate effective in computing the average value of c_t during an erosion event.

For cohesive sediments, c will be less than c_t, because of the greater energy per unit mass required to remove sediment, and in order to quantify this greater resistance, an erodibility parameter, β, was introduced by Rose (1993) where:

$$\bar{c} = \bar{c}_t^{\beta} \tag{17.7}$$

The magnitude of β is ≤ 1, unless erosion processes other than those due to overland flow enhance $\bar{c}$, and β decreases as soil strength increases (Misra and Rose, 1995). Considering a non-rilled situation for simplicity, Fig. 17.1 illustrates the form of relationship between c (or $\bar{c}$) and stream power, Ω (defined as the rate of working of the shear stress between the soil and flow over it: Rose, 1993). The evaluation of β using a program called GUEST is discussed in Rose *et al.* (1997a).

Finally, the total mass of soil lost during an erosion event, M, is then given by:

$$\begin{aligned} M &= \bar{c}\Sigma Q \\ &= c_t^{\beta}\Sigma Q \\ &= k^{\beta} Q_e^{0.4\beta}\Sigma Q \end{aligned} \tag{17.8}$$

Equation 17.8 shows that, to reduce soil loss due to flow-driven erosion, it is necessary to reduce the total runoff amount and the rate of runoff per unit area, Q, or the velocity of runoff flow, which is related to Q. As will be illustrated in the results, contact cover provided by surface mulch, for example, can be effective in achieving this objective, as well as having other soil conserving effects.

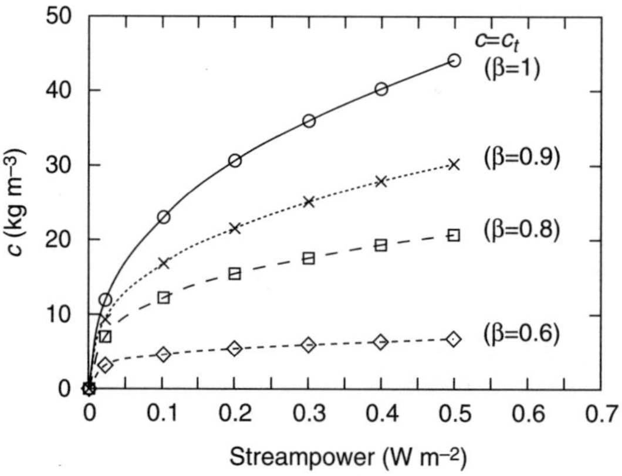

Fig. 17.1. The relationship between sediment concentration c and c_t (upper curve) and stream power, assuming sheet flow and various values of β (from Rose, 1994).

Plot-scale Runoff Modelling for Soil Loss Prediction

Theory given in the earlier section defines the crucial role of rate of runoff in determining soil loss. In the Water Erosion Prediction Project (WEPP) (Foster and Lane, 1987), infiltration is modelled using the Green–Ampt equation, and the peak runoff rate obtained using the kinematic flow approximation (both described in Marshall *et al.*, 1996) is assumed as the steady-state runoff rate for erosion computations.

The research projects reported in this chapter provided a vast body of data on runoff rate, possibly the largest body of data collected worldwide on runoff rate at a 1-min time scale. Apparent infiltration rate at all sites is closely related to rainfall rate, and is essentially independent of cumulative infiltration amount, features not in accord with commonly used infiltration equations such as the Green–Ampt equation. There is a body of literature which supports the consistent finding of a positive relationship between infiltration rate and rainfall rate. There is also growing consensus that the reason for this relationship, so apparent in natural storm events when rainfall rate varies greatly over time, is due to spatial variability in infiltration rate across the field. The exponential decay in infiltration rate over time, predicted by classic infiltration theory, holds when rainfall rate is constant, an uncommon feature of natural rainfall. However, even at the plot scale, this effect, observable at a point in the field, is completely overshadowed under natural rainfall by rate-correlated apparent infiltration variation due to the interaction of variable rainfall rate and spatial variation in infiltration rate. Yu *et al.* (1997a) have shown that a simple, one-parameter, infiltration rate equation, which is the simplest way of modelling the effect of spatial variation, can provide a very good description of the entire ACIAR project database. This model relates the actual infiltration rate, I, to the rainfall rate. The model for I is:

$$I = I_{\mathrm{m}}\left(1 \bullet \mathrm{e}^{\bullet P - I_{\mathrm{m}}}\right), \tag{17.9}$$

where P is the rainfall rate in mm h^{-1} and I_m is the only parameter, which is the mean maximum infiltration rate in mm h^{-1} over the whole plot or field, which would occur if the entire plot was saturated and generated runoff.

In general, there is an initial amount of infiltration of rainfall before any runoff occurs, F_o, and this is a second model parameter in characterizing the relation between rainfall and runoff described by Yu *et al.* (1997b).

Following the initial removal of amount F_o, the rainfall excess, given by $P-I$, needs to be routed to the plot outlet. Yu *et al.* (1997b) have shown that, for plots of modest size, a simplified solution of the kinematic wave approximation provides a realistic description, involving only a single dimensionless parameter, α, which depends on the lag in overland flow and the time interval of measurement.

Thus an efficient three-parameter model of runoff rate has been found to provide very good agreement between observed and modelled runoff rate for all of the ACIAR sites. This model is called SSRRM for Small-Scale Runoff-Routing Model.

This modelling approach allows prediction of Q as a function of time from measured rainfall rate P (once the appropriate value of the three parameters I_m, F_o and α have been determined), as is illustrated in Fig. 17.2.

Experimental Methodology

The experimental methodology involved collection of water and sediment from experimental plots in a settling basin of low slope in which coarser sediment (hereafter referred to as the 'bedload') was deposited. Because of its importance in the theory (e.g. Eqn 17.5), the runoff rate of water was then measured and recorded as a function of time on an electronic data logger, together with rainfall rate, for subsequent data checking and analysis using a suite of computer programs. The concentration of the finer sediment (or 'suspended load') was also measured, enabling determination of total soil loss for the erosion event. Other factors affected by management, such as surface contact cover, rilling and crop yield, were recorded manually. Nutrient characteristics of surface soil and of eroded sediment were measured in some major runoff events, but not on all occasions. Hydraulic roughness and the settling velocity characteristics of the soil (involved in the theory mobilized by program GUEST) were separately determined. More detail on methodology is given by Ciesiolka *et al.* (1995a) and Ciesiolka and Rose (Chapter 18, this volume).

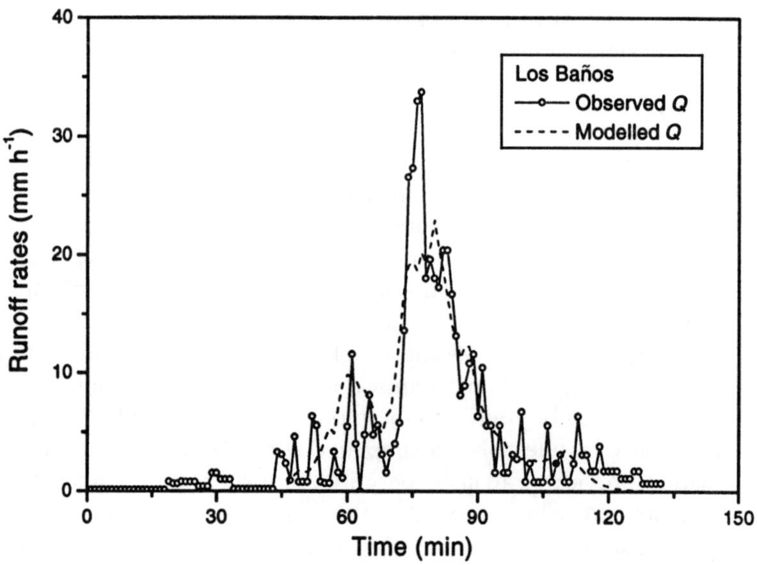

Fig. 17.2. A comparison of runoff rate measured at the Los Baños site with tipping-bucket technology and modelled runoff rate referred to in the text (from Yu *et al.*, 1997b).

Review of Results

This section of the chapter reviews the results of this ACIAR project, which are given in more detail in Coughlan and Rose (1997). Emphasis is given to those major outcomes which have some generality of application as indicated by data from two sites in the Philippines and Australia, and one in Thailand and Malaysia, located as shown in Fig. 17.3. Results of a related prior ACIAR project are given in a special issue of *Soil Technology* (Anon., 1995).

Soil and water loss

Table 17.1 provides a summary of average annual runoff, runoff coefficient, soil loss and average sediment concentration for the five sites involved in the reported research. Results for each site are given first for a reference plot kept essentially bare by weeding, followed in descending order by results for the local conventional farmer practice, and an 'improved practice' believed likely to be acceptable and effective in conserving soil. Soil texture and soil description for each site are also given in the table.

Very high soil losses of well over 100 t ha^{-1} $year^{-1}$ from bare plots were measured at all sites with two exceptions: at Nan, Thailand, where there was a drought and little runoff for the single year of measurement, and at VISCA (the Visayas State College of Agriculture), the Philippines, where sediment concentration was extremely high, but runoff was limited because of a low runoff coefficient. Table 17.1 shows that conventional farmer practice did not reduce these high losses to tolerable levels. Improved practices varied with site, but these reduced the rate of soil loss to less than 20 t ha^{-1} $year^{-1}$ at Kemaman, Malaysia, and less than 10 t ha^{-1} $year^{-1}$ at the other sites; rates which are probably tolerable, given expected higher potential rates of soil

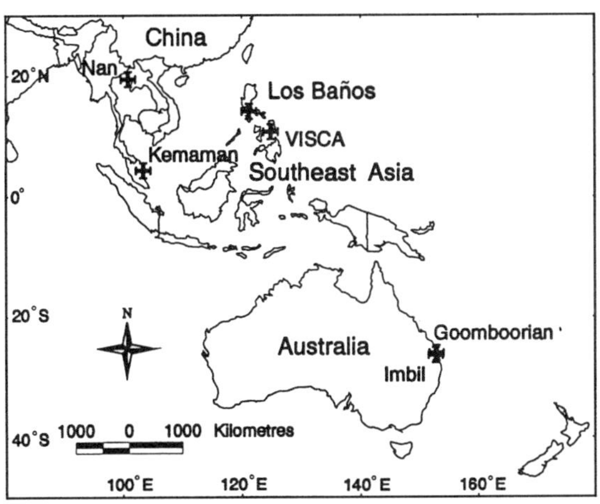

Fig. 17.3. Location of project sites in South-East Asia and Australia discussed in this paper.

Table 17.1. Average annual runoff, runoff coefficient, soil loss and sediment concentration from plots at five sites.

Site[a]	Treatments (soil type)	Runoff (mm)	Runoff coefficient[b]	Soil loss (t ha^{-1})	Sediment concentration (kg m^{-3})
Kemaman, Malaysia 4.5 years Sandy clay loam 17% slope Average annual rainfall = 3638 mm	Bare plot (orthoxic tropudult)	2245	0.62	127	5.7
	No living ground cover	1287	0.35	90	7.0
	Grass and legume ground cover	413	0.11	17	4.1
Nan, Thailand 1 year Clay Average slope = 30% Annual rainfall 1993 = 1886 mm	Bare plot (oxic paleustult)	42	0.02	7.2	17.1
	Clean cultivated farmers practice	10	<0.01	0.6	6.0
	Tephrosia hedgerows	10	<0.01	0.4	4.0
Los Baños, the Philippines 6 years Clay Average slope = 18% Average annual rainfall = 2037 mm	Bare plot (typic tropudalf)	393	0.19	184	47
	Clean cultivated farmers practice	387	0.19	119	31
	Alley cropping and mulching	114	0.06	6	5.3
VISCA, the Philippines 2 years Clay 50% slope plots Average annual rainfall = 2800 mm	Bare plot (oxic dystropept)	55	0.02	60	125
	Clean cultivated furrows up-and-down slope	84	0.03	38	45
	Alley cropping and mulching	16	<0.01	3	19
Goomboorian, Gympie, Australia 3 years Loamy sand Slope: Landslope = 14% Furrow slope = < 6% Average annual rainfall = 1045 mm	Bare plot (typic eutropept)	286	0.27	216	76
	Conventional plot, no surface contact cover	213	0.20	51	24
	Improved practice – furrow mulching	150	0.14	3	2

[a] Information on length of experimental period, soil type, slope, and average annual rainfall over the experimental period is given.
[b] Runoff coefficient, R_c = average annual runoff/average annual rainfall.

formation in the humid tropics (Brady, 1974). All improved practices were agronomic in nature, made use of the potential for biomass production in the humid tropics, and involved the use of cover sufficiently close to the ground surface to impede overland flow – here referred to as 'surface contact cover'. When soil surface contact cover is measured using small quadrats or in other ways, only the fractional area covered by vegetation or stone sufficiently close to the soil surface that it will interfere with overland flow should be included. This type of cover may be contrasted with 'canopy cover' provided by vegetation not in such intimate contact with the soil surface.

The great effectiveness of surface contact cover in reducing soil loss is illustrated by comparing results for both non-bare-soil treatments at the Kemaman site (Table 17.1), for which canopy cover by the cocoa and companion shade trees was similar. However, the living grass–legume surface contact cover of the improved practice reduced average annual soil loss by a factor of over seven, whereas soil loss from the crop treatment with no ground cover reduced soil loss only by a factor of 1.4.

At both the Los Baños and VISCA sites in the Philippines, the improved practice for which data is given included the introduction of leguminous hedgerows whose trimmings were a component of mulch added to the cropped alley between contour hedgerows. Though the relative improvement due to hedgerow or mulch cannot be ascertained from these data, the reduction in soil loss is greater than would be expected in terms of reduced runoff alone (Table 17.1). Though hedgerows can fail under extreme typhoon events, causing a dramatic increase in soil loss akin to contour bank failure, intact hedgerows reduce soil loss more than would be expected in terms of reduced runoff, evidently due to net deposition as sediment-laden water moves through the half-metre or so of hedgerow. A better confirmation of the mechanisms involved in this reduction in sediment concentration could be useful.

The finer fraction of soil loss, termed the 'suspended load', is often enriched in nutrients, and is an indicator of off-site pollution potential, which was highest for the Kemaman and Goomboorian sites (Hashim *et al.*, 1995). The fraction of soil lost as suspended load increased with surface contact cover as found by Palis *et al.* (1990), and this fraction was higher for lighter texture soils, despite reduction in the absolute amount.

Total runoff from plots with conventional up-and-down slope cultivation was little less than from bare soil, or even greater in the case of the VISCA site, where weed control in the bare plot by hand hoe under wet conditions appeared to produce large stable aggregates and enhance infiltration. However, soil loss was reduced by conventional practices compared with bare soil, mostly due to a reduction in sediment concentration (Table 17.1).

The improved practice at each site resulted in a substantial reduction in runoff, notably at the Kemaman site, though more data are needed for the Nan site. For soils of light texture (at Goomboorian in Australia and Khon Kaen in Thailand; Sombatpanit *et al.*, 1995), the improved practice had a larger effect on sediment concentration than on runoff.

The runoff coefficient, R_c, the ratio of average annual runoff to rainfall, is shown in Table 17.1 to be very variable. However, the range in R_c for lighter texture soils (0.27 to 0.62) is higher than the range for clay soils (0.02 to 0.19), presumably due to the better water-stable aggregation of the clay soils, or to associated higher biotic activity.

Soil erodibility

To provide an indication of the relative importance of rainfall-driven versus overland flow-driven erosion mechanisms, the sediment concentration from detachment trays of soil at low slope was compared with the sediment concentration of plot runoff. These investigations confirmed the dominance of flow-driven erosion, even for plots of 12 m length, when slopes were more than some 15%. In such situations, the erodibility parameter β, defined in Eqn 17.7, is meaningfully related to the energy per unit mass required for flow to erode soil (Rose, 1993).

The value of β, calculated from bare plot data using program GUEST, varied in a manner dependent on soil type, cultivation and time since last cultivation, the method of weed control, and probably on factors such as soil strength and consolidation of soil through time. At Goomboorian, preparation for the pineapple crop is preceded by intense cultivation, with no subsequent cultivation, since weeds are controlled chemically. Over the 3-year period of experimentation, β initially had values fluctuating from 1.2 to 1.0, then declined steadily, probably due to soil consolidation, to values between 1.1 and 0.9, with minor fluctuation possibly due to pulsing of sediment through the 36-m-long ridge and furrow system (Rose *et al.*, 1997a). Since furrow slope was only about 5%, the reason for β being greater than 1.0 initially could be due to a rainfall-driven contribution to sediment concentration.

At both Philippine sites (Los Baños and VISCA), soils were clay in texture and the value of β fluctuated more than for the loamy sand at the Goomboorian site, apparently due to soil weakening associated with cultivation used for weed control (see Rose, 1994; Paningbatan *et al.*, 1995). Also, the average value of β was less than for the Goomboorian soil, perhaps due to the soils being more cohesive in nature.

Cover effects in soil conservation

The sensitivity of runoff from the research sites to either surface contact or canopy cover was less than commonly reported for example from some of the agriculturally important cropping locations in south-east Queensland (Rose and Freebairn, 1985) or New South Wales (Lang, personal communication). However, there is much greater similarity in the response to cover of both sediment concentration and soil loss between the data for research sites reported here and for data from sites elsewhere, in Australia and overseas. In Fig. 17.4 for Los Baños, this response is expressed as the relationship between the ratio c/c_b and C_s, where c is the average sediment concentration of runoff from plots with a range of surface contact cover percentages, C_s, and c_b is the average sediment concentration in runoff from the bare plot.

For the pineapple crop at the Goomboorian site, cover is dominantly by

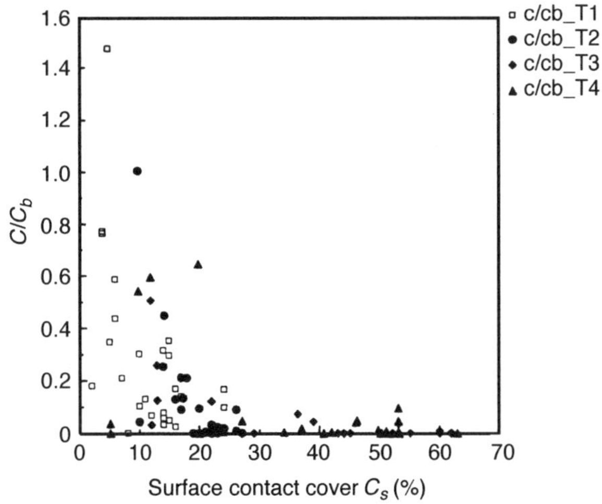

Fig. 17.4. The ratio of the sediment concentration from any particular treatment to that from the bare soil plot as a function of the surface contact cover C_s (%) for each treatment at the Los Baños site. (T1 refers to farmer's practice of up-and-down slope cultivation; T2 to hedgerow/alley cropping with no mulch cover; T3 as for T2 but with added mulch; T4 as for T3 with zero tillage) (from Rose *et al.*, 1997b).

leaves close to the soil surface, but recorded as canopy cover. Fig. 17.5 shows the ratio of event soil loss (conventional treatment/bare soil) versus canopy cover for this site. The relationship is similar if contact cover is added to give a total cover percentage.

Cover for the Los Baños site (Fig. 17.4) consists of both a canopy and surface contact component. The fact that there is a (somewhat complex) relationship between these two forms of cover makes it impossible to separate unambiguously the contribution of either type of cover to the reduction in sediment concentration. There is support given by the Los Baños data for conclusions from the literature that c/c_b is approximately linearly related to canopy cover, and strongly non-linearly related to surface contact cover (Rose *et al.*, 1997b).

As discussed more fully in Rose *et al.* (1997b), surface contact cover expressed as a fraction, C_s, can be mathematically related to its effect in reducing the measured sediment, c, relative to that measured from a bare soil plot, c_b. Despite considerable scatter (see Fig. 17.4), no doubt in part due to the great variety in form of contact cover and consequent estimation difficulties, c/c_b can be related to C_s by

$$c / c_b = \exp\left(-k_s C_s\right), \tag{17.10}$$

where k_s is non-dimensional, and can vary over a range, but $k_s = 10$ fits some data, such as that in Fig. 17.4.

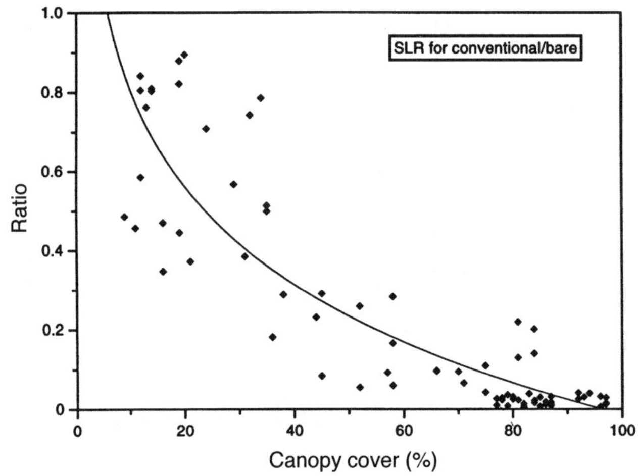

Fig. 17.5. The ratio of event soil loss (conventional cultivation of pineapples to bare soil), as a function of canopy cover (%) for the Goomboorian site (slope 5.5%) (from Rose *et al.*, 1997b).

Soil erosion and nutrient loss

There are important links between soil erosion and nutrient loss which are explored in Anon. (1995) and by Hashim *et al.* (Chapter 14, this volume). Enrichment of nutrients in eroded sediment, especially for soils of lighter texture (at the Goomboorian and Khon Kaen sites for example), complicates the relationship between soil and nutrient loss. Also, nutrient loss in soluble form can be important, especially if fertilizer applications are substantial, as was the case for pineapple production at Goomboorian.

With clay soils and limited fertilizer application, there is limited enrichment and the relation between soil and nutrient loss is much more direct. These characteristics applied to the Los Baños site in the Philippines, and because of the full database including yield and economic factors collected by Paningbatan, that site was chosen for an investigation of the long-term effects of land management on soil erosion, crop yield, and on-farm economics in that country.

Runoff and Soil Loss Prediction

So far, this chapter has outlined a methodology for interpreting experimental data on the dynamics of overland flow measured at the plot scale, and the consequences of this in terms of soil lost from the plot through erosion processes. The soil erosion theory used, employing GUEST methodology, showed a dependence of sediment concentration on runoff rate, indicating the desirability of measuring that rate – a measurement not required in USLE methodology.

Given experimental data of the type collected in the reported research, the parameters in the hydrological model SSRRM and the GUEST soil erosion model can be determined. Models of exactly the same theoretical structure as embodied in SSRRM and GUEST can be used for purposes of prediction, provided the parameters in these models are known or can be estimated. Thus the theory employed is identical, whether the model is used to analyse data in order to determine model parameters, or when the model is used for prediction, in which case the value of model parameters may be available from prior experience or might have to be predicted from surrogate information. One of the reasons for using process-type models is the hope that the parameters in these models have some physical meaning, and hence, when direct data is lacking, may be estimated, however approximately, from surrogate information that is either available or may be readily obtained. Thus one context where the need for prediction arises is where experimental information allowing adequate determination of predictive model parameters is not available or is inadequate. Another context for prediction is to extend greatly the time period of information in a situation where only a limited period of adequate measurement has been made.

Yu *et al.* (1997d) consider the range of possible prediction scenarios, and how these may be addressed. Such scenarios can be defined by type and availability of data required for predictive purposes. From Eqn 17.8, it follows that the hydrological predictions required for GUEST are the effective runoff rate, Q_e, and the total event runoff amount, ΣQ. In Eqn 17.8, the power on ΣQ is higher than on Q_e, making it the more important of the two quantities for prediction. A flow chart for soil erosion prediction using GUEST is shown in Fig. 17.6.

The prediction of runoff amount (ΣQ), given information on rainfall, requires a continuous water-balance accounting model of the type described by Yu *et al.* (1997c) and Yu *et al.* (1997d). This keeps track of water stored in the upper layer of the soil, since antecedent water content has a substantial effect on the amount of runoff for any given rainfall amount. Yu *et al.* (1997d) showed that good prediction of ΣQ could be obtained with a two-parameter model involving an apparent water storage capacity of the soil and any mulch layer if present, together with the runoff coefficient for excess rainfall. Some data at the site of interest is currently required in order to evaluate these two parameters.

The second required hydrological variable, the effective runoff rate, Q_e, has been modelled in different ways, depending on data availability. Because of the widespread influence of USLE methodology, there is a considerable body of collected data where rainfall rate and runoff amounts have been measured (in addition to soil loss). Yu (1997) has developed a program, GOSH (Generation Of Synthetic Hydrograph), which produces a synthetic runoff rate, given the above type of data. Program GOSH codifies three one-parameter infiltration models, including the exponential model given in Eqn 17.9 which is the most effective of alternatives tested.

However, even if the only data available are rainfall rates, Yu *et al.* (1997c, d) have shown that a one-parameter model using peak rainfall

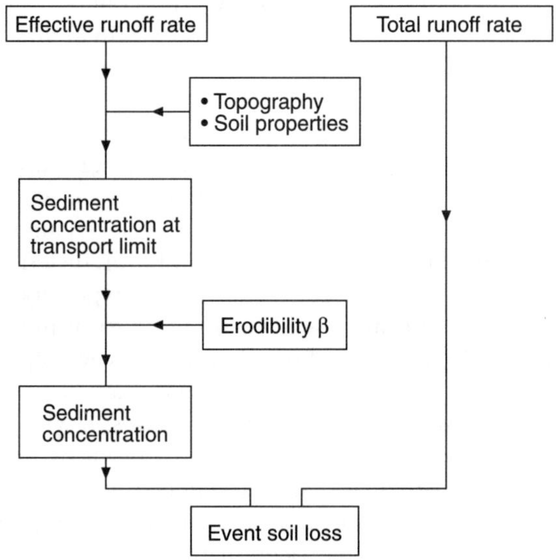

Fig. 17.6. Flow chart for event soil-loss prediction using GUEST methodology (from Yu *et al.*, 1997d).

intensities at 6 or 15 min can be a useful estimator of Q_e. Again, however, data on runoff rates are needed to ascertain accurately values for the two parameters involved.

A program that both allows calculation of the soil erodibility parameter, β, and prediction of the amount of soil loss using GUEST methodology is now available, and is called GUEPS (Griffith University Erosion Prediction System) (Yu and Rose, 1997). As well as providing both an analysis facility (yielding β) and a predictive capacity (when β is known), GUEPS relieves the requirement of GUEST that runoff rate must be supplied at a fixed constant time interval.

A flowchart illustrating the GUEPS methodology is given in Fig. 17.7.

Discussion and Conclusions

This chapter outlines a theoretically driven field experimental methodology different from that traditionally used with the type of interpretation based on the USLE, where rainfall rate was seen to be the driving variable, to experimentation where rate of runoff is included as a variable with potency to erode in its own right. Electronic forms of data recording, and the use of computer programs for data checking and analysis, have also been helpful to this development (Ciesiolka *et al.*, 1995a). This soil conservation research methodology has been applied to a multicountry study of alternative and locally decided soil-conserving systems in tropical steeplands.

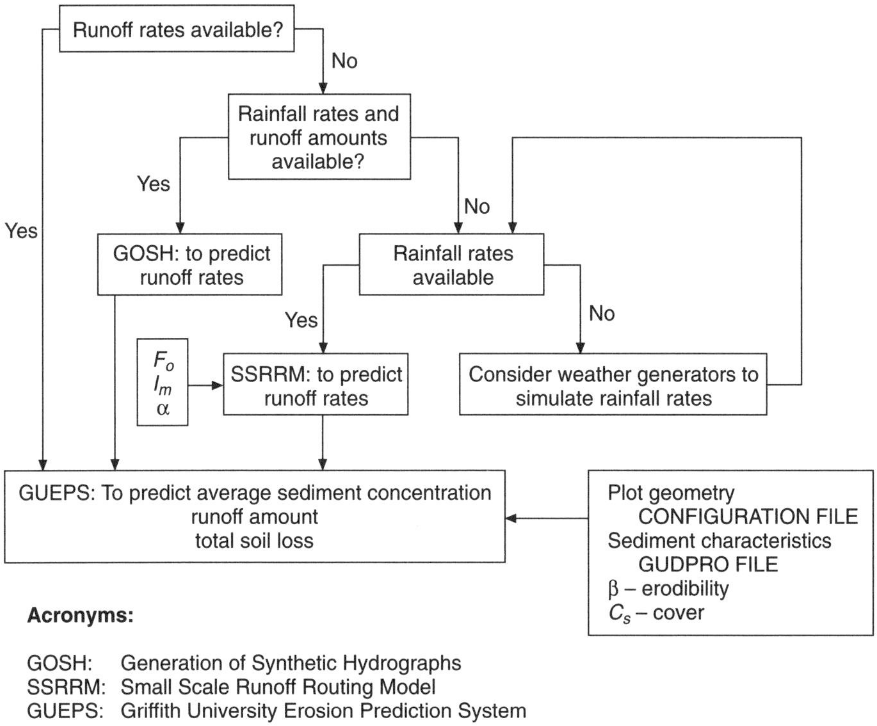

Fig. 17.7. Flow chart for event soil loss and runoff prediction using GUEPS methodology.

In general, there is even less data available on runoff rate than on rainfall rate. It is partly for this reason that the reported research sought a better understanding of the quantitative connection between these processes. The research has expanded experience in evaluating soil erodibility and its changes with management. Whilst the erodibility factor β appears to have theoretical and practical advantages over its equivalent in the USLE methodology, namely the *K*-factor, the relationship between β and more readily measured soil characteristics is not simple, despite the development of locally valid relationships (Ciesiolka *et al.*, 1995b; Misra and Rose, 1995).

Outcomes of the reported research emphasize the important role and effectiveness of agronomically based methods of soil conservation, as distinct from structural methods which require significant soil movement. In particular, it has been shown that use of mulch to provide effective levels of cover in intimate contact with the soil surface reduces soil loss to very low levels, even on steep slopes. Whilst the use of leguminous hedgerows adds further protection and stability to the system, and encourages contour cultivation, the establishment cost in particular can be an economic disincentive.

The intensive data collected in the research projects reviewed herein have provided an invaluable platform for developing and testing predictive models of runoff and soil loss at the plot scale when less intensive but more commonly collected data are all that is available. Progress in this area of research is reviewed, with emphasis on hydrological aspects.

Acknowledgements

The chapter draws heavily on the outcomes of Project PN 9201 and, to a lesser extent, PN 8551, programmes supported financially by the Australian Centre for International Agricultural Research (ACIAR). These ACIAR projects involved collaboration of a great number of organizations, scientists and support staff in all the countries involved. However, particular thanks are due to C.A.A. Ciesiolka of the Queensland Department of Primary Industries, G.M. Hashim of the Malaysian Agricultural Research Institute, E.P. Paningbatan of the University of the Philippines at Los Baños, A.L. Presbitero of the Visayas State College of Agriculture, Leyte, C. Anecksamphant of the Department of Land Development, Thailand, and B. Fentie who provided us with valuable research support.

References

Anon. (1995) *Soil Technology. Special Issue: Soil Erosion and Conservation* 8(3), 258 pp.

Brady, N.C. (1974) *The Nature and Properties of Soils*, 8th edn. Macmillan Publishing Co. Inc., New York.

Ciesiolka, C.A., Coughlan, K.J., Rose, C.W., Escalante, M.C., Hashim, G.M., Paningbatan, E.P. and Sombatpanit, S. (1995a) Methodology for a multi-country study of soil erosion management. *Soil Technology* 8, 179–192.

Ciesioka, C.A.A., Coughlan, K.J., Rose, C.W. and Smith, G.D. (1995b) Erosion and hydrology of steeplands under commercial pineapple production. *Soil Technology*, 8, 243–258.

Coughlan, K.J. and Rose, C.W. (eds) (1997) *A New Soil Conservation Methodology and Applications to Cropping Systems in Tropical Steeplands. ACIAR Technical Report No. 40.* ACIAR, Canberra, ACT, Australia.

Foster, G.R. (1982) Modelling the erosion process. In: Hann, C.T. (ed.) *Hydrologic Modelling of Small Watersheds. American Society of Agricultural Engineering Monograph No. 5.* St. Joseph, Michigan, pp. 297–379.

Foster, G.R. and Lane, L.J. (1987) *User requirements: USDA – Water Erosion Prediction Project (WEPP). NSERL Report No. 1, Sept. 1,* USDA-ARS National Soil Erosion Research Laboratory, Purdue University, West Lafayette, Ind.

Govers, G. and Poesen, J. (1988) Assessment of the interrill and rill contributions to total soil loss from an upland field plot. *Geomorphology* 1, 343–354.

Hairsine, P.B. and Rose, C.W. (1991) Rainfall detachment and deposition: sediment transport in the absence of flow-driven processes. *Soil Science Society of America Journal* 55, 320–324.

Hairsine, P.B. and Rose, C.W. (1992a) Modelling water erosion due to overland flow using physical principles: I. Uniform flow. *Water Resources Research* 28, 237–243.

Hairsine, P.B. and Rose, C.W. (1992b) Modelling water erosion due to overland flow using physical principles: II. Rill flow. *Water Resources Research* 28, 245–250.

Hashim, G.M., Ciesiolka, C.A., Yusoff, W.A., Nafis, A.W., Mispan, M.R., Rose, C.W. and Coughlan, K.J. (1995) Soil erosion processes in sloping land in the east coast of Peninsular Malaysia. *Soil Technology* 8, 215–233.

Lisle, I.G., Coughlan, K.J. and Rose, C.W. (1996) *GUDPRO 3.1: A Program for Calculating Particle Size and Settling Velocity Characteristics. User Guide and Reference Manual. ENS Working Paper* 7/96. Faculty of Environmental Sciences, Griffith University, Nathan, Queensland, Australia.

Lisle, I.G., Rose, C.W., Hogarth, W.L., Hairsine, P.B., Sander, G.C. and Parlange, J-Y. (1998) Stochastic sediment transport in soil erosion. *Journal of Hydrology* 204, 217–230.

Marshall, T.J., Holmes, J.W. and Rose, C.W. (1996) *Soil Physics*, 3rd edn. Cambridge University Press, Cambridge.

Misra, R.K. and Rose, C.W. (1990) *Manual for Use of Program GUEST.* Monograph, Division of Australian Environmental Studies, Griffith University, Brisbane, Australia.

Misra, R.K. and Rose, C.W. (1995) An examination of the relationship between erodibility parameters and soil strength. *Australian Journal of Soil Research* 33, 715–732.

Misra, R.K. and Rose, C.W. (1996) Application and sensitivity analysis of process-based erosion model GUEST. *European Journal of Soil Science* 47, 593–604.

Morgan, R.P.C., Quenton, J.N. and Rickson, R.J. (1992) *EUROSEM: Documentation Manual.* Silsoe College, Silsoe, UK.

Nearing, M.A., Foster, G.R., Lane, L.J. and Finkner, S.C. (1989) A process based erosion model for USDA water erosion prediction project technology. *Transactions of the American Society of Agricultural Engineers* 32, 1587–1593.

Palis, R.G., Okwach, G., Rose, C.W. and Saffigna, P.G. (1990) Soil erosion processes and nutrient loss II. The effect of surface contact cover and erosion processes on enrichment ratio and nutrient loss in eroded sediment. *Australian Journal of Soil Research* 28, 641–658.

Paningbatan, E.P., Ciesiolka, C.A., Coughlan, K.J. and Rose, C.W. (1995) Alley cropping for managing soil erosion of hilly lands in the Philippines. *Soil Technology* 8, 193–204.

Rose, C.W. (1993) Erosion and sedimentation. In: Bonell, M., Hufschmidt, M.M. and Gladwell, J.S. (eds) *Hydrology and Water Management in the Humid Tropics – Hydrological Research Issues and Strategies for Water Management.* Cambridge University Press, Cambridge, pp. 301–343.

Rose, C.W. (1994) Research progress on soil erosion processes and a basis for soil conservation practices. In: Lal, R. (ed.) *Soil Erosion Research Methods.* Soil and Water Conservation Society, Ankeny, Iowa, pp. 159–180.

Rose, C.W. and Freebairn, D.M. (1985) A new mathematical model of soil erosion and deposition processes with applications to field data. In: El Swaify, S.A., Moldenhauer, W.C. and Lo, A. (eds) *Soil Erosion and Conservation.* Soil Conservation Society of America, Iowa, pp. 549–557.

Rose, C.W. and Hairsine, P.B. (1988) Processes of water erosion. In: Steffen, W.L. and Denmead, O.T. (eds) *Flow and Transport in the Natural Environment.* Springer-Verlag, Berlin, pp. 312–316.

Rose, C.W., Coughlan, K.J., Ciesiolka, C.A.A. and Fentie, B. (1997a) Program GUEST (Griffith University Erosion System Template). In: Coughlan, K.J. and Rose, C.W. (eds) *A New Soil Conservation Methodology and Applications to Cropping Systems in Tropical Steeplands. ACIAR Technical Report No. 40.* ACIAR, Canberra, ACT, Australia, pp. 34–58.

Rose, C.W., Coughlan, K.J., Ciesiolka, C.A. and Fentie, B. (1997b) The role of cover in soil conservation. In: Coughlan, K.J. and Rose, C.W. (eds) *A New Soil Conservation Methodology and Applications to Cropping Systems in Tropical*

Steeplands. ACIAR Technical Report No. 40. ACIAR, Canberra, ACT, Australia, pp. 59–78.

Sombatpanit, S., Rose, C.W., Ciesiolka, C.A.A. and Coughlan, K.J. (1995) Soil and nutrient loss under rozelle (*Hibiscus subdariffa* L. var. *altissima*) at Khon Kaen, Thailand. *Soil Technology* 8, 235–241.

Yu, B. (1997) *GOSH: A Program for Calculating Runoff Rates Given Rainfall Rates and Runoff Amount. User Guide and Reference Manual. ENS Working Paper 2/97.* Faculty of Environmental Sciences, Griffith University, Nathan Campus, Brisbane, Queensland, Australia.

Yu, B. and Rose, C.W. (1997) *GUEPS 1.2: A Program for Calculating the Soil Erodibility Parameter β and Predicting the Amount of Soil Loss Using GUEST Methodology. User Guide and Reference Manual. ENS Working Paper 3/97.* Faculty of Environmental Sciences, Griffith University, Nathan Campus, Brisbane, Queensland, Australia.

Yu, B., Rose, C.W., Coughlan, K.J. and Fentie, B. (1997a) Plot-scale runoff modelling for soil loss predictions. In: Coughlan, K.J. and Rose, C.W. (eds) *A New Soil Conservation Methodology and Applications to Cropping Systems in Tropical Steeplands. ACIAR Technical Report No. 40.* ACIAR, Canberra, ACT, Australia, pp. 24–33.

Yu, B., Rose, C.W., Coughlan, K.J. and Fentie, B. (1997b) Plot-scale rainfall-runoff characteristics and modelling at six sites in Australia and Southeast Asia. *Transactions of the American Society of Agricultural Engineers* 40, 1295–1303.

Yu, B., Rose, C.W., Coughlan, K.J. and Fentie, B. (1997c) Runoff and soil loss prediction. In: Coughlan, K.J. and Rose, C.W. (eds) *A New Soil Conservation Methodology and Applications to Cropping Systems in Tropical Steeplands. ACIAR Technical Report No. 40.* ACIAR, Canberra, ACT, Australia, pp. 101–110.

Yu, B., Rose, C.W., Ciesiolka, C.A.A., Coughlan, K.J. and Fentie, B. (1997d) Toward a framework for runoff and soil loss prediction using GUEST technology. *Australian Journal of Soil Research* 35, 1191–1212.

18

The Measurement of Soil Erosion

C.A.A. Ciesiolka[1] and C.W. Rose[2]

[1]*Department of Natural Resources, 203 Tor Street, Toowoomba, Queensland 4350, Australia;*
[2]*Faculty of Environmental Sciences, Griffith University Nathan, Queensland 4111, Australia*

The objectives and measurement challenges in soil erosion studies are described for the plot, subcatchment and catchment scales. The selection of suitable sites at all three scales is discussed. Studies at the plot scale provide the opportunity to evaluate quantitatively the effectiveness of alternative land management treatments in terms of soil, water and nutrient loss. The use of tipping-bucket technology, which is appropriate at this scale, provides a most effective method for separating the bedload sediment complement from the suspended load which has the capability of transporting fine sediment and chemicals over long distances to receiving water bodies. Such knowledge obtained at this scale provides the opportunity for effective control when problems recognized at larger scales indicate the need for remedial action.

Studies at the subcatchment and catchment scale indicate the integrated consequences of the complex combination of land characteristics and land management contained within the larger unit.

The chapter discusses the range of measurement methods appropriate to this range of scales, problems associated with the installation and use of equipment, and the principles involved in data logging and data conversion. Particular attention or bias is toward the use of economical and reliable sensors and measurement methods suitable for field studies in any part of the world.

Introduction

There has been an amplification of objectives in soil erosion studies in the last decade giving more emphasis to water quality issues than to soil loss *per se.* This development has been accompanied by recognition of the need to move up in scale from the plot scale through subcatchment to the catchment side. The plot scale of investigation typically is based on the scale of local farmer operation, and this is reflected for instance in ACIAR projects such as PN9201, where the plot scale investigated ranged from, 70 to 1000 m^2. Investigations at the subcatchment scale could range from 0.5 to 20 ha. Catchment-scale investigations typically can involve hundreds of hectares or more.

This chapter describes the outcome of extensive experience in the design, installation and operation of measurement systems for studies of hydrology and soil loss at the three scales described earlier ranging from plots to catchments.

Project Objectives

In general as the scale of investigation progresses from plot to catchment, objectives of the studies change from predominantly on-site to off-site concerns.

Objectives of plot studies

Studies at the plot scale are typically concerned with the effect of management options on soil and water loss and productivity. Such options would include current local farmer practice, but also explore options deemed to be feasible, acceptable, innovative but also based on the best understanding available of soil conserving principles.

At the plot scale it is feasible to investigate the hydrological and soil erosion processes, and how processes are modified by the management options selected for investigation. Study of processes at this relatively small scale carries with it the desire to follow process dynamics at small time scales of the order of minutes. Recent advances in the understanding of soil erosion processes have indicated the need to measure the rate of overland flow as well as rainfall (Ciesiolka *et al.*, 1995). The development of electronic logging equipment has made feasible the measurement of rates at this time scale.

Recent studies give more emphasis to the loss of soluble nutrients and agricultural chemicals such as herbicides and insecticides. Sampling to allow reliable measurement of such soluble components presents particular problems currently not well addressed by commercial sampling equipment.

Objectives at the sub-catchment scale

At this scale both hydrology and sediment transport are affected by spatial variability in factors such as land management, cover, soil type and infiltration characteristics. Thus an objective of studies at this scale is to assess the realism of extrapolation in space from the smaller-scale plot studies. Study at this scale should indicate the effect of processes not evident or expressed at the plot scale, such as rilling, channellized flow and hillslope throughflow.

Objectives at the catchment scale

At this scale the commercial interest in the quantity and quality of the water resource plays a predominant role in study objectives. Communal concerns include such issues as potability of water, and suitability for multiple use, for example recreation and fish culture. Sediment delivery ratios vary downstream according to the stream's ability to transport its load, consequently affecting the distribution of off-site effects. Engineering principles can be brought to bear on the management of downstream hydraulics to translate a flood or spread its flow in order to increase soil moisture across a floodplain.

Measurement at the Plot Scale

Studies of soil erosion in the field have traditionally been largely based on the standard plot length of 22 m adopted in the universal soil loss equation (USLE) experimentation (Wischmeir and Smith, 1965, 1978). More recently, theoretical development has allowed plot length to be chosen to represent farmer practice. In plot studies plot dimensions are strongly influenced by cultivation methods, in particular whether or not these are mechanized. Whilst it is desirable for plot length to represent farmer field practice, the choice of width of experimental plot may be strongly determined by available financial resources; financial limitations may enforce plots of modest width, but these can still provide realistic data, and cost obviously increases with scale of plot.

At the plot scale it is desirable to choose plot layout so that it occupies a linear hillslope element of the landscape. Such choice allows for much simpler interpretation, and relieves the need to quantify within-plot areas of net deposition. This choice thus separates the effect of management from the effects of more complex land surfaces on erosion/deposition.

For soils which are known to consolidate strongly, and are thus less likely to form deep rills, an option is to define the upper and side boundaries of the rectangular plot by strips of galvanized iron embedded some 20 cm into the soil, or to reach the clay subsoil, whichever is less. The purpose of this is to isolate the plot hydrologically.

For soils that do not consolidate and thus are likely to form rills, the use of galvanized iron strips for plot definition is not recommended. Rather, formed soil bunds of some 20 cm height are recommended, and their stability can be assisted by planting with grass or provision of a deep mulch cover.

The plot exit boundary should be designed in such a way that it provides no impediment to the water and soil flowing off the plot. It has been found extremely useful to incorporate a trough of low slope to catch the soil-laden water as it leaves the plot, ensuring that the coarser fraction of the sediment drops out through net deposition in this trough, to form what is termed the *bedload*. One advantage of this prior bedload separation of sediment is that the ongoing flow of water contains only suspended sediment which will not deposit in subsequent flow-measuring devices, and thus not affect their calibration. A second advantage of bedload separation is that subsampling of suspended load is more accurate and less expensive than seeking to subsample the entire sediment flux. A third advantage is that this innovative method provides an ongoing *suspended load* which is a direct measurement of sediment and associated organic and chemical constituents. The suspended load can be transported over substantial distances to water storages, thus providing a direct indication of pollution potential.

Monitoring of soil moisture and water-table levels has improved with the use of electronic equipment. Installation of equipment causes some plot disturbance initially, but is favoured over weekly sampling which causes regular disturbance. Spacing of throughflow lines in hillsides makes unsystematic measurement of soil moisture and water table elevations somewhat meaningless as the results cannot be extrapolated across the whole landscape. Consequently, measurements of soil moisture and water level at a minimum of three points in an outside buffer strip of each plot will provide a satisfactory set of data for modelling soil water. It is recommended that plots be hydrologically confined with a membrane to the depth of the clay so that shallow subsurface water can be intercepted and measured.

Hydrological instrumentation

Pluviometry

Provided the sensitivity of the instrument is adequate, there are distinct advantages in a tipping-bucket type pluviometer for the measurement of rainfall rate. The sensitivity of this type of pluviometer depends on the ratio between its catchment area and the volume of water required to cause the collection bucket to tip. For example, a 300-mm diameter rain gauge with a bucket volume of 7 ml results in an acceptable accuracy (Yu *et al.*, 1997). It is desirable to have rainfall rate measured by at least one pluviometer, and the possible spatial variation in rainfall amount can be ascertained by strategically locating standard 200-mm totalizing rain gauges. Data from such rain gauges should also be used to check the validity of the logged pluviometer data.

There are a number of handbooks available from Meteorological Offices throughout the world that describe location and maintenance of such instruments (WMO, 1965; Rodda, 1969).

Rain gauges should be positioned so that no vegetation taller than 20 cm below the gauge exists for the surrounding 30 m if possible (SCS National Engineering Handbook 224). Thus, if located within a growing crop, the instrument stand needs to be adjustable in height to suit crop stage.

Flow measuring devices

As a general guide, flow can be best measured by tipping-bucket flow devices when the area of runoff plots is less than about 600 m^2. For areas greater than this some form of flume is preferable, since there are upper limits to the feasibility of flow measurement using tipping-bucket devices. For example, a bucket tipping rate approaching the high rate of once per second produces instability in performance, and there are structural difficulties in constructing very large buckets. The maximum desirable length of bucket is about 1 m, and maximum bucket capacity is about 20 l per tip.

To assist in the design of tipping buckets, use can be made of the theory (Yu *et al.*, 1997) that gives the standard error of measurement (σ) as:

$$\sigma = V / A\Delta t\sqrt{6} \tag{18.1}$$

where V is the bucket volume, A the catchment area of the plot, and Δt is the period over which the number of tips is recorded.

For example:

$$\text{If } A = 480\ \text{m}^2,\ V = 16\ \text{l},\ \Delta t = 1\ \text{min},$$

$$\text{then } \sigma = \frac{16 \times 10^6\text{mm}^3}{480 \times 10^6\text{mm}^2 \times \left(60^{-1}\text{h}\right)\sqrt{6}}$$

$$= 0.816\ \text{mm h}^{-1}$$

Thus, if the runoff rate is 10 mm h^{-1} the relative error is $(0.816/10) \times 100 = 8\%$.

A suitable design of a tipping bucket is given in Fig. 18.1. Construction of the tipping bucket can be in moulded PVC or rust-proofed metal. The location of the axis of support for the bucket must be such that when one side of

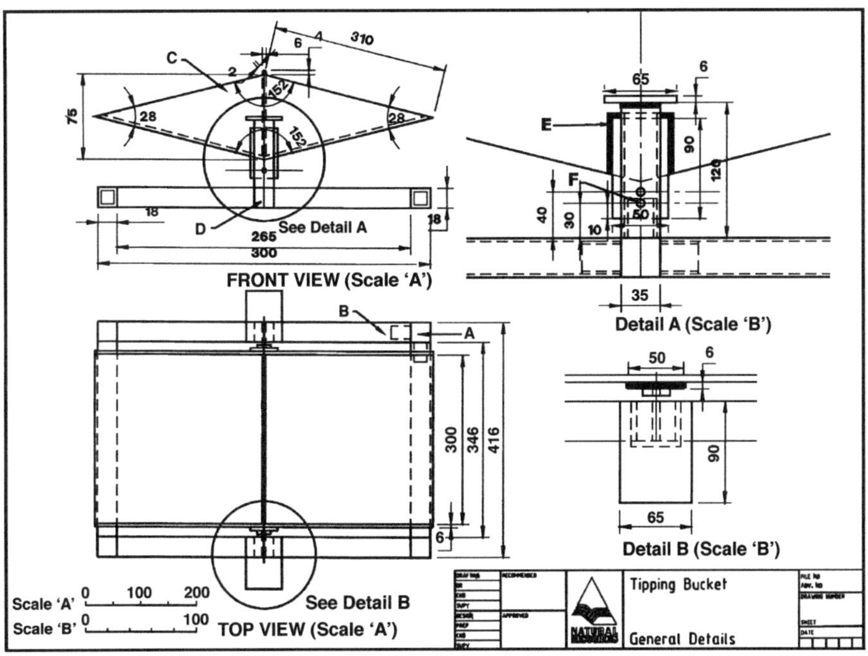

Fig. 18.1. Working drawing for a tipping bucket of width 300 mm.

the bucket fills with water to within 2 cm of its top (under static conditions), the bucket will tip to expose the opposite bucket. For self cleaning of the bucket it is desirable that at rest the lower lip of the bucket is inclined at a slope of 8%.

Prior to field installation, tipping-bucket devices need calibration in the laboratory at a number of rates. It is desirable to calibrate over the range from 1 tip in 5 min to 50 tips per min. Rotameters or a 30° V-notch weir can be used for accurate measurement of the calibrating flow rate. Such a calibration is not sensitively dependent on how level the bucket is when installed.

If estimates of tipping rate indicate a possible tipping rate of once per second or greater, then some type of flume should be used for flow measurement. The choice of the most desirable form of flume depends on estimated sediment concentrations. If the sediment concentration is expected to be high, then a San Dimas type of flume is recommended (Bermel, 1950), despite its accuracy in the measurement of low flows possibly being poorer than the alternative type – a Parshall flume (Bos, 1989). Parshall and trapezoidal flumes have been commonly used in this context (Hudson, 1973). However, field experience includes the possibility of sediment deposition within such flumes, and this will affect their discharge rating curve.

In the detailed design of any flume, its geometry should be such that a flow depth of at least 3 cm should correspond to a runoff rate of 1 mm h^{-1} or less from the flume's catchment area. For example, if the catchment area is 1000 m^2, a runoff rate of 1 mm h^{-1} will lead to a volumetric flow rate of 0.28 l s^{-1}, and this will imply a certain flume geometry, depending on its rating curve.

If inexperienced in flume design, it is strongly recommended that Bermel (1950) be consulted.

Sediment measuring devices

The type of low-slope modified Gerlach trough in which bedload is deposited is illustrated in Fig. 18.2.

The slope of the trough should be less than 1%. The loss by flow of bedload which deposits in the trough is restricted by placing a number of wooden cross-pieces along the length of the trough. Following the erosion event, these cross-pieces are removed, a number of subsamples of bedload are taken for laboratory measurement of water content, and then the entire remaining bedload is shovelled into 30-litre drums for weighing in the field.

The size of the trough should be such that it will store an amount of sediment corresponding to a loss of 50 t ha^{-1} per event, or if slopes exceed 30% or rilling is extensive, a figure of at least 70 t ha^{-1} per event should be used in trough design. It may be necessary to use more than one trough, one overflowing laterally into the next, if the amount of sediment is very high. If the cost of trough construction cannot be met, then a reduction in plot width is indicated.

Suspended load is measured by extracting a flow-weighted small propor-

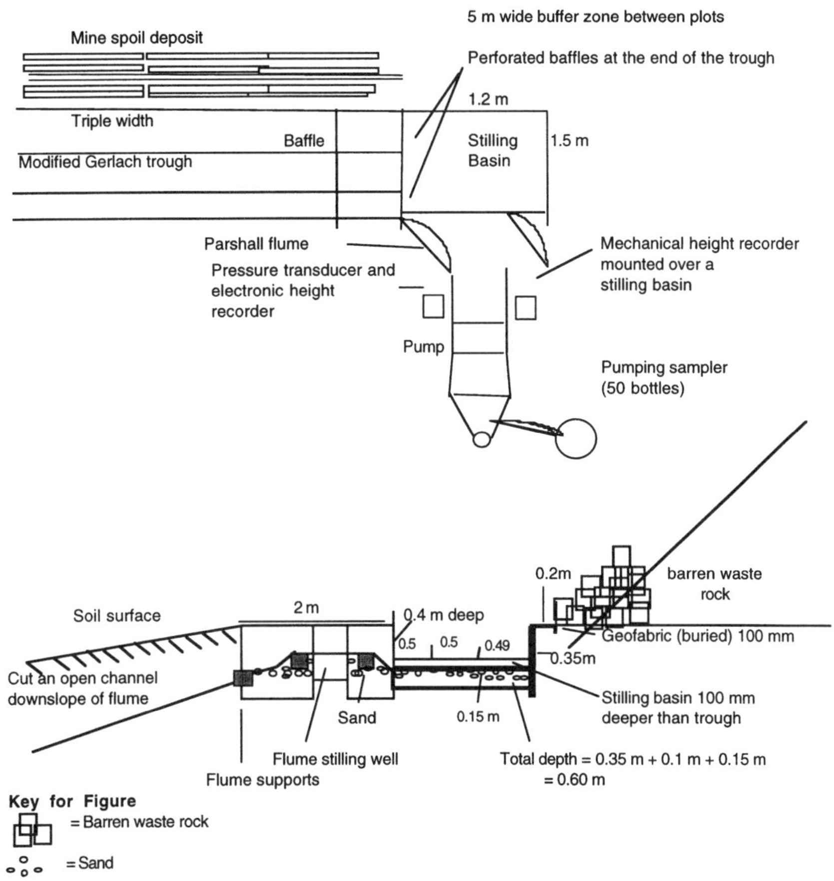

Fig. 18.2. Schematic plan and side views of a modified Gerlach trough installation.

tion of the total flow which has exited the flow measuring device. If this measuring device is a tipping bucket, then an accurate subsampling technique is to direct the flow shed by each bucket on to a plastic pipe with 3-mm-wide slots exposed to the flow at about every 10 cm along the pipe. This device can be called a flow splitter. The water collected in the pipe is drained into a container located lower than the pipe. This entire suspended sediment sample is allowed or encouraged to settle (e.g. by cooling in a refrigerator), the supernatant water decanted carefully, and the remaining sediment oven-dried at the same temperature as that used for oven-drying the bedload.

The quantity required for purposes of analysis is the sediment concentration, which is calculated as follows:

$$\text{Sediment concentration} = \frac{\text{oven-dry mass of sediment}}{\text{volume of (sediment + water)}} \quad (18.2)$$

It is desirable to obtain, by calibration, the fraction of the flow which is collected by the flow splitter. With this fraction and the volume collected known, the total flow can be calculated. This estimate can be valuable in data recovery should the main flow measurement equipment fail.

When flumes are used for flow measurement, pumping samplers to extract samples can be used for determination of sediment concentration. For determining total soil loss this sample extraction needs to take place at a site well upstream of a Parshall flume, since the flume causes deposition to occur in its upstream section. This sediment concentration, when multiplied by flow rate and the product integrated over the duration of the event, provides an estimation of total sediment flux, and thus of soil loss.

It should be noted that such total sediment concentration sampling does not allow convenient separation of sediment into bedload and suspended load as is achieved with tipping-bucket devices.

Field installation of flow measuring devices

Troughs

When study objectives include the sampling of subsurface flow through the plot, it is convenient to excavate a trench exposing the subsoil. This trench is lined with an impermeable membrane to collect seepage, and the flow rate of such seepage can be measured with its own (smaller) tipping bucket. If such an installation is made, then the bedload trough can be installed above the trench.

To avoid the trough edge becoming an impediment to sediment loss from the plot, its elevation needs to be adjusted. In up-and-down slope furrowed crops, for example, the upper edge of the trough should align with the bed of the furrows.

Troughs can be constructed in galvanized iron. Concrete construction is also a possibility, though this can affect water quality.

Tipping-bucket manifold

In order to produce a controlled inflow into the tipping-bucket measuring system, a flow-controlling device called a *manifold* is used. The manifold accepts water from the trough, and directs flow with suitable characteristics for entry on to the tipping bucket. Where plant debris may be sufficient to affect measurement, a coarse screen can be fitted to the manifold upstream of the slot through which the water falls on to the tipping bucket.

Tipping bucket

The frame supporting the tipping bucket needs stable support in a horizontal position. The frame needs to be positioned so that the axis of the tipping-bucket system is aligned vertically below the slot in the manifold through which collected water falls.

As shown in Fig. 18.1, the tipping bucket is supported from the base frame so that the bucket axis is 10 mm above the axle on which rotation of the buckets can occur. The axle is supported with low friction material (lubron). The distance from the slot in the manifold to the top of the arc of rotation of the buckets should not exceed 2 cm in order to minimize splash loss of water falling into the bucket.

The angle of rotation of the tipping bucket is controlled by shock-absorbent and UV-resistant plastic attached to the frame, the outfall angle reaching 8%. The suspended sediment splitter should be located so that water shed by the tipping bucket splashes directly on to it.

Flume

The common choice in soil erosion studies is either the Parshall or San Dimas flume, and both have a similar installation. There exists a danger that subsurface flow could undermine the installed flume structure. To avoid this possibility, subsurface flow is obstructed as completely as possible by digging a trench to bedrock (if it exists) or as deep as practical, and the trench is then either filled with concrete or a galvanized iron sheet fitted. Immediately downslope of this cut-off trench, the concrete form of the flume commences. This concrete flume shape is laid on top of a 10-cm thick layer of coarse sand, which in turn overlays a 30-cm layer of coarse gravel, providing a stable foundation.

Shapes typical of a San Dimas flume discussed above are illustrated in Fig. 18.3. The vertical profile of a base for a flume installation in the field is shown in Fig. 18.4.

Wings at the intake of a flume are constructed of corrugated iron with corrugations in the vertical direction. This corrugated iron sheet is bolted to vertical steel posts driven into the soil and rust-proofed.

Data Logging and Data Conversion

Higher-order data manipulation can be carried out in the laboratory using computer facilities. Thus there is no real necessity to in-build such capacity into data loggers for field data capture. Reducing the complexity of field data logging equipment has the following advantages:

- reduced cost of data recording,
- reduced size, which allows improved instrument security because of the capacity for unobtrusive installation.

In studies of soil erosion, a field visit following each erosion event is required because either bedload must be measured and removed (in tipping-bucket installations), or sample bottles and charts must be changed (in flume installations). Thus there is no necessity for data loggers in this context to have extensive memory capacity.

It is most desirable on every site visit for the validity of electronically logged data to be examined. This can be readily achieved using a portable computer with suitable data acquisition and verification procedures. Consonance in time between pluviometer and surface runoff measurement is an important and easily checked feature of the data. A check that runoff totals

SAN DIMAS FLUME

NOTES	(600 mm depth flume) Throat width (mm)						(1000 mm depth flume)		
	75	100	150	250	300	450	750	900	2500
1. Radius of wings (mm)	75	100	150	250	300	350	750	900	1500
2. Inlet width (mm)	225	300	450	750	800	1150	2150	2700	7500
3. Distance to stilling well inlet (mm) NB Changes with flume depth	600	600	600	700	800	900	1500	1800	2000
4. Flume tail length = two times flow depth at measuring point (mm)	600	600	600	600	600	600	1500	1800	2000
5. Riser (mm)	36	36	36	39	42	45	90	108	1200
6. Contributing area (ha)	0.07	0.1	0.15	0.2	0.3	0.5	3.0	5.0	20.0

450 mm deep flume of 150 mm width
(for 600 mm and 1000 mm depth flumes refer to notes above)

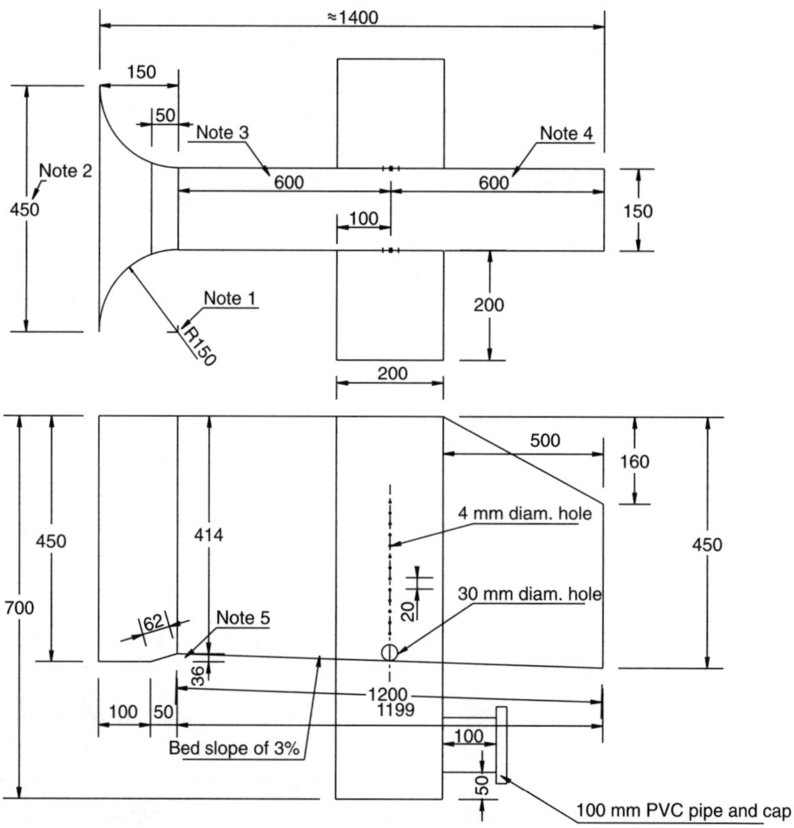

Not to scale

Fig. 18.3. San Dimas Flume.

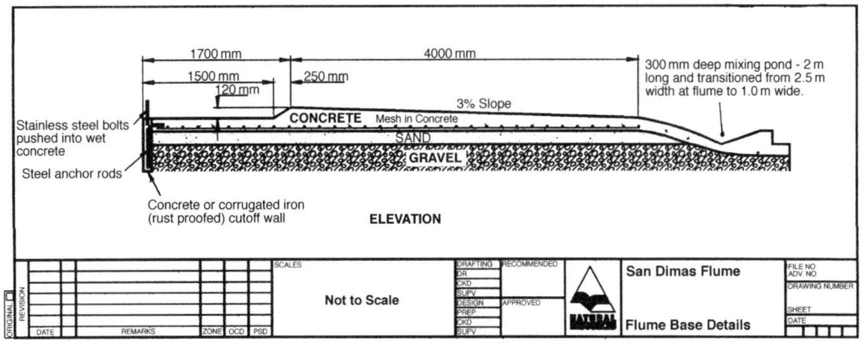

Fig. 18.4. Cross-sectional view of a base used in field installation of flumes.

and rates do not exceed rainfall totals and rates provides another check. The relative comparison of results with treatment characteristics can also provide a useful and more testing check on data reliability.

On-site identification of possible data incongruities allows an immediate check of field equipment and site characteristics. This on-site identification of possible problems with data, followed by investigation, is the most effective way of ensuring good data quality.

Measurement at the Subcatchment Scale

At this larger scale, complexity is added by variety of landscape form, the possibility of channellized flow due to erosion, roadways or other cultural features, and the possible emergence of subsurface flow.

Within a subcatchment it is common for a region dominated by net loss of soil to form a first-order flow path whose characteristics are associated with the first evidence of substantial net deposition. If so, it is of value in understanding the subcatchment process for measurement to be made just prior to the location in the stream of such net deposition.

Especially in more humid regions, it is also possible to find locations within the subcatchment where base flow emerges, either within a channel or as seepages on surrounding slopes. This may occur downstream of the confluence of more than one first-order stream. Measurement at a site suitable to capture both base flow and surface flow allows the possibility of being able to separate the quantity and water quality associated with these two types of flow. There is particular concern with the water quality of base flow, as this can be the source of well water.

With increase in scale the delivery ratios of sediment drop, and the effects of individual land treatments become less separable.

The measurement of flow height in stream channels and in flumes can be either by electronic recording of static water pressure or by flotation devices recording on a chart. Battery recharge for either type of recording can be by solar panel. Chart recorder drums can be driven either by clock mechanisms or electronic stepping motors.

The height of flow measured at a standard location within a flume is converted to a flow rate via the calibration curve relating flow height to discharge.

Samples for the measurement of water quality or sediment concentration can be taken manually, but this is not often feasible. Hence pumping samplers are commonly employed to obtain water samples at times related to flow characteristics. Since the type and duration of storm events is variable, there is difficulty in ensuring an adequate number of samples to characterize fully the runoff event. Because they sample a more consistent flow regime, most commercially available pumping samplers have an inadequate number of collection bottles to meet this need.

Another limitation of commercially available sampling equipment in soil erosion studies is that the size of sample collected is too small to provide an adequate mass of solids for a complete range of chemical analyses. The efficiency of sampling in the dynamically varying context of storm erosion events can be improved by altering the sampling frequency in a manner proportional to flow rate using computer control. An uncertainty in the design of such sample control programs is the likely total sampling duration required to describe the event, and this time period becomes increasingly uncertain as the scale of catchment increases.

Measurement at the Catchment Scale

A contributing area could be defined as a catchment when the characteristics of stream flow resulting from the confluence of two or more depression lines result in a flow path that has measurable hydraulic characteristics. Even accepting this criterion, the problem remains of site selection on the stream for suitable measurement of flow and water quality. At least in upper catchments, more linear stream segments may be found between meanders, and such a linear or riffle segment is a preferable measurement site since such segments can possess the following desirable characteristics:

- stable bank and bed materials;
- simpler flow dynamics that is typical of meanders;
- closer balance between sediment deposition and entrainment, thus providing a more typical example of sediment transport in the stream;
- for smaller flows, sampling by wading may be possible; and
- bridges are commonly located over such linear stream segments, and these existing structures can be utilized to support stream gauging and sampling equipment, thus avoiding the necessity of providing an alternative sampling support structure.

There is a well-developed suite of traditional flow velocity measurement and suspended sampling equipment. The appropriate technology for measuring bedload transport is more varied, reflecting the great variety of bedload materials. A useful reference on the above methods is Goudie *et al.* (1981).

However, it should be noted that recent advances in electronic equipment have been employed in the development of new types of flow velocity and depth measurement. An example of this is use of the measurement of Doppler signal shift for indicating flow velocity profiles and flow depth (Atkinson, 1976). Water velocity is measured acoustically by recording the Doppler shift using particles and microscopic air bubbles carried in water and corrected for the effects of water temperature. This equipment is attached to the stream bed, and thus will generally require some protection from burial by bedload, or damage from larger bedload components.

Both in streams and in flumes, water depth measurement can be efficiently and accurately made by measuring static water pressure using the electrical output of a pressure transducer coupled to a 32 bit logger. Problems due to clogging of inlet ports to the stilling well above the pressure transducer can be avoided by having multiple ports along the stilling well tube. More recent water depth measurement techniques include laser equipment.

Traditional methods of sediment sampling in streams require knowledge of the stream velocity profile, and vertical movement of the sampling device to provide a representative flow-weighted sample of sediment concentration. This involves a team of people, and is thus expensive, as well as presenting logistical problems to ensure personnel presence during the sampling event. An alternative, more automotive method of stream sampling during a storm event is to have a perforated sampling tube supported at its lower end by a hinge to the stream bed, and connected at its upper end to a float. When the float height exceeds some predetermined limit, samples can be pumped from the array of ports along the sampling tube; this array is designed to provide a good approximation to a flow-weighted sample. Pumped samples can be collected in bottles, and the sample-timing computer controlled. The sample-storing equipment can be refrigerated to preserve more transient chemical elements for later laboratory analysis.

Conclusions

Rising community interest in the quality of potable water from streams or groundwater wells has stimulated the interest of governments to support scientific studies of processes which control such water quality. Recognition of the longer-range transport of chemicals with suspended load sediments led to the desirability of separating bedload from the total sediment load; this is done most effectively using the recently developed tipping-bucket technology described earlier in this chapter.

A complementary source of stimulation to such field study of sediment and chemical transport has been recent developments in the theory of such processes, in particular drawing attention to the fact that it is the rates as well as the amounts which are important in interpreting and managing these processes. It is also now evident that such an understanding of processes provides a sound basis for suggesting a range of alternative practical management options for farmer evaluation. An important example of this flow-on from principles to processes to changing practices is in the pineapple industry in Queensland.

The advent of solid, static, computer-controlled, data acquisition equipment has made more feasible the remote and frequent logging of hydrological, sediment and chemical transport processes. These advances have also improved the quality and reliability of sensors under the trying field conditions typical of erosion studies in the field. All these types of development have greatly improved the feasibility of data acquisition at time scales related to actual physical processes, providing greater opportunity to test the realism of models seeking to describe such processes.

These improvements in data acquisition capacity have reduced the cost of labour and equipment in such studies, and removed the necessity of personnel presence during dangerous storm events.

A general reason why non-commercial development of equipment and computer programs has been so necessary and important in soil erosion studies is that commercial equipment has been designed for the higher-volume market, where there is less variation in time and variety of measurement type than is involved in studies of water, sediment and chemical transport. The comments made earlier concerning commercial water sampling equipment provide an example of the validity of this comment.

In order to develop sound and effective integrated catchment management strategies it is desirable to understand hydrological, sediment and chemical transport processes at a number of landscape scales. Hence follows the desirability of what is commonly referred to as 'nested' studies in which measurements are carried out at all the three scales discussed in this chapter. Study at each of the three scales – plot, subcatchment and catchment – provides synergistic information useful in the interpretation of processes at related scales, and thus to landscape and water quality management.

Acknowledgements

The authors gratefully acknowledge financial assistance from the Queensland Department of Primary Industries and the Australian Centre for International Agricultural Research in the development of the equipment described in this chapter.

References

Atkinson, P. (1976) A fundamental interpretation of ultrasonic velocity meters. *Ultrasound in Medicine and Biology*, Vol. 2, 288 pp., Perigan Press, London.

Bermel, K.J. (1950) Hydraulic influence of modifications to the San Dimas critical-depth measuring flume. *Transactions, American Geophysical Union*, 31(5), 763–770.

Bos, M.G. (1989) *Discharge Measurement Structures*, 3rd edn. *ILRI Publication No. 20*, Wageningen, the Netherlands.

Ciesiolka, C.A, Coughlan, K.J., Rose, C.W., Escalante, M.C., Mohd, G., Hashim, E.P. Paningbatan Jr. and Sombatpanit, S. (1995) Methodology for a multi-country study of soil erosion management. *Soil Technology*, 8(3), 179–192.

Goudie, A., Anderson, M., Bunt, T., Lewin, J., Richards, K., Whally, B. and Worsley, P. (1981) *Geomorphological Techniques.* George Allen & Unwin, London, pp. 203–205.

Hudson, N. (1973) *Soil Conservation.* Batsford, London.

Rodda, J.C. (1969) The assessment of precipitation. In: Chorley, R.J. (ed.) *Water, Earth and Man.* Methuen, London.

SCS National Engineering Handbook (1972) *Hydrology Section 4.* USDA, Washington, DC.

Wischmeier, W.H. and Smith, D.D. (1965) *Predicting rainfall-erosion losses for cropland east of the Rocky Mountains – Guide for selecting of practices for soil and water conservation. Agricultural Handbook No. 282,* USDA, Washington, DC.

Wischmeier, W.H. and Smith, D.D. (1978) *Predicting rainfall-erosion losses. A guide to conservation planning. Agricultural Handbook No. 537,* USDA, Washington, DC.

WMO (1965) Guide to Hydrometeorological Practices. World Meteorological Organisation, Geneva, 168,TP 82.

Yu, B., Ciesiolka, C.A.A., Rose, C.W. and Coughlan, K.J. (1997) A note on sampling errors in the rainfall and runoff data collected using tipping bucket technology. *Transactions of the American Society of Agricultural Engineers,* 40(5), 1305–1305.

19

The Use of GIS and Remote Sensing Techniques to Predict Erosion in the Nepal Middle Hills

R. Collins[1], A. Dickinson[2] and R. Johnson[1]
[1] *Institute of Hydrology, Crowmarsh Gifford, Wallingford, Oxon OX10 8BB, UK;* [2] *HR Wallingford, Howbery Park, Wallingford, Oxon OX10 8BA, UK*

Empirical relationships derived from erosion plot data are incorporated into a GIS framework to predict soil loss in the Middle Hills, Nepal. The region experiences high natural rates of soil loss exacerbated by the expansion of terraced agriculture and thinning of forests. Estimates of erosion are undertaken for individual rainfall events and quantify soil loss on a grid cell basis throughout a whole catchment. Material detached on the hillslope is subsequently routed towards the channel network using flow direction data in conjunction with a simple runoff velocity equation. The GIS framework enables the impact of land use and climatic change upon erosion to be assessed. The potential for using data derived from satellite imagery in erosion studies is explored; simple image processing techniques have discriminated the broad land-use types of the Middle Hills region and broad variations in soil erodibility can potentially be determined using imagery recorded in the dry season.

Introduction

In response to increasing population in Nepal, it is estimated that agricultural production will have to more than double by the year 2000 even though virtually all land suitable for cultivation is being used (HMG Nepal, Ministry of Forests and Soil Conservation, 1988). This population pressure is greatest in the Middle Hills region and has caused the expansion of cultivated land on to steep and marginal slopes. This has increased the destruction of forest cover to make way for agricultural terraces and to meet the increased demand for forest products for fuel and fodder. Associated with the land-use changes are a greater exposure of bare land, reduced infiltration rates and a loss of interception storage. As a result, agricultural terraces are susceptible to soil detachment by rain splash, particularly before each crop is fully established. The

(eds F.W.T. Penning de Vries, F. Agus and J. Kerr)

greater volume of runoff increases the eroding and transporting capacity of overland flow. Consequently, there is widespread concern that the naturally high erosion rates of the region are now exacerbated by agricultural practices which accelerate the loss of topsoil at an unsustainable rate. This may result in a decline in soil fertility and hence a decrease in crop productivity. In addition, increased hillslope erosion is likely to increase the sediment load of rivers, enhancing the potential for siltation of reservoirs and decreasing their operational life.

A significant increase in the application of inorganic fertilizers since the early 1960s has occurred throughout the region. There is evidence that large and possibly inappropriate fertilizer applications have increased the leaching of nutrients to catchment drainage waters, increasing the potential for eutrophication. A significant proportion of this nutrient flux is likely to be attached to sediments, increasing the importance of erosion processes in determining the quality of the water resource.

In response to these concerns, the UK ODA funded fieldwork that aimed, in part, to quantify soil erosion in the Nepal Middle Hills. The results from this study were utilized by the Institute of Hydrology to develop a GIS-based catchment management tool with which to predict soil erosion and sediment yield. In contrast to long-term approaches to predicting erosion, for example CALSITE, simulations are undertaken for individual storm events and are able to capture the seasonal dynamics of soil loss apparent in this region. A menu interface allows the user to define the rainfall characteristics of each application, enabling the impact of extreme events to be assessed.

An initial evaluation indicated that the USLE was unsuitable for prediction of soil loss from terraced agriculture in the Nepal Middle Hills. Specific land management practices are undertaken upon the terraces and these are not readily encompassed by the management and conservation factors of the universal soil loss equation (USLE). In addition, the narrow terraces are generally characterized by lateral drainage ditches and steep high risers. This morphology makes application of the slope length factor inappropriate. As a consequence of the limitations of the USLE, empirical field-derived relationships were used to predict erosion. These are site specific and application of the tool is limited to analogous regions characterized by a similar range of terrace types. This contrasts with CALSITE which, as a consequence of its incorporation of the USLE, is generally widely applicable with respect to annual soil loss prediction.

Erosion Plot Studies

In order to evaluate soil loss, 24 erosion plots were established under different land-use types in the Likhu Khola watershed, located 10 km to the north of Kathmandu. Full details of this work are provided by Gardner *et al.* (1995). The plots were located upon: rainfed and irrigated agricultural terraces; forestry, both established and degraded; grazing and scrub land; and bare degraded land. The plots encompassed a range of slope angles and two broad soil types.

On rainfed terraces, ditches beneath each terrace face remove runoff and eroded soil, directing it to gullies at the side of each terrace set. This practice is undertaken by the farmers to reduce the build up of overland flow on each terrace thereby reducing the potential for rill development. These lateral ditches were utilized to collect runoff and eroded soil, avoiding the use of artificial bounding. The flooded rice terraces precluded the use of conventional erosion plots. Instead the sediment flux in irrigation canals both feeding and draining terrace sets was monitored and it was established that these terrace systems trap soil. Discussions with local farmers revealed that in some areas of each catchment, rice terraces can trap up to a metre of soil over a 6-year period. The management of irrigated terraces therefore clearly aids soil conservation. On all other land uses, the erosion plots were fully bounded by metal sheets which were designed to follow the natural topography and drainage of the land.

Each plot or terrace was instrumented with 2 × 200-litre collecting drums with a 1/10 divisor between the drums, and in some cases an additional overflow container. The use of drum divisors enabled extreme storm events to be sampled; however, calibration was required to determine the exact fraction of runoff and sediment (*c.* 10%) passing to the second drum.

Runoff was measured, and samples of the thoroughly mixed suspended sediment collected from each drum after each monitored rainfall event. Samples were subsequently filtered, dried and weighed, and multiplied up by the volume of runoff to give estimated total soil loss for each event. All plots were located within 300 m of an automatic rain gauge, and most within 50 m. Daily manual rainfall collection was also undertaken.

In addition to sampling soil loss and runoff, soil moisture was monitored upon the rainfed terraces, forestry and scrub plots using tensiometers. These recorded the seasonal change in soil moisture and enabled processes of runoff generation to be inferred. Infiltration rates were also determined on each plot. These indicated that the lower rainfall intensities at which runoff was observed were often lower than the derived infiltration capacities. This discrepancy is explained by the formation of surface crusts under the impact of rainfall, a process not accounted for by the use of the double ring infiltrometer.

Seasonal variation

Seasonal differences in soil loss were apparent under terraced, grazing, scrub and degraded land. These reflect the lack of a fully established vegetative cover during the premonsoon months (April to June), when high erosivity rainfall can occur. By early July, the growth of crops, ground cover and algal mats provides significant soil protection during the high intensity monsoon storms. However, the harvesting of maize and planting of millet in mid-August renders the rainfed terraces susceptible to high soil loss over this period. In this respect, the practice of turning over soil and organic residues in preparation for the planting of millet is likely to be beneficial; it is undertaken by hand, leaving large aggregates intact and creating an irregular surface with a large volume of depression storage.

Changes in soil moisture also influence the seasonal variation in soil loss. Soil moisture levels are very low during the dry season and runoff is only generated by infiltration excess rainfall. The horizons gradually increase in moisture, from the surface downwards, over the premonsoon season. During the monsoon, tensiometer data indicate that the soil under all land-use types is at, or close to, saturation and runoff may be generated by both infiltration excess rainfall and the exceedance of the soil moisture storage capacity.

Soil type

Although a comprehensive soil mapping programme was undertaken in the Likhu Khola watershed, the soils data input to the GIS consists of just two broad categories. This generalization approach was taken as detailed soil maps in other areas of the Middle Hills region are unlikely to be available. In addition, logistical constraints prevented erosion plots from encompassing the full range of soil variability. Red, iron-rich soils are found extensively in each catchment. These are fine textured and relatively erodible with a low organic matter content. Buff-coloured soils are characterized by highly erodible, saprolitic material and are found in areas where mass movement has removed the surface soil horizons. Plots for each land-use type were located on both soil types.

Slope angle

Topographical mapping shows that approximately 80% of non-terraced land in the watershed lies upon 11–45° slopes. Slope angles on rainfed terraces are generally 5–10°. At this angle, runoff generated upon each terrace drains off, preventing flow from concentrating and so decreasing the potential for rills to develop.

The influence of season, soil type and slope angle upon erosion is accounted for in the suite of empirical relationships established between soil loss and rainfall intensity for each land-use type. Some of these relationships are illustrated in Table 19.1, and enable any combination of factors throughout a catchment to be represented by one of the loss equations. The relationships are derived for four ranges of storm magnitude (2–10 mm, 10–20 mm, 20–40 mm and >40 mm).

Model Development

Estimation of soil detachment

The development of the GIS-based modelling tool was undertaken using Arc/Info and involved the rasterization of land use, soil and topography data for each study catchment. Land use was digitized from 1:50,000 scale maps derived by field survey and analysis of aerial photography. Topography was digitized from 1-m interval contour maps and modified to derive slope angle.

Table 19.1. Soil loss relationships.

Land use	Slope any (%)	Season	Soil	Rainfall (cm)	Relationship (Sl in g m^{-2})
Rainfed terrace	5–10	Monsoon	Red	10–20	Sl = exp(10.4 × int.)
Rainfed terrace	5–10	Premonsoon	Red	10–20	Sl = 50.5 × int.
Rainfed terrace	5–10	Dry	Red	10–20	Sl = 12.5 × int.
Rainfed terrace	10–15	Monsoon	Red	>40	Sl = 125 × int.
Rainfed terrace	10–15	Monsoon	Red	20–40	Sl = exp(10.4 × int.)
Grassland	>30	N/a	Red	2–10	Sl = 0.1 × int.
Scrub	15–25	Monsoon	Red	10–20	Sl = 7.2 × int.
Scrub	15–25	Premonsoon	Red	10–20	Sl = 5.0 × int.
Scrub	15–25	Premonsoon	Red	>40	Sl = 0.5 *(exp(8.2 × int.)
Degraded forest	15–25	N/a	Buff	>40	Sl = exp(11.77 × int.)
Degraded forest	15–25	N/a	Buff	20–40	Sl = 111 × int.
Forest	15–25	N/a	Red	20–40	Sl = 22 × int.
Forest	15–25	N/a	Red	>40	Sl = 1.5 × (exp(5.756 × int.)

Sl, soil loss.

The detailed soils maps were digitized and reclassified and each grid cell represented by one of two soil types.

During a simulation, the land use, soil and slope grids are queried in an automated process using the Arc Macro Language (AML). This determines the attribute values of each grid cell enabling the appropriate loss equation to be applied. A menu interface was developed to enable the user to define the season and rainfall characteristics (intensity and magnitude) of each simulation. The rainfall values are currently applied uniformly across each catchment, although there is potential to incorporate spatial variability into the model. Both rainfall characteristics and season are used to solve each loss equation, enabling erosion to be quantified on a grid cell basis throughout a whole catchment. By the nature of the field sampling procedure these erosion values represent a combination of soil detachment and transport within the boundary of an erosion plot. In the subsequent sediment routing analysis the values represent a grid cell value of detached soil which has the potential to be transported.

The spatially distributed modelling approach provides a useful tool with which to identify those areas within a catchment that are particularly susceptible to soil loss. Once the location and severity of soil loss is quantified, remedial land management practices can be implemented. The GIS framework enables the impact of environmental change scenarios to be assessed as land use and soil data are easily modified by reclassification of each attribute grid.

Hillslope transport

Calculation of flow direction

Material detached on the hillslope is routed towards the channel network using rasterized flow direction data in conjunction with a simple runoff velocity equation. The incorporation of sediment routing within the model enables the original plot data to be scaled up. This allows crude quantification of the total material, derived only from hillslope erosion, which reaches the channel network during a given storm. It does not account for material derived from mass movement or bank erosion. Equal proportions of the total soil loss in a grid cell are transported in a number of time steps to simulate the movement of sediment during a storm. Future model development will enable the intensity and magnitude of rainfall to vary within each time step, providing more accurate representation of the transport process.

The initial step in calculation of flow direction involves the creation of a hydrologically correct surface from the altitude grid. This process enforces surface continuity by removing sinks that trap the simulated flow of water. The modified altitude grid is then utilized to compute the direction of flow from every cell to its steepest downslope neighbour. If the descent to all eight adjacent cells is the same, the neighbourhood is enlarged until a solution is found. During a simulation, sediment is moved from a cell to its steepest downslope neighbour if the transport velocity in that time step is such that the material is able to travel the distance of one grid cell. The total material potentially available for transport from each cell is the sum of deposition from neighbouring cells and the soil detached *in situ*. Each cell is potentially able to provide and receive sediment during the same time step, simulating both erosion and deposition.

Calculation of sediment transport velocity

The erosion plot studies collected runoff and sediment from rainfed terraces at the end of each lateral riser ditch. These ditches drain into ravines and gullies which run downslope, flanking each terrace set. Consequently, predicted soil loss from rainfed terraces, derived from the plot studies, is assumed to have reached the channel network by the end of each storm. Once in the channel, this material is not subject to further transport during a simulation; however, in-stream transport equations can be incorporated subsequently to route the material down each channel to the main river.

Soil eroded from non-terraced land is routed downslope using a simple sheetwash velocity equation. Transport occurs until a grid cell is reached in which a channel is located. The material is then assumed to enter the channel and sheetwash transport ceases. The limited number of erosion plots precluded the establishment of empirical relationships between rainfall and runoff velocity for an extensive range of slope angles. Consequently, the potential to incorporate velocity/transport equations from commonly used erosion models was assessed, for example:

1. AGNPS

$$V = 10^{0.5\times\log10 \quad S\times100 \quad SCC} \tag{19.1}$$

where V = velocity of overland flow (m s^{-1}); S = land slope; and SCC = surface condition constant (roughness parameter).

2. Manning equation

$$V = D^{0.66}\ S^{0.5} / n \tag{19.2}$$

where V = overland flow velocity (m s^{-1}); D = depth of flow (m); S = slope (m m^{-1}); and n = surface roughness.

3. Beasley *et al.* (1980)

$$TF = 161 \times S \times Q^{0.5} \qquad \text{if } Q \Leftarrow 0.046\ \text{m}^2\ \text{min}^{-1} \tag{19.3}$$

$$TF = 16320 \times S \times Q^{2} \qquad \text{if } Q > 0.046\ \text{m}^2\ \text{min}^{-1} \tag{19.4}$$

where TF = potential transport rate of sediment (kg min^{-1} m^{-1}); S = slope; and Q = runoff flow rate (m^2 min^{-1}).

The relationships of Beasley *et al.* (1980) were chosen to compute the sediment transport capacity of overland flow. These relationships utilize slope angle and a crude estimate of runoff flow rate (derived from the plot studies) to calculate the velocity of sheetwash prior to its concentration in channels. The first equation is valid for laminar flow and the second for turbulent flow, the former being more applicable to the sheetwash evident on the erosion plots. These relationships are based upon an equation by Yalin (1963), who assumes that the mechanism of sediment transport by a shallow flow (sheetwash) is similar to the mechanism of bedload transport in channels and that a critical shear stress exists acting on the soil at the beginning of sediment transport. However, use of this approach introduces error into the GIS estimates of transport because the loss relationships are derived from soil that has been transported to a collecting drum and therefore has already overcome the initial shear stress. However, all these simple transport equations necessitate gross assumptions, for example, runoff volume may not necessarily be incorporated into the calculation of velocity, and all particles, regardless of size, are assumed to be entrained. In addition, roughness coefficients necessitate each grid cell to be represented by one lumped, literature-derived value.

Soil loss calculated from the plot-derived relationships represents material detached by both rain splash and overland flow within the confines of the erosion plot. It is assumed that during subsequent movement by sheetwash, represented by movement to a neighbouring grid cell, no further soil detachment occurs.

Although a very simple representation of sediment transport is incorporated into the GIS, it is only applicable to sheetwash over non-terraced land. In addition, transport only occurs over relatively short distances as the extensive drainage network soon concentrates flow. The simple representation is advantageous with respect to data requirements; only two input variables are required, enabling the model to be potentially widely applicable throughout the region.

Application to the Langtang Catchment

Although the predictive tool was initially developed using soil loss relationships established in the Likhu Khola watershed, the relatively uniform landuse, geology and soil type of the Middle Hills region suggests that more widespread application may be possible. Subsequent application has focused upon the Langtang catchment about 60 km north-west of Kathmandu. The Langtang is currently the focus of an Institute of Hydrology research programme which aims to improve water resource assessment in Himalayan rivers. The research aims to develop a methodology with which to characterize and quantify fluvial sediments. This includes the quantification of each sediment source, which necessitates the prediction of hillslope erosion. With the exception of ice and rock in its upper reaches, the Langtang has a very similar range of land-use types to that of the Likhu Khola. The underlying geology of each catchment is similar, dominated by gneisses, and both catchments are dominated by the same two broad categories of soil type. Consequently, the soil loss relationships developed in the Likhu Khola were assumed to be generally applicable to the Langtang.

Figure 19.1 illustrates the predicted soil detachment for the Langtang catchment for each 500 × 500 m grid cell during a typical monsoon storm. The values reflect a simulation with a mean rainfall intensity of 0.2 mm min^{-1} and magnitude of 20–40 mm. Soil detachment is greatest from areas of badly degraded forest, up to 350 kg ha^{-1}. Steeply sloping scrub land and some rainfed terraces are subject to relatively high losses, up to 50 kg ha^{-1}.

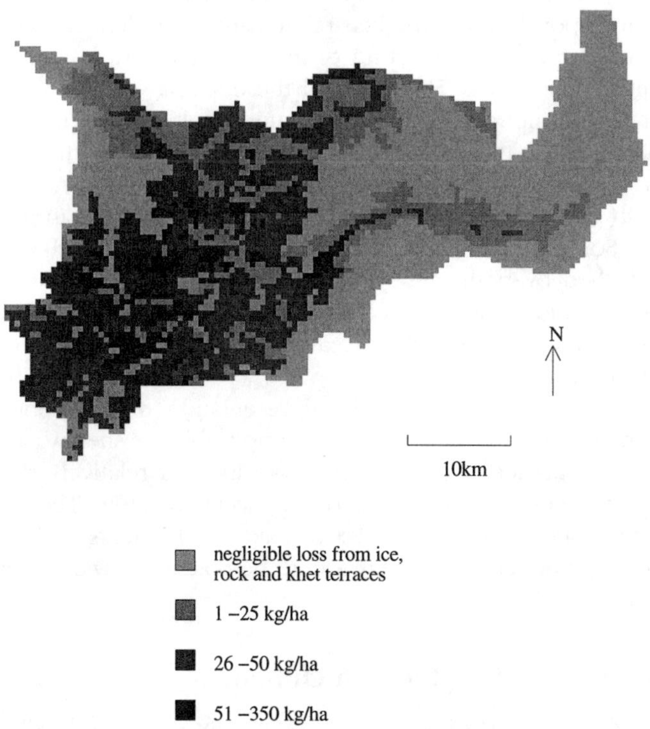

Fig. 19.1. Hillslope sediment detachment – typical monsoon storm.

Extrapolation of soil loss from each land-use type over a typical wet season (premonsoon and monsoon) rainfall pattern indicates that erosion ranges from less than 0.01 t ha^{-1} from forestry and grassland to 20 t ha^{-1} from degraded forest and bare land.

Figure 19.2 shows the results of routing soil, detached during an extreme monsoon event, down the hillslope through four time steps. Sediment gradually

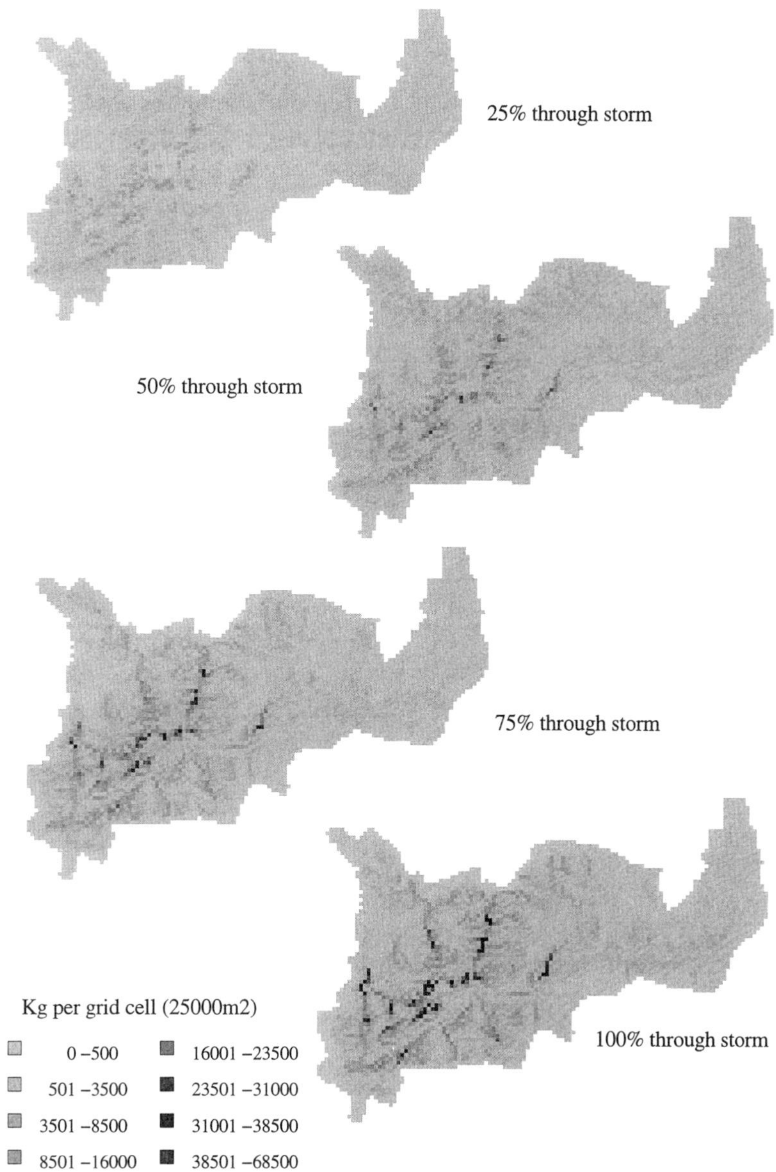

Fig. 19.2. Hillslope sediment detachment – Langtang catchment.

concentrates within the rills, gullies, ravines and streams of the catchment. Although the velocity equations lead to only a crude estimation of sediment transport, their inclusion in the GIS provides a useful qualitative assessment of areas of erosion and deposition within a catchment. Input files, defining the rainfall characteristics of a series of storm events, can be read into the GIS, enabling assessment of erosion and sediment redistribution over the course of a season.

Validation

Although the single event application enables the impact of rainfall variability upon erosion to be assessed, validation of the hillslope sediment predicted to reach the channel network during that event is particularly difficult. Stream sediment derived from the hillslope source is not easily discriminated from material derived from landslides or bankside erosion. In rivers with glaciated headwaters, the heterogeneous sediment supply may entirely mask any diagnostic property that each sediment source may possess. The difficulty in quantifying sediment sources is unfortunate since it limits the effectiveness of conservation measures and land management strategies.

Long-term monitoring programmes can establish a sediment delivery ratio; however, quantification over individual events requires specialist techniques. Fallout radionuclide concentrations offer considerable potential in this respect, the source 'fingerprints' are independent of lithology and soil type and can clearly distinguish surficial and channel sources. The method has been successful in identifying inter- and intra-event variations in sediment source (Walling and Woodward, 1992).

The Application of Remote Sensing to Erosion Studies

There is potential for regional application of erosion models through the use of land-use, topographical and soils data derived from remote sensing imagery. Acquisition of such data may be particularly valuable in areas where logistical and analytical constraints preclude the use of field-derived data.

Simple image processing techniques have been relatively successful in identifying the major land-use types in the Nepal Middle Hills. Unsupervised classification of a Landsat Thematic Mapper image (recorded in January 1989) was undertaken and the resulting classes identified by calculating a normalized difference vegetation index (NDVI), (Rouse *et al.*, 1973). The NDVI is the difference between radiances in the near infrared and the red region of the spectrum. Green vegetation has a high reflectance in the infrared and a low reflectance in the red; for non-vegetated surfaces the reverse is true. This process clearly discriminates forest, vegetation and bare earth (Fig. 19.3) and the patterns broadly reflect the land use mapped from field survey; forestry is confined to high-altitude interfluve areas, vegetation to the steep valley sides and the bare earth areas to the valley bottoms. At the time of year the image was recorded, little or no rice was grown and a mustard crop was only grown on a limited number of rainfed terraces. The bare earth areas therefore reflect

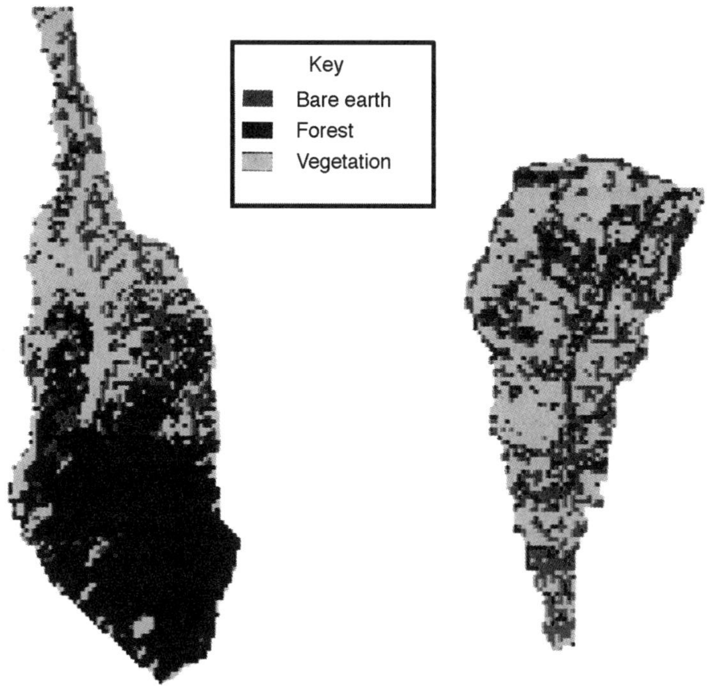

Fig. 19.3. Classification of land use from satellite imagery, Middle Hills, Nepal.

non-vegetated terraces. Imagery recorded in April or May is potentially better able to distinguish between rainfed and irrigated terraces as the respective maize and rice crops become established. Such discrimination is essential if imagery is to be used to parameterize erosion models in regions with similar types of terracing.

The dry season imagery reveals much of the floodplain sediment that subsequently becomes transported during monsoon flows. Image processing and ground truth surveys have enabled the discrimination of sediment size at the confluence of the Likhu and Tadi rivers, north-west of Kathmandu. Coarse boulder/pebble size material is heaped between braided channels, whilst sand and soil deposits mark the banks of the monsoon flow. This smaller material is often associated with temporary dry season terraces. Initiation of downstream transport will occur at different flow thresholds for each sediment size. Consequently, the imagery may be used to parameterize in-stream sediment transport models and enable crude estimation of sediment load for broad flow ranges.

The Earth resources satellite, SPOT, incorporates a mirror in the optical path of its sensors enabling stereoscopic coverage to be obtained by viewing one area in successive orbits. Topography can therefore be accurately mapped from the overlapped region of the two images and, subsequently, digital terrain models and flow direction derived. Detailed topography can discriminate slope form and morphology, factors which strongly influence land failure and hence sediment supply.

The identification of soil properties

Remote sensing offers the potential to determine broad variations in a number of soil physical and chemical properties. However, with the exception of images derived from radar, this requires images to be recorded on relatively cloud-free days from areas where the soil spectral response is not extensively masked by vegetation. Soil reflectance generally increases with wavelength, although significant differences may exist between soil types due to variation in the key factors that influence reflectance properties, such as moisture content, organic content, texture, structure and iron oxide content (Hoffer, 1978).

The structure of clay soils encourages moisture retention giving rise to a relatively low reflectance. In contrast, the higher porosity of sandy soil aids drainage, decreasing relative moisture content and resulting in a high reflectance. In visible, near and middle infrared wavelengths, the presence of soil moisture reduces the surface reflectance of soil. This is accentuated by water- and hydroxyl-absorbing wavebands that absorb at 0.9, 1.4, 1.9, 2.2 and 2.7 mm. This absorption is marked in clay soils as these can have structurally bound water and very strong hydroxyl absorption properties. Soil moisture variations can therefore be used indirectly to derive soil structure, a key factor in determining the erodibility of a soil. The acquisition of imagery at a suitable temporal scale can identify diurnal and seasonal changes in soil moisture. This can be used to determine available soil moisture capacity, an important factor in determining the time taken for runoff generation to occur.

Soil organic matter decreases soil reflectance. An increase in organic matter from 1 to 5% has been shown to decrease reflectance from 50 to 25% (Page, 1974). The identification of soil structure, texture and organic matter content from images indicates that broad variations in soil erodibility can be determined, providing data with which to parameterize regional scale erosion models.

In the Likhu Khola watershed, soil colour generally reflects broad variations in erodibility; the red soils have a higher clay and organic matter content and lower erodibility than the buff soils. Iron oxide is the principal cause of the red coloration and this may therefore be used to identify soil erodibility. Iron oxide selectively reflects red light and absorbs green, hence a ratio image of the red to green reflectance can discriminate variations in soil iron oxide content. Field sampling and ground truth is then necessary to derive a relationship between iron oxide content and erodibility. Ratio images, cross-referenced to soils maps and chemical analyses, are currently being used to assess the potential application of this technique to the soils of the Nepal Middle Hills.

Conclusions

A GIS-based predictive tool has been developed with which to assess soil erosion in the Middle Hills of Nepal. This region has high natural rates of erosion, currently exacerbated by the thinning of forests and an expansion of agricultural terraces in response to population growth. Erosion plot studies established empirical relationships between soil loss, land use, rainfall characteristics and soil type. These are incorporated into a GIS framework enabling quantification of soil loss throughout a catchment during one or several rainfall events. This process identifies those areas within a catchment that are particularly susceptible to soil loss, allowing appropriate conservation measures and land management strategies to be undertaken. A menu interface allows the user to define the rainfall characteristics of each application and hence the impact of extreme events can be assessed. In addition, the GIS framework enables the effect of land-use change upon erosion to be determined. Material detached on the hillslope is routed towards the river channel using a simple runoff velocity equation in conjunction with a flow direction grid. This enables the redistribution of sediment throughout a catchment to be modelled.

Data derived from satellite images provide the potential to apply erosion models upon a regional scale. Simple image processing techniques have identified the major land-use types in the Nepal Middle Hills and current research is attempting to estimate soil erodibility. The discrimination of different floodplain sediment deposits suggests that the imagery can also provide data for in-stream sediment transport models.

References

Beasley, D.B., Huggins, L.F. and Monke, E.J. (1980) ANSWERS: A model for watershed planning. *Transactions of the American Society of Agricultural Engineers* 23(4), 938–944.

Gardner, R., Thapa, K. and Tripathi, R. (1995) Soil erosion from overland flow on hillslopes. In: Gardner, R. and Jenkins, A. (eds). *Land Use, Soil Conservation and Water Resource Management in the Nepal Middle Hills.* Commissioned report for ODA.

HMG Nepal, Ministry of Forests and Soil Conservation (1988) *Master Plan for the Forestry Sector.* HMG Nepal, Kathmandu.

Hoffer, R.M. (1978) Biological and physical considerations in applying computer-aided analysis techniques to remote sensing data. In: Swain, P.H and Davis, S.M. (eds) *Remote Sensing the Quantitative Approach.* McGraw Hill, pp. 227–289.

Page, N.R. (1974) Estimation of organic matter in Atlantic coastal plain soils with a colour-difference meter. *Agronomy Journal* 66, 652–653.

Rouse, J.W. Jr., Haas, R.H., Schell, J.A. and Deering, D.W. (1973) Monitoring vegetation systems in the Great Plains with ERTS. *Earth Res. Techn. Satellite-1 Symp., Goddard Space Flight Center, Washington DC,* pp. 309–317.

Walling, D.E. and Woodward, J.C. (1992) Erosion and sediment transport monitoring programmes in river basins. *Proceedings of the International Symposium, Oslo, Norway, 24–28 August 1992.* International Association of Hydrological Sciences, Wallingford, UK, 210.

Yalin, S. (1963) An expression for bed-load transportation. *Journal of Hydraulic Engineering,* 89(3), 221–250.

20

Predicting Erosion and Sediment Yield at the Catchment Scale

A. Dickinson[1] and R. Collins[2]

[1]HR Wallingford, Howbery Park, Wallingford, Oxon OX10 8BA, UK;

[2]Institute of Hydrology, Crowmarsh Gifford, Wallingford, Oxon OX10 8BB, UK

The development of geographical information systems (GIS) has facilitated the prediction of soil loss in a spatially distributed manner. A range of data types can be incorporated to provide an integrated approach to catchment studies. In addition, digital terrain models can be utilized to determine hydrological pathways, enabling sediment hillslope transport to be incorporated into erosion models. This chapter describes a GIS-based approach to predicting erosion and sediment yield at the catchment scale; CALSITE, a user-friendly software package has been designed to predict soil erosion rates and sediment yields from limited data, enabling it to be applied widely.

Introduction

CALSITE, (**cal**ibrated **si**mulation of **t**ransported **e**rosion) is a catchment planning tool for identifying the main sources of soil erosion and sediment yielded to downstream structures such as reservoir or irrigation off-takes, and for predicting the effects of land-use change on soil erosion and sediment yield. CALSITE uses a geographical information system (GIS) approach to analyse a catchment on a grid cell basis. It has been developed in response to the need to improve sediment yield predictions and to provide answers to the following questions: (i) why is it that large catchments produce lower sediment yields than subcatchments, (ii) where is the sediment coming from within the catchment, and (iii) what will be the effect of land-use changes in the catchment on future sediment yield?

(eds F.W.T. Penning de Vries, F. Agus and J. Kerr)

Measuring Sediment Production from Subcatchments Compared with Plots

Much research work on erosion measurement has been carried out using plots, which provide valuable data to compare erosion rates between different crop covers and soil types. Plots are typically rectangular, being about 20 m long and 2 or 3 m wide. Unfortunately, measurements at this scale cannot be 'scaled up' to evaluate erosion for a whole catchment. There are many other effects, which are evident at the larger scale, that cannot be determined from plot data; for example, it is impossible to assess gully and stream bank erosion on plots. Care must also be taken to ensure that the scale of the parameter being tested is small enough for the experiment to be valid; there is no point, for example, in trying to identify the effects of tree spacing if there are only one or two trees to the plot.

To evaluate subcatchment erosion, catchments of 20–30 ha in area are monitored. These are generally characterized by one natural stream outlet where a gauging structure may be constructed. This scale of measurement is used to gain information about sediment yield, crop yields, runoff rates and nutrient loss. Subcatchments may also provide valuable data on the effects of various catchment management practices, such as tree spacing and contour terraces.

At the single, well-defined, catchment outlet channel it is also possible to construct a rating structure (e.g. a structure in which the discharge is directly related to the depth of flow), allowing reasonably accurate determination of runoff and sediment yield. Because sediment monitoring is also required, extreme care must be taken to ensure that the design of structure chosen will allow all the sediment to pass through while still maintaining a known discharge rating throughout a wide range of flows.

HR Wallingford uses a modified 'H' flume with sloping bed (in order to keep the structure self-flushing) and a downstream drop structure to create a turbulent zone (Fig. 20.1), which provides an ideal location to collect pumped samples of water and sediment (Crickmore and Aked, 1975; Amphlett and Dickinson, 1989). The automatic sediment sampler developed by HR Wallingford allows sampling of flood events at any time. A pump mounted in the turbulent drop zone between two wing walls ensures samples can be collected from most storms including those with low discharge. Discrete water and sediment samples are pumped into a maximum of 24 numbered, half-litre bottles at preprogrammed time intervals (e.g. from 1 min to several hours), the whole sampling sequence being triggered when water reaches a predetermined level in the flume. The bottles are checked daily and any samples collected returned to the laboratory for determination of sediment concentration. If these data are to be converted into sediment flux (g $m^{-2}s^{-1}$), the instantaneous discharge at the time each sample is collected must be known. This can be done by automatically logging the sample time on the water level logger.

Fig. 20.1. HR Wallingford pumping sampler at an HL flume in the Philippines.

Why is it Large Catchments Produce Lower Sediment Yields than Subcatchments?

Studies of large catchments are necessary to provide information for basin management projects and to assess the capability and life of possible reservoirs. At this scale, it is necessary to monitor river discharges and sediment loads from a catchment of more than 10 km^2, and often greater than 200 km^2. In these cases, construction of a flow gauging structure is unlikely to be practical due to cost, head, levels of sediment and discharge, and scale of the catchment outlet.

Large catchment measurements

Measurements of sediment yield from plots or small catchments cannot be directly extrapolated to large catchments, since the effect of the sediment delivery ratio (see Sediment delivery effects below) is not easily quantifiable. There is also no reliable means of estimating delivery ratios from climatic and catchment characteristics other than those previously studied. To measure at this scale a different set of techniques has been devised, which require quantification of two main parameters – stream discharge and sediment flux.

There are a number of tried and tested methods for measuring river discharges (such as International Standards Organisation ISO 748, 2nd Edn, 1979), which all demand a careful selection of the monitoring site. This site should be on a straight and stable reach with a 'control' section to provide a fixed relationship between discharge and stage (water level). In practice, it is often difficult to find a suitable site and, frequently, velocity and cross-section measurements must be undertaken for the duration of the sediment study. Current metering can be used to carry out these measurements independently or in conjunction with sediment sampling. Velocity data can be collected by using a current meter suspended from a boat or cableway throughout the whole range of flows. Because of problems in monitoring high discharges, extrapolation of the measurement to higher values is often necessary. Once the stage/discharge relationship has been established, a simple record of the variation of water level (stage) with time will provide the required discharge data. The sediment monitoring technique employed is another form of pump sampling technique (Fig. 20.2). Since the samples are to be collected from a natural river, the means whereby a representative sample (or set of samples) may be obtained must be considered carefully. Previous research studies have clearly shown the need to consider bed material load and wash material load separately when collecting and analysing samples. This is because different hydraulic factors govern their transport rate in natural rivers. Wash load is expected to be uniformly distributed through the cross-section and is thus readily calculated as the product of a single measured concentration and mean discharge at the time of sampling. Suspended bed material particles are not uniformly distributed throughout the depth, as the concentration normally decreases with height above the bed. Measurements are thus required at a

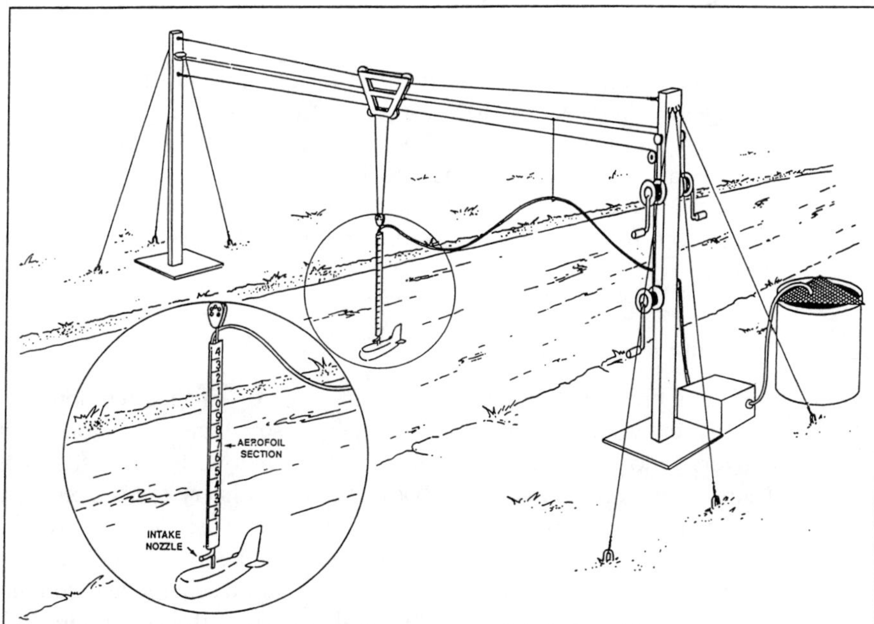

Fig. 20.2. Cableway manual pump sampling system.

number of discrete points in a vertical, in order to define the sediment concentration profile. The distinction between wash load and bed material load is not well defined, an arbitrary split being made at 63 μm (Dickinson and Bolton, 1992).

Many practitioners have tried using sediment ratings to extend the periods of measurement to estimate for periods of missing data. Inaccuracies in the predictions may occur since sediment flow relationships exhibit hysteresis effects and large scatter (Lawrence, 1996).

Even when successful measurement has been undertaken for a number of years, the annual yields determined may still be considerably far from the long-term average annual yields.

Sediment delivery effects

The quantity of sediment that is exported from a catchment is generally much smaller than the amount indicated by erosion rates measured or estimated from contributing small land areas. This is due to sediment storage in the catchment; deposition on the land surface, at the base of slopes, on valley floors, and on the beds and banks of channels.

The sediment delivery ratio is defined as the sediment yield at a catchment outlet divided by the summed erosion rates from all contributing areas. There are many formulas in the literature that predict sediment delivery from one or more parameters, mostly based on data from the United States. Walling (1983) reproduces a number of delivery functions based on catchment area, which are appropriate to different rivers and this is reproduced in Fig. 20.3. Functions derived from data collected in different regions produce very different relationships. This is not surprising as the processes of erosion, transport, deposition and re-erosion that occur as sediment moves through the fluvial system to a catchment outlet are both complex and controlled by site-specific factors.

The most important of these is the size range of the sediment being eroded and transported. Delivery ratio will be large for the finest sediments, and low or zero for the very coarse sediments. The sediment mixture measured in a flood flow in a river may thus consist of fine sediments eroded from the land and transported to the catchment outlet during the storm event being measured, and coarser sediments that were first eroded from the land surface or from channel banks decades or even centuries earlier, and have been deposited and re-eroded many times before reaching the catchment outlet. Delivery is thus dominated by size-related factors that are not accounted for in simple delivery relationships (Atkinson, 1992).

Many other factors influence delivery ratio such as catchment area, catchment steepness, basin length, maximum elevation difference and stream bifurcation ratio. Many of these parameters are not readily available.

Nutrient enrichment

The problems of sediment measurement and delivery can be further compounded when it comes to look at nutrient losses from a catchment. Work

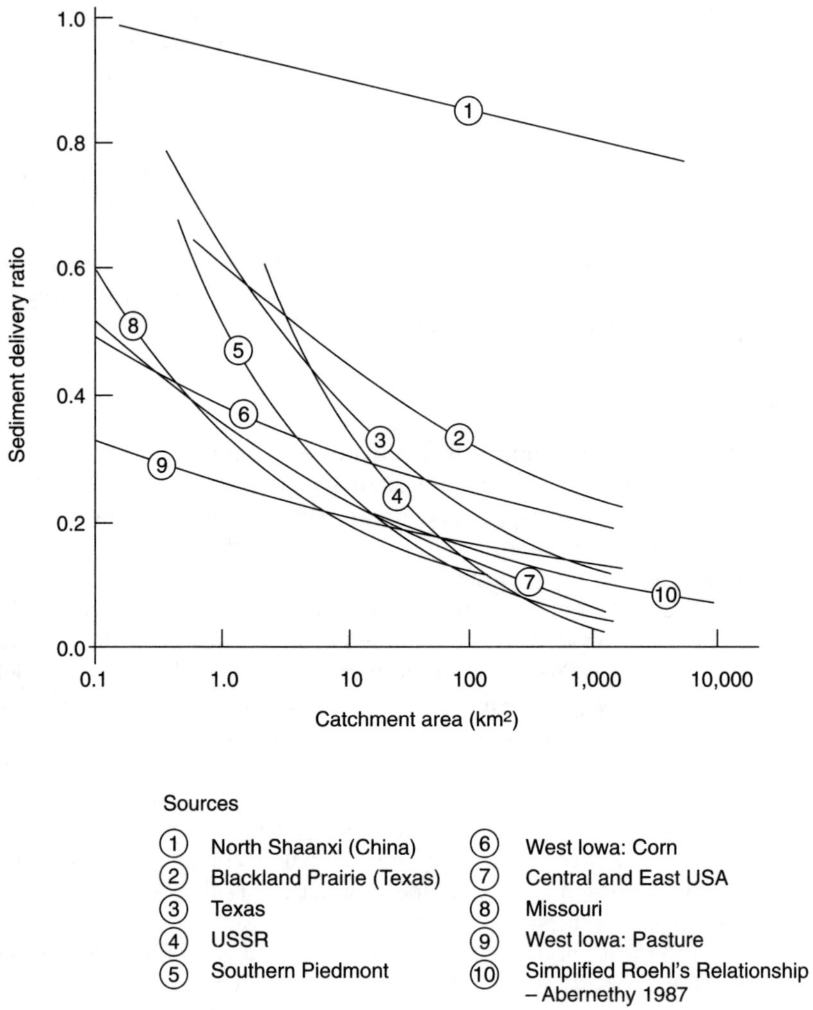

Fig. 20.3. Sediment delivery ratios.

in southern Africa has shown that an enrichment can occur whereby the concentrations of a given nutrient yielded by the catchment, in a given volume of sediment, are much greater than those found in an equivalent volume of non-eroded soil, *in situ*, in the catchment (Stocking, 1988). This relative increase, often termed the nutrient enrichment, can result in a need to know the relative yields of soil with high enrichment ratios to those of low enrichment.

Sediment yield determination

Thus, long-term estimates of sediment yield in large catchments can be highly inaccurate due to problems in determining long-term yield, because year to year variation can be highly significant (Lawrence, 1996). These problems may be compounded by changes in land use, which will also affect the long-term average rates. The situation may be further complicated by the need to know potential enrichment ratios of certain chemical elements such as N, P and K, of transported material.

Reservoir surveys

A further method with which to estimate catchment-wide, mean sediment yield is to run a series of comparative surveys of the capacity of a suitable reservoir every few years after impoundment. The accuracy of this method is dependent on the number of range lines surveyed and the method used to calculate the deposited sediment volume. HR Wallingford has considerable experience in such work and has developed the RESservoir Survey Analysis and Sedimentation Simulation (RESSASS) to aid in accurate volume and sediment yield calculation and long-term accretion simulation.

Where Does the Material Measured Come From?

If attempts are to be made to reduce sediment yields, it is important to identify and quantify sources of erosion. CALSITE was specifically developed to do this.

Many models attempt to estimate location and erosion rate. These range from complex physically based simulation models to the application of simple empirical predictors. At present the use of complex simulation models is restricted to research catchments as the data required to set up and run the models is unavailable elsewhere. At the other extreme, simple empirical methods, which can be applied with limited data, do not account for complex processes such as sediment or plant nutrient loss (N, P, K), transport, deposition and re-erosion, and have a very low accuracy when conditions differ from those for which the methods are calibrated. Simulation of such processes is important as they can have significant impacts on the redistribution of soil and nutrients. CALSITE does not model all processes but gets around this problem to some extent by using calibration data based on measurements which encompass the cumulative effects of these processes.

Since erosion rates will vary from place to place, it is necessary to repeat the estimation process many times over the complete catchment area. For many areas in the developing world, little or no information is available and hence for general application a limited data set must be used. For the design of CALSITE it was felt most appropriate to develop a tool offering a consistent technique applicable to any location in the world. The aim of the method is to model some of the processes described above through spatial calculation, visual presentation, and with appropriate but minimal input data.

This is where GIS-based techniques offer great power in speed of spatial calculation and benefit in easy visualization of spatial data. GISs come in two main forms, vector-based systems and raster-based systems. Vector-based systems consist of a list of coordinates which define lines or polygons as shown below. This approach offers high spatial accuracy and low data storage. However, they are complex to use, slow in computation speed and generally more costly. Raster-based systems store data in a grid cell format (pixel format) as shown in Fig. 20.4, so that points, lines and enclosed areas (polygons) within a map need to be converted into a collection of pixels (grid cells) to represent the point, line or area. Each pixel in a particular area is given the same identification number. Generally, remotely sensed data from satellites or aerial photography are stored in a raster format. This format is normally of lower spatial accuracy, simpler to use and faster to process on low-cost computers. It does, however, require greater data storage space.

CALSITE uses a raster-based GIS, which requires minimal input data, and because of this assumption, the output data does not need to be highly spatially accurate. This also minimizes the cost of the system, allowing it to be more widely applied. CALSITE utilizes the low-cost IDRISI GIS package for the preparation of input map data and for map display and has been designed to run on PCs of moderate capability (Eastman, 1992a,b, 1993). Map information on rainfall, soils, land use and topography is digitized within IDRISI then used within CALSITE to determine soil erosion for each pixel.

To determine soil erosion, two existing models have been included in CALSITE. These are the USLE (universal soil loss equation) and SLEMSA (soil

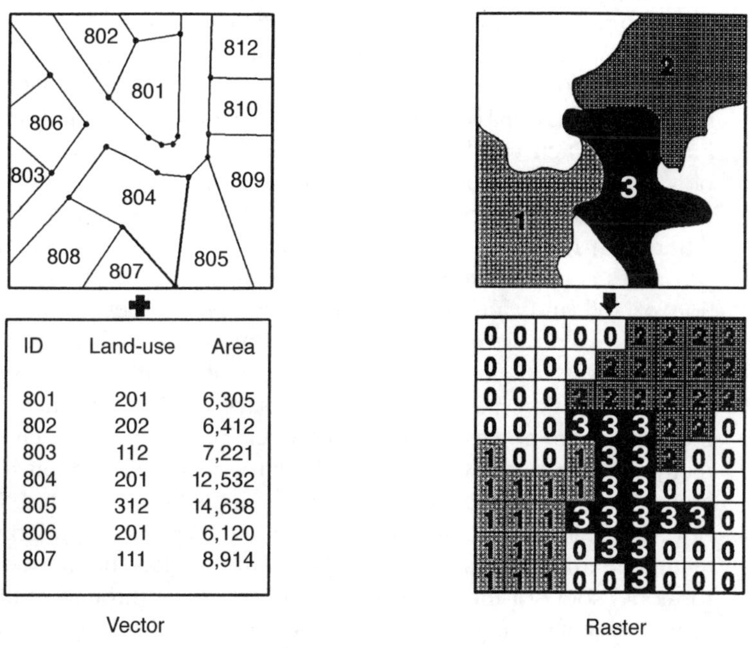

Fig. 20.4. Comparison of vector- and raster-based data (reproduced from Eastman, 1992b).

loss estimator for southern Africa) (Fig. 20.5). The latter is only applicable to catchments in southern Africa, being originally developed for use in Zimbabwe. The USLE model was developed for use in the USA but many of the empirical factors used have been derived for other countries throughout the world.

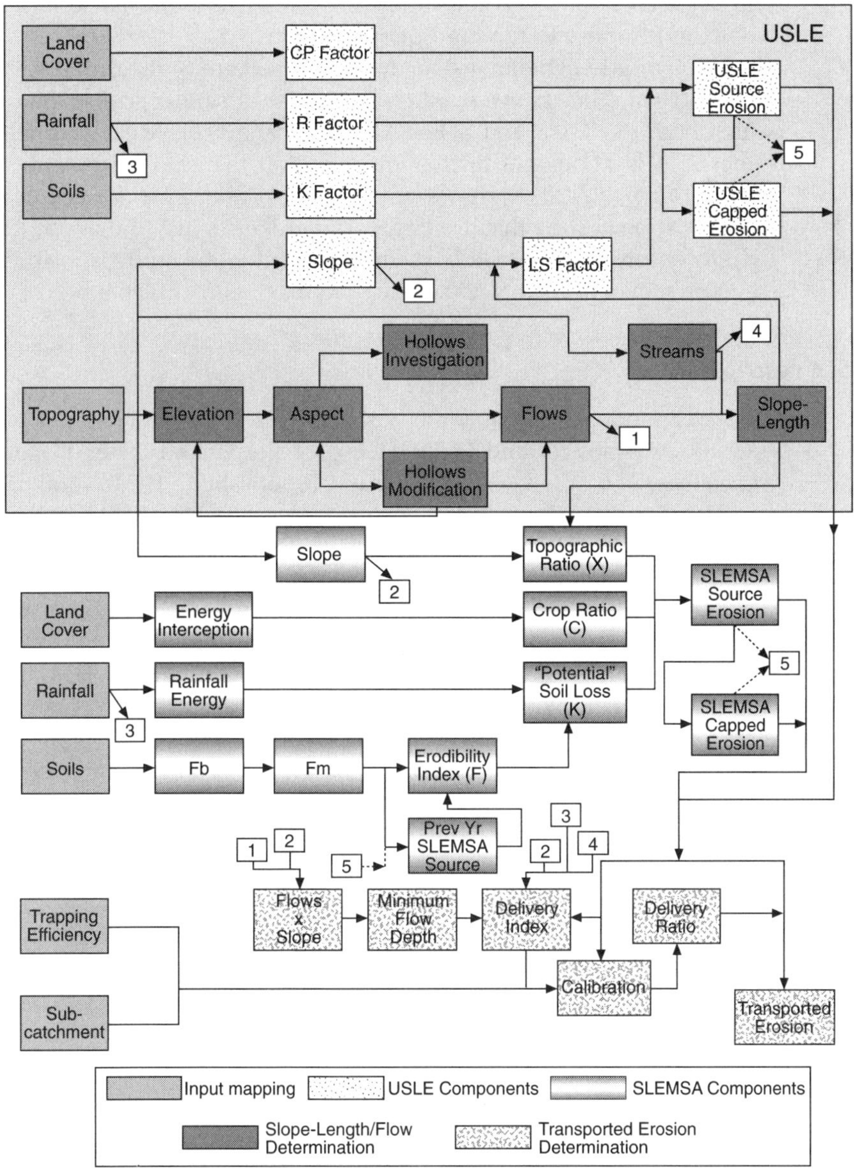

Fig. 20.5. The USLE and SLEMSA models incorporated within CALSITE.

CALSITE combines these soil erosion models with sediment routing techniques in order to predict annual sediment loads. The equations are calibrated by comparing estimated values with measured values at one or more catchment or subcatchment outlets. Two key outputs are produced by CALSITE, first, a map depicting the amount of soil loss from each grid cell in the catchment, and second, a map of sediment yield is produced showing the smaller quantity of soil from each cell which reaches the catchment outlet. The latter takes into account the sediment delivery and is explained further below, in response to the third question.

Pixel, or grid cell dimensions typically used are of the order of 25 × 25 m to 50 × 50 m. These have been chosen to be of similar size to the plots from which both the USLE and SLEMSA model parameters were determined. The upper size is about the maximum definition that can be determined on a 1:50,000 map. CALSITE allows any size of pixel to be used, but error will increase at pixel sizes that are larger. Larger values may, however, have to be used to ensure that the whole catchment used in the study is included within the approximately 1000 × 1000 pixel grid maximum available size of CALSITE.

USLE description

The USLE model is probably the best-known soil loss estimator. It was developed by Wischmeier and Smith (1965) of the United States Department of Agriculture, and was later updated and republished (1978). Modifications to the underlying methods have occurred, resulting in the MUSLE (modified USLE) and other versions. It has been widely applied throughout the world following research in many countries to determine local parameter values.

The approach is a statistical correlation type with relationships being derived from research studies to fit data from almost 10,000 plot-years of experiments on sites east of the Rocky Mountains in the USA. Data were restricted to small plots, and to low slopes where cultivation is permissible in the USA.

The output from the equation is an estimate of long-term, average annual sheet and rill erosion. The equation cannot be used with confidence for individual storms or even individual years with hope of success.

The formula is

$$E_a = R \times K \times LS \times C_m \times C_p \qquad (20.1)$$

where:

E_a = estimated average annual soil loss (t ha^{-1})

R = rainfall erosivity factor (MJ ha^{-1} × mm h^{-1}) = EI_{30}

where:

E = kinetic energy of the storm (MJ ha^{-1})

I_{30} = maximum 30-min rainfall intensity (mm h^{-1})
(Note : We are interested in the average long-term value)

K = soil erodibility factor (t.h(MJ.mm)$^{-1}$)
(Soil loss per unit of rainfall erosivity index, from bare fallow ground on a 9% slope of length 22.1 m)

LS = slope length and steepness factor (dimensionless)
where:
L = slope length (m)
m = an exponent
S = angle of slope

Recommended values for m (Wischmeier and Smith, 1978) are:
m = 0.5 if slope > 5%
m = 0.4 if 5% > slope > 3%
m = 0.3 if 3% > slope > 1%
m = 0.2 if slope < 1%

C_m = cropping and management factor (dimensionless)
(The ratio between soil loss under specified crop-cover management conditions and that under the worst conditions, i.e. cultivated fallow land). Commonly referred to as the *C* factor.

C_p = conservation practice factor (or erosion-control-supporting-practice factor) (dimensionless)
(The ratio which compares soil loss with that from a field with no conservation practice, i.e. ploughed up and down the slope. With no conservation methods C_p = 1.0) Commonly referred to as the *P* factor.

Several modifications have been made to the USLE, usually involving a change to just one of the five factors. Two of the most widely known modifications are by Onstad and Foster (1975) and Williams (1975, 1981).

The range of conditions to which the USLE can be applied with confidence is well defined and potential users must ensure that their particular application does not cross these boundaries (Wischmeier, 1977). A problem may arise in that the name 'universal' suggests that the model can be used in any situation, anywhere in the world. Wischmeier (1972) defends this name, saying that extensive research in the USA on the interrelationships and form of the five parameters in the USLE suggest that, if they are properly calibrated for an area, they should be applicable anywhere. However, this does not provide a complete answer as there have been problems in applying the USLE in parts of the USA (McDole and Reinersten, 1983), as well as elsewhere (Wendelaar, 1978), even after careful calibration has been carried out. It appears that some of the factors (especially *R*) are not applicable to areas with prevailing conditions different to those present during development of the model. See Hudson (1971: p. 65) and Elwell (1978, 1980a,b).

It follows that application of the model outside the conditions defined when the USLE was being developed has to be done with care and with due regard to the sensitivity and impact that extrapolation or modification can induce. Particular care should be taken when implementing the rainfall (*R*), slope and length of slope (*SL*), cropping/management factor and conservation practice factor (*CP*) components of the USLE. Research carried out in individual countries can provide values for these factors. Examples are provided in Roose (1977).

SLEMSA described

The SLEMSA has been designed to predict mean annual soil loss arising from sheet erosion on the area of arable land between two adjacent contour ridges. Its application to situations other than these will, in many circumstances, require the inclusion of controlling variables not accounted for within the SLEMSA, and this important point should be borne in mind by potential users.

Extensive testing of the USLE had proved its unreliability for prediction on the high veldt of Zimbabwe (Wendelarr, 1978). For the USLE, poor correlations were obtained for EI_{30} (on which R is based) with soil loss and difficulty was found in relating K and C values from the Zimbabwe conditions to those in the USA used to derive the USLE. The SLEMSA was thus developed by Elwell (1978, 1980a,b) and a team at the Agricultural Engineering Institute, Harare, Zimbabwe, to replace the USLE under these high veldt conditions. The applicability of the model throughout southern Africa has been tested in many of the countries contributing to SARCCUS (Southern Africa Regional Commission on Conservation and Utilisation of Soils).

The SLEMSA is similar to the USLE, including similar parameters, but some of these have been adapted to fit African conditions better. The model consists of three submodels K, X and C to account for soil loss estimation from bare soil, slope steepness and length, and crop types and cropping practices, respectively. These are related to the annual soil loss by :

$$E_a = K \times X \times C \tag{20.2}$$

where:

E_a = estimated average annual soil loss (t ha^{-1} $year^{-1}$)

K = mean annual soil loss from a standard, conventionally tilled field plot of size 30 × 10 m at a 4.5% slope, for a known soil erodibility, F, under a weed-free, bare, fallow management (t ha^{-1} $year^{-1}$) (F is related to K through a function dependent on rainfall energy)

X = the ratio of soil loss from a plot of length L metres and 5% slope, to that lost from the standard plot

C = the ratio of soil loss from a cropped plot to that under bare fallow management.

Figure 20.5 provides a schematic diagram of CALSITE incorporating the above three subcomponents. For any location the initial stage is to produce mapped information on four physical systems: climate, soil, crop and topography. In the SLEMSA these are treated as separate entities and this is an advantage over the USLE where interactions between model components can cause problems. A number of control variables have been identified and interpolating from these produced a considerable reduction to the number of field-plot measurements required.

Five control variables have been identified for estimating sheet erosion losses from arable lands. These are:

- seasonal rainfall energy (E),
- the amount of rainfall energy intercepted by the vegetation (I),

- the soil erodibility (F),
- slope length (L), and
- slope percentage (S).

The control variables, expressed in terms of soil loss, are combined in the three principal subcomponents. The K component combines the influences of mean seasonal energy E and soil erodibility to give an estimate of the mean annual soil loss from conventionally tilled bare soil at 4.5% slopes, 30 m long. The C component is a ratio which corrects the soil loss from bare fallow to that for the crop grown; and the X component, again a ratio, adjusts for changes of slope steepness and length.

In CALSITE the model components as developed for the Zimbabwean high veldt have been used, which is where the SLEMSA was developed and tested. The following proviso should be noted, Elwell (1980a) states, referring to the SLEMSA's subcomponents, that 'although some small progress has been made towards rationale, they remain essentially empirical and cannot be used with any certainty outside the area for which they have been developed. Furthermore, the Estimator is by no means fully developed as a predictor of high veldt soil losses'.

How CALSITE uses the model to produce a soil erosion estimate map

CALSITE determines the soil erosion through processing and creating a series of images that represent stages in the above models. For example the USLE process involves creating images that represent the various model factors of *R*, *K*, *CP* and *SL*. Digitized maps of each of the above four factors are produced from existing paper maps or by processing digital data, either using the IDRISI GIS or other GIS package. Figure 20.6 shows a map of land cover for the upper Magat catchment in the Philippines, which was used to provide the *CP* factor of the USLE (Bradbury *et al.*, 1993).

K and *CP* factors

Given catchments with minimal available data it is often necessary to be able to modify these factors easily, enabling multiple runs to test the sensitivity of the impact of the factor on the final result. To ensure this process is fairly straightforward, CALSITE makes use of a series of files called Look Up Tables (LUT). For the example of soil erodibility (*K*), the LUT contains the *K* factor value that is required for each of the different soil classes that occur in the catchment. Thus, if the user is unsure exactly what the correct factor for a given soil type is, a number of CALSITE runs can be undertaken to assess the impact of varying the selected value for that soil. Such LUTs are used for the *K* and *CP* factors.

R factor

Reliable information on rainfall intensity is rarely available and it is often necessary to estimate rainfall erosivity from total annual rainfall using empirical

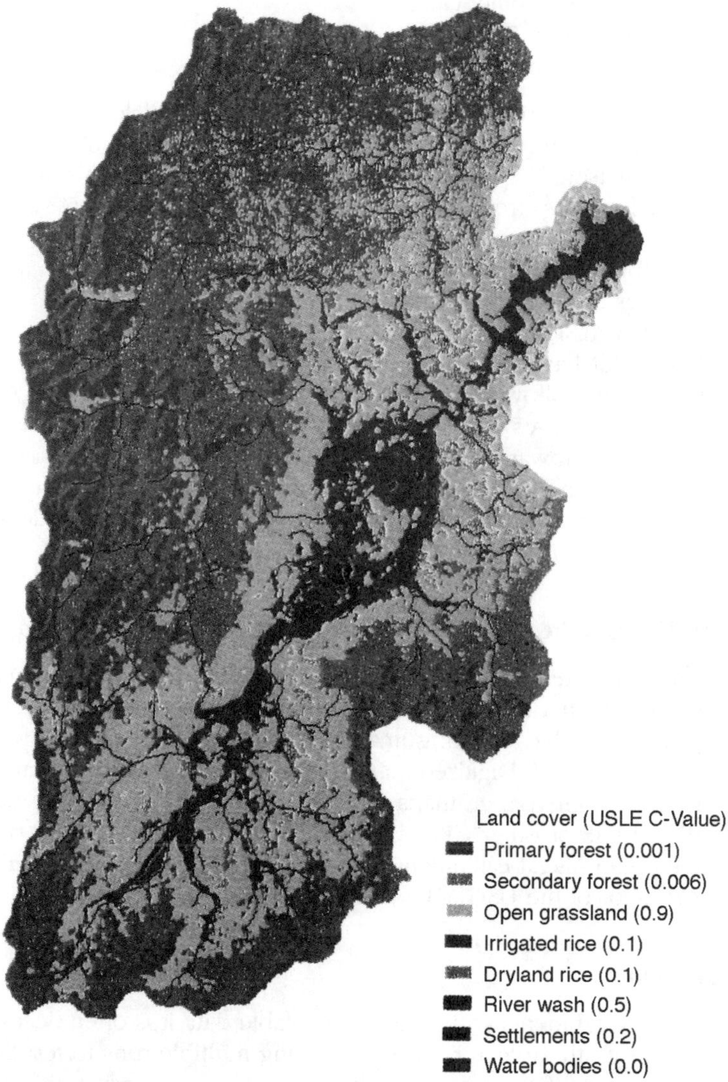

Fig. 20.6. A land cover image for the Magat catchment, North Central Luzon, the Philippines.

relationships. In CALSITE the following equation is provided which is based upon research carried out in Indonesia by Bols (1978). The equation has also been used in the example from the Philippines presented below.

$$R = 2.5P_a^2 / (0.073P_a + 0.73) \tag{20.3}$$

where R is rainfall erosivity and P_a is the annual rainfall in mm.

In generating the annual rainfall image it is necessary to overcome three major problems. There will be (i) a spatial variation of rainfall across the catchment, (ii) a possible further variation of rainfall related to elevation, and (iii) a variation of rainfall with time. As previously stated, the aim is to determine a long-term average annual rainfall. In the Magat catchment in North Central Luzon, the Philippines, considerable variation in rainfall was apparent reflecting the large elevation range of the catchment. Unfortunately, the majority of the rain gauges were located in the valley bottoms as access to the surrounding mountains was difficult. Different climatic zones were expected to influence different parts of the catchment, but rainfall records were not able to mirror this.

The *LS* factor

The process to determine the *LS* factor is the most complex of all the factors in the USLE. An almost identical procedure is used in the SLEMSA to determine the topographic ratio (*X*). Initially, the contours on a topographical map of a rectangular area encompassing the catchment are digitized. This is then converted to an elevation surface using a surface-fitting algorithm. IDRISI provides such an algorithm, which is a modification of the CONSURF algorithm developed by David Douglas of the University of Ottawa, Canada.

A slope image is also produced using another IDRISI function called SURFACE. The resulting image contains units of slope gradient in percentages representing the tangent of the angle multiplied by 100.

The aspect image is derived from the surface elevation image. The aspect of each pixel is based upon the elevation of the four corners of the pixel, using an algorithm defined in Lea (1992), and is given in units of radians. Zero radians is due west with a pixel value of 128 and, as the pixel value increases from 34 to 222, the aspect changes from east to north to west to south then back to east.

To determine the flows through the catchment, CALSITE uses the aspect of a pixel to determine the point of exit of flow along one of the pixel's four sides. The program then repetitively calculates entry points for each pixel over which the flow subsequently passes. This algorithm again is described in Lea (1992). The algorithm sums the number of converging flow paths and uses the total number of flow paths passing through a pixel as a measure of flow. As the number of converging flow paths increases, there will become a point where it can be said that the count represents a stream and in CALSITE this is regarded as being at a pixel value of 255. It is possible for flow paths to terminate in a hollow and CALSITE incorporates methods to filter these out.

The 'stream' image, digitized from the original topographical map, and the 'flows' image are combined to produce a 'length of slope' image. Estimates of length of slope are based upon the size of the contributing area. The values in the flow accumulation image show the number of pixels whose flow path's converge on that point. Thus a value of 50 indicates that the size of the contributing area is 50 pixels. By assuming the contributing area to be of uniform shape, the length of the contributing area from the furthest contributing pixel to the pixel under study can be estimated. This will be the longest possible length of slope and in reality is likely to be an overestimate of the length of slope contributing to overland flow since it does not represent an average flow path or take account of infiltration of overland flow.

In order to produce more realistic length of slope estimates, it has been assumed that the shape of the contributing area can be treated as circular rather than elliptical. The radius can then be used as the estimate of length of slope. The premise is that most overland flow will be due to pixels in the lower part of an elliptical contributing area and that the radius of the circular contributing area closely approximates the average length of slope of the contributing area.

Thus the formula for calculating the slope length L is

$$L = \sqrt{(F\,X\,Y / \pi)} \tag{20.4}$$

where L = length of slope in m
F = number of contributing flow paths
X = pixel's east–west dimensions in m
Y = pixel's north–south dimensions in m

The user can limit the length of slope. It is unlikely that water will flow as overland flow for long distances as it will inevitably converge to form a stream or infiltrate in a hollow.

The final determination of the LS factor in the USLE is based upon combining the length of slope (L) and slope images (S). CALSITE offers an alternative of a constant length of slope so only the slope image is used to determine the LS factor image. A maximum slope percentage may be set in order to exclude very steep slopes. For the USLE the LS factor is calculated using the equations of McCool *et al.* (1993).

Determination of soil erosion

Soil erosion is finally determined by multiplying together the four factors, R, K, LS, and CP. Figure 20.7 presents an example for a catchment in Thailand showing the soil erosion for 1979 and 1994.

What will be the Effect of Land-use Changes in the Catchment on Future Sediment Yield?

CALSITE attempts to assess this through use of transported erosion functions using the 'delivery ratio' concept. Given the above discussion of delivery ratio, CALSITE's relatively simple approach can still provide considerable information over the medium to long term. Here the 'delivery ratio' (DR) represents the proportion of predicted soil erosion which reaches an outlet, i.e.

$$DR = \text{sediment at catchment outlet/total soil erosion} \tag{20.5}$$

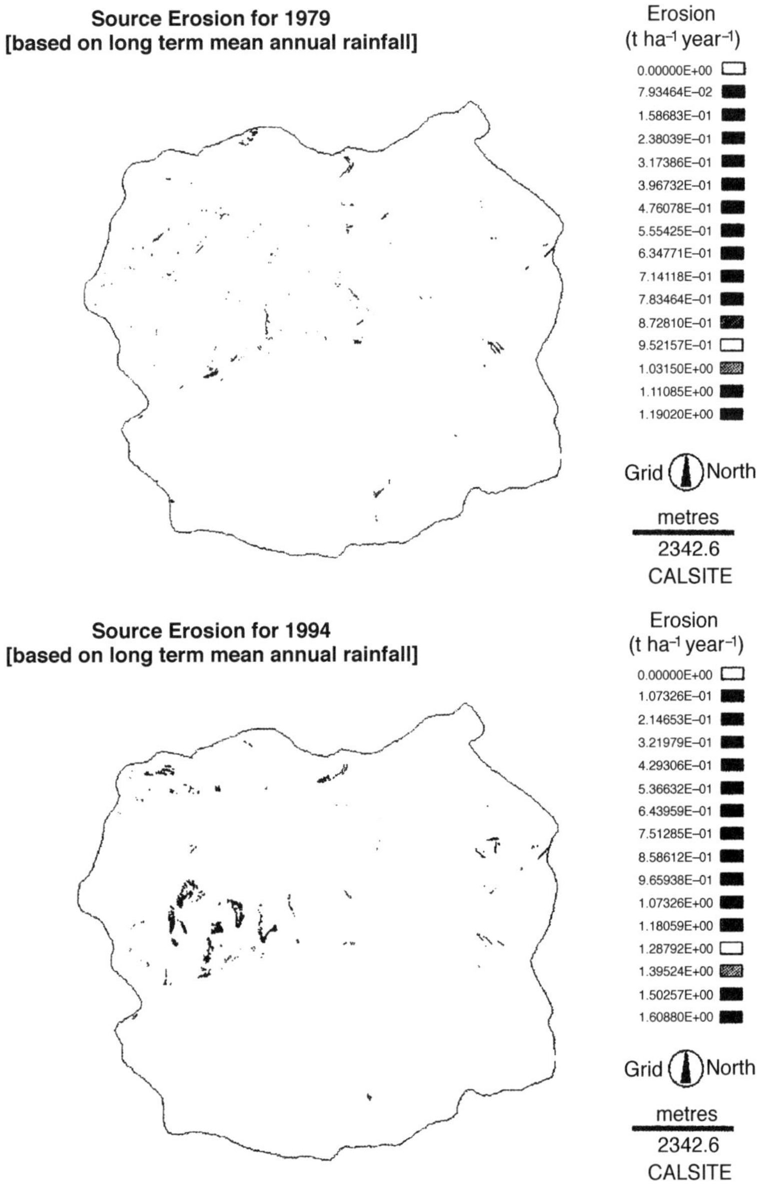

Fig. 20.7. Soil erosion images for Huai Ka Kang catchment, North Eastern Region, Thailand.

Thus a delivery ratio of 0.5 means that only 50% of the predicted erosion from a pixel reaches a reservoir or other outlet; the rest of the soil erosion is redeposited along the flow path. Within CALSITE the term 'delivery index'

(*DI*) is used to refer to an arbitrary value from 0 to 255, which is an estimate of the limiting transporting capacity down the flow path, and is related to delivery ratio. The delivery index value is determined for each pixel, where:

$$DI = \text{transport capacity/sediment supply} \tag{20.6}$$

A high *DI* value indicates that most sediment will be transported and a low value shows that there is insufficient transport capacity to transport all the sediment.

The equation used to determine delivery index is defined, as follows:

$$DI = \frac{Lc}{SE/Q} \tag{20.7}$$

where Lc = minimum limiting sediment concentration transporting capacity of flow path
SE = soil erosion
Q = potential runoff

The reason for selecting the minimum value is that this value will limit sediment transport and cause sediment deposition. Thus, if flow along a flow path traverses from a steep slope to an area of more gentle slope then sediment will begin to drop out around this point, since the available energy to transport is less. The location of minimum energy for transport thus limits the sediment transport.

Transporting capacity may be defined using equations based upon the laboratory studies of Govers (1990) as:

$$L = kQ^{1.66}S^{1.44} \tag{20.8}$$

where L = transport capacity
k = a constant
Q = discharge ($m^3\ s^{-1}$)
S = slope (%)

This minimum transporting capacity, *Lc*, can then be derived by assuming that discharge is proportional to runoff, which is determined from the contributing area:

$$Lc = k' P_a^{0.92} \min\left[F^{0.66} S^{1.44}\right] \tag{20.9}$$

where P_a = annual rainfall in mm
F = number of converging flow paths
S = slope (%)
min = minimum value down flow path
k' = a second constant

In Eqn 20.8, *Q*, the discharge, is estimated based upon empirical studies by Amphlett and Dickinson (1989) using the equation:

$$Q \propto P_a^{1.4} \quad (20.10)$$

Thus rewriting Eqn 20.7 using Eqns 20.9 and 20.10, the full equation becomes:

$$DI = P_a^{2.32} \min\left[F^{0.66}\ S^{1.44}\right] / SE \quad (20.11)$$

Although the exponents of P_a, F and S have values that are intended for general use in tropical catchments, editing facilities have been provided. For example, the exponents of F and S can be changed if it is known that in the project area the (D_{50}) median sediment diameter of eroding material is significantly greater than the 0 to 0.1 mm range assumed by CALSITE. Likewise, the exponent of P_a can also be changed if a power relationship has been established for the region under study relating runoff to annual precipitation.

Determination of minimum value down flow path

This operation determines the minimum $F^{0.66}\ S^{1.44}$ value down the flow path of each pixel. The routine is similar to the flow path determination routine except that a record is kept of the minimum value down the flow path of each pixel and this value is written to the 'Route Min' image file. A stream is assumed to have been reached if the 'flow' image has a value of 200 or more. Digitized stream locations are copied to the 'Route Min' image file and given a value of 255.

Converting delivery index to delivery ratio

In general terms, delivery ratio, *DR*, can be expected to increase with delivery index, until the limit $DR = 1$ is reached. The overall shape of the curve that converts delivery index to delivery ratio values is defined from consideration of the range of sediment sizes in transport. A multiplying constant is used that can be defined in terms of the *DR* value for $DI = 100$.

Calibrating the sediment yield estimates

Once the soil erosion and delivery index images have been created, the calibration procedure is run before determining sediment yield on a pixel by pixel basis. The calibration is performed using measurements of annual sediment load for the different subcatchments under study and where possible for the catchment as a whole. Since the sediment yield will represent a long-term average of say 25 years, it is important that data used for calibration are also consistent with that time frame. The application of CALSITE to catchments where an impounded reservoir exists and for which at least two reservoir surveys have been carried out can provide an alternative longer-term mean annual sediment yield. The trapping efficiencies of such reservoirs should be taken into account when the calibration is performed. CALSITE also allows the inclusion of trapping efficiencies for reservoirs within the catchment.

In the absence of any reliable information on sediment loads, the calibration may be based upon an estimated overall delivery ratio for the whole catchment. Such estimates may be based upon empirical equations derived from catchment characteristics such as area, length and average slope (Walling, 1983).

For the best calibrations of sediment yield estimates, average annual sediment load values (converted to tonnes per hectare) should be used for at least three areas representing different soil erosion and sediment delivery conditions. One of these areas should be the whole of the study area.

Performing the calibration

The calibration is performed by varying the estimated delivery ratio value for the delivery index of 100 between 0 and 1 until a good fit between observed and predicted sediment yields is obtained.

Once the calibration has been performed, the shape of the delivery index to delivery ratio graph which gives the best fit is displayed, in a similar format to Fig. 20.8, alongside details of the observed and predicted sediment yield values for the catchment. Also included at the bottom of the table is the modelled overall delivery ratio for the whole catchment and the 'critical' delivery ratio, which is the *DR* value corresponding to a delivery index value of 100. A chi squared (χ^2) value is included in the table to assess the goodness of fit of the calibration.

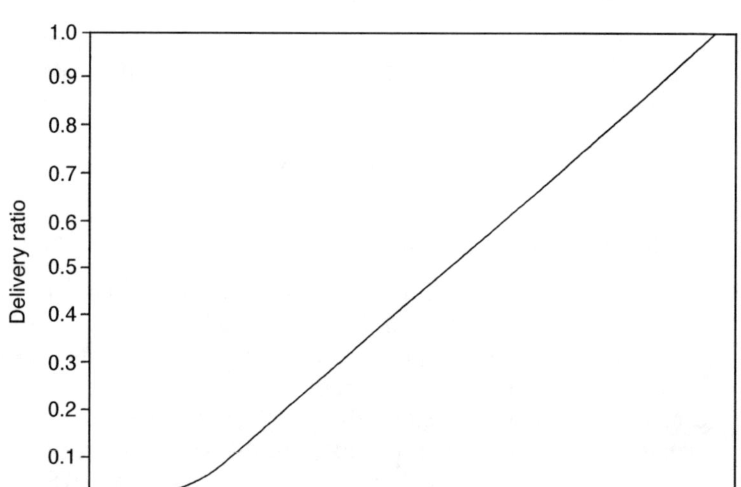

Fig. 20.8. An example of a delivery function.

Determination of sediment yield

Finally, the sediment yield is calculated for each pixel within the image based on the equation:

$$SY = DR \times SE \tag{20.12}$$

where SY = sediment yield
SE = soil erosion
DR = delivery ratio, a calibrated value from 0 to 1 based upon the delivery index image and a LUT file.

Comparison of soil erosion and sediment yield images

When both images have been created, it is possible to identify areas of high erosion potential. Areas that provide erosion products causing the greatest impact on downstream structures in the long term can be identified. There is a growing debate about whether it is better to target, for remedial action, older highly eroded areas from which material is already in transport and those at presenteroding, both of which do (or will in the near future) have an impact on downstream structures, or whether to target potential remedial action to areas that at present do not contribute to downstream impact but may do so in the future. In catchments with low delivery ratios, there is already a large amount of material stored which could be in transport for many years. In this case remedial action is effectively too late. Targeting potential high-erosion areas, on the other hand, may reduce sediment yields in the long term.

CALSITE offers the planner the opportunity to plan remedial actions or land-use change scenarios and rapidly see their medium- to long-term impact on the catchment erosion rates and sediment yield. It also allows comparison of options for catchment management at a variety of investment levels.

CALSITE Applications

The following sections are designed to give examples of some problems where CALSITE can provide useful information.

An application of CALSITE in the Philippines

CALSITE was applied to the Magat catchment to determine the source of erosion and sediment yield that impacts upon the Magat reservoir. The catchment covered an area of 4000 km^2 and had poor and out-of-date land-use mapping. To overcome this problem a Landsat 4 Multi-Spectral Scanner (MSS) image was used. Ground truth data for a number of known areas in the catchment were then used to perform a 'supervised' classification of the image into eight land-use categories. Since such imagery has a ground resolution of approximately 80 m, a pixel size of 50 × 50 m was used.

To overcome the earlier mentioned poor rain gauge cover for the catchment, a solution was derived using a linear function relating rainfall to annual rainfall at a particular rain gauge. This presents a problem: how does any particular year relate to the long-term average annual rainfall? To illustrate this, Table 20.1 shows the results from running CALSITE using the 1986 annual rainfall and that year's measured soil erosion data collected at three catchments nested inside one another, as calibration points.

Table 20.1 shows that CALSITE's calibration process produces a reasonable fit between the observed and predicted sediment yields. However, use of this calibration becomes dubious when the next year's results are taken into consideration. The full record for 1986 to 1991 indicates that 1986 represents an average year in terms of sediment yield and rainfall (Bradbury *et al.*, 1995).

An application of CALSITE in Thailand

As part of a study to quantify the amount of sediment that had entered the Huai Ka Kang Reservoir, in north-east Thailand, CALSITE was applied to the reservoir catchment to gain an understanding of the long-term erosion rates. The catchment drains the sandy plains of the north-eastern region, covers an area of 80 km^2 and has a change in elevation of only 40 m. Available topographical mapping only provided contours at 10 m intervals.

The pixel density used was 2 × 25 m and a long-term average annual rainfall was produced using a 35-year record from the only two rain gauges near the catchment. A long-term calibration was not undertaken as it was apparent that sources of erosion other than sheet erosion, which are not included in CALSITE, predominated, thus only the soil erosion component of CALSITE was run. The initial run used the long-term average rainfall and the only available land use based on a 1979 survey of the catchment. A fast transact land-use survey of the catchment was undertaken to incorporate any significant changes in land use between 1979 and 1994. CALSITE was rerun incorporating this data set along with the same long-term average annual rainfall to assess sediment yield rates at the reservoir.

Table 20.2 provides a summary of soil erosion information for both the 1979 and 1994 land-use images. Although considerable changes in land use had occurred between the two land-use map dates, there was little change in predicted erosion rate. The mean erosion rate determined is low (less than

Table 20.1. CALSITE results from the Magat catchment, Luzon, the Philippines.

Subcatchment	Area (km^2)	Sediment yield (1986) Predicted (t ha^{-1} year^{-1})	Sediment yield (1986) Observed (t ha^{-1} year^{-1})
Baretbet	2149	19.93	25.25
Aritao	144	23.17	22.00
Santa Fe	15	41.87	39.70

Table 20.2. Huai Ka Kang – summary of soil erosion information.[a]

Year	Mean pixel soil erosion ($t\ ha^{-1}\ year^{-1}$)	Maximum pixel soil erosion ($t\ ha^{-1}\ year^{-1}$)
1979	0.007	1.044
1994	0.009	1.694

[a] Based on long-term mean annual rainfall, not rainfall for the year in question.

0.1 $t\ ha^{-1}\ year^{-1}$) suggesting erosion is not of great significance. This result is supported by a very low mean sediment rate determined for the reservoir over 38 years of operation (Dickinson, 1995).

CALSITE experience in southern Africa

At the initial stages of incorporation of the SLEMSA into CALSITE there were studies, carried out by the Division of Water, Environment and Forestry Technology, CSIR, South Africa, which compared CALSITE with a number of assessments of soil erosion and sediment yield for the Mgeni catchment. This catchment is approximately 4353 km^2, rising to an altitude of 2000 m above sea level and drains eastwards for 257 km into the Indian Ocean in Kwazula-Natal. Rainfall is variable in the catchment, ranging from in excess of 1400 to 600 mm.

For this catchment, Table 20.3 shows results from CALSITE for a number of subcatchments. This study (Donald, 1997) compared CALSITE with a number of methods in the Mgeni catchment and also in the Sabie catchment, located on the border of South Africa with Mozambique, in the Kruger National Park. The methods compared with CALSITE were: the ACRU (Agricultural Catchments Research Unit) model developed by the Department

Table 20.3. Results from the application of CALSITE to the Mgeni catchment, South Africa.

Catchment and period simulated	Observed sediment yield ($t\ km^{-2}\ year^{-1}$)	Predicted sediment yield ($t\ km^{-2}\ year^{-1}$)	Area (km^2)
Inanda (1988–1998)	118.2	82.9	4082
Henley (1960–1987)	47.0	39.0	220
Albert Falls (1974–1983)	13.7	12.7	1642
Midmar (1965–1983)	10.0	11.3	927

of Agricultural Engineering of the University of Natal, Pietermaritzburg, South Africa; Rooseboom *et al.* (1992) using a regionalized regression technique: and finally, Niekerk and Heritage's (1994) method (see Donald, 1997, for description of technique), which used a lumped vector approach.

The conclusion from these comparisons was that although both the Rooseboom, and Niekerk and Heritage methods were easy to apply, the results were 'coarse'. The ACRU approach was believed to be the 'most detailed', including being event based. However, this meant that the inputs for the model were 'often not available' and CALSITE was described as 'a useful and reasonably accurate tool for predicting long term, average annual sediment yields'.

Conclusions

The development of geographical information systems has facilitated the prediction of soil loss in a spatially distributed manner; a range of data types can be incorporated to provide an integrated approach to catchment erosion studies. The identification and quantification of erosion sources is required in order to select appropriate conservation measures and land management strategies.

CALSITE has been designed to predict soil erosion rates and sediment yields from limited data on soils, land use, sediment yields, rainfall and topography. The required input data is generally available in most developing countries enabling the software to be widely applicable. CALSITE is based on simple empirical soil erosion predictors, the USLE and SLEMSA, which are applied to land units of a similar size to the plots from which erosion data are derived. Predicted annual soil erosion rates can be adjusted to account for deficiencies in the soil predictors. The software carries out simple sediment routing computations. These enable sediment yields, which in large catchments represent a proportion of the soil erosion, to be estimated. Once calibrated, the impact of various scenarios over the medium to long term can be rapidly compared. Successful applications to catchments in Thailand, the Philippines and South Africa have been undertaken.

References

Abernethy, C.L. (1987) Soil erosion and sediment yield: A review of data on river-borne sediment, and their use to indicate catchment scale soil erosion rates. Unpublished report for FAO, November 1987.

Amphlett, M.B. and Dickinson, A. (1989) *Dallao Soil Erosion Study, Magat, Philippines. Summary Report (1984–1987). Report No. OD 111*, HR Wallingford, in collaboration with the National Irrigation Administration, Government of the Philippines.

Atkinson, E. (1992) *Sediment Delivery in River Systems. OD/TN 48.* Overseas Development Unit, HR Wallingford, UK.

Bols, P.L. (1978) *The Iso-Erodent Map of Java and Madura.* Belgian Technical Assistance Project ATA 105, Soil Research Institute, Bogor, Indonesia.

Bradbury, P.A., Lea, N.J. and Bolton, P. (1993) *Estimating Catchment Sediment Yield:*

Development of the GIS-based CALSITE model. Report No. OD 125, April 1993, HR Wallingford, UK.

Bradbury, P.A., Lawrence, P. and Atkinson, E. (1995) *Estimating Catchment Sediment Yield. Refinement of the GIS-based CALSITE model.* HR Wallingford, UK.

Crickmore, M.J. and Aked, R.F. (1975) Pumped samplers for measuring sand transport in tidal waters. *IERE Conference on Instrumentation in Oceanography, Bangor, North Wales.*

Dickinson, A. (1995) *The Huai Ka Kang Catchment Sedimentation Study. Report OD/ITM 53*, HR Wallingford, UK.

Dickinson, A. and Bolton, P. (1992). A programme of monitoring sediment transport in north central Luzon, the Philippines. In: *Erosion and Sediment Transport Monitoring Programmes in River Basins, Proceedings of the Oslo Symposium, August 1992.* IAHS Publ. No. 210.

Donald, P. (1997) *GIS Modelling of Erosion and Sediment Yield in a Semi-Arid Environment. Report No. ENV/P/I 97089*, Division of Water, Environment and Forestry Technology, CSIR, Pretoria, South Africa.

Eastman, J.R. (1992a) *IDRISI Technical Reference Version 4.0.* IDRISI Project, School of Geography, Clark University, Worcester, Massachusetts, USA.

Eastman, J.R. (1992b). *IDRISI User's Guide Version 4.0.* IDRISI Project, School of Geography, Clark University, Worcester, Massachusetts, USA.

Eastman, J.R. (1993). *IDRISI Version 4.1 Update Manual.* IDRISI Project, School of Geography, Clark University, Worcester, Massachusetts, USA.

Elwell, H.A. (1978) *Soil Loss Estimation – Compiled Works of the Rhodesian Multi-Disciplinary Team on Soil Loss Estimation.* Department of Conservation and Extension, Zimbabwe, August 1978.

Elwell, H.A. (1980a) *Design of Safe Rotational Systems. For the Compiled Works of the Multi-Disciplinary Team on Soil Loss Estimation.* Department of Conservation and Extension, Zimbabwe.

Elwell, H.A. (1980b) A soil loss estimation technique for Southern Africa. In: *Proceedings of 'Conservation 80', Silsoe College, Silsoe, Bedford, UK.*

Govers, G. (1990) Empirical relationships for the transport capacity of overland flow. In: *Erosion Transport and Deposition Processes, Proceedings of the Jerusalem Workshop, March–April, 1987.* IAHS Publ. No. 189.

Hudson, N. (1971) *Soil Conservation.* Botsford Academic & Educational Ltd, London, UK.

Lawrence, P. (1996) *Guidelines on Field Measurement Procedures for Quantifying Rates of Soil Erosion and Catchment Sediment Yield. Report OD/TN 77*, HR Wallingford, UK.

Lea, N.J. (1992) An aspect-driven routing algorithm. In: Pearsons, A.J. and Abrahams, A.D. (eds) *Overland Flow: Hydraulics and Erosion Mechanics.* UCL Press, London, UK.

McCool, D.K., George, G.O., Freckleton, M., Douglas, C.L. Jr. and Papendick, R.I. (1993) Topographic effect on erosion from cropland in the North-western wheat region. *Transactions of the American Society of Agricultural Engineers* 36(s), 771–775.

McDole, R.E. and Reinersten, S.A. (1983) STEEP – an inter-agency, multidisciplinary approach to soil conservation. *Journal of Soil and Water Conservation*, May/June, 235–244

Onstad, C.A. and Foster, G.R. (1975) Erosion modelling on a watershed. *Transactions of the American Society of Agricultural Engineers* 18(2), 288–292.

Roose, E. (1977) Use of the Universal Soil Loss Equation to reduce erosion in West

Africa. In: *Proceedings of the National Conference on Soil Erosion, Soil Conservation Society of America, Ankeney, Iowa, USA.*

Rooseboom, A., Verster, E., Zietsman, H.L. and Lotriet, H.H. (1992) *The Development of the New Sediment Yield Map of Southern Africa. Water Research Commission Report No. 297/2/92,* South Africa.

Stocking, M. (1988) Quantifying the on-site impact of soil erosion. *Paper to 5th International Soil Conservation Conference, Bangkok, Thailand, January 18–2.*

Walling, D.E. (1983) The sediment delivery problem. *Journal of Hydrology* 65, 209–237.

Wendelarr, F.E. (1978) *Applying the Universal Soil Loss Equation in Rhodesia.* Unreferenced report, Institute of Agricultural Engineers, Harare, Zimbabwe.

Williams, J.R. (1975) Sediment routing for agricultural watersheds. *Water Resources Bulletin* 11(5), 965–974.

Williams, J.R. (1981) Testing the modified universal soil loss equation. In: *Proceedings of the Workshop on Estimating Erosion and Sediment Yield on Rangelands,. USDA, ARS, ARM-W-26, Tuscon, Arizona, USA.*

Wischmeier, W.H. (1972) Upslope erosion analysis. In: *Environmental Impacts of Rivers.* Water Resources Publication, Colorado, USA.

Wischmeier, W.H. (1977) Use and misuse of the USLE. In: *Soil Erosion Prediction and Control. Soil Conservation Society of America, Special Publication No. 21,* Ankeney, Iowa, USA.

Wischmeier, W.H. and Smith, D.D. (1965) *Predicting Rainfall-erosion Losses from Cropland East of the Rocky Mountains. USDA Agricultural Handbook No. 282,* Washington, DC.

Wischmeier, W.H. and Smith, D.D. (1978) *Predicting Rainfall-erosion Losses: a Guide to Conservation Planning. USDA-SEA Agricultural Handbook 537.* Agr. Res. Serv., USDA, Washington DC, 58 pp.

21

Modelling Catchment Erosion, Sediment and Nutrient Transport in Large Basins

A.J. Jakeman[1], T.R. Green[2], L. Zhang[1], S.G. Beavis[1], J.P. Evans[1], C.R. Dietrich[3] and B. Barnes[4]

[1]Centre for Resource and Environmental Studies and Centre for Integrated Catchment Assessment and Management, The Australian National University, Canberra, ACT 0200, Australia; [2]CSIRO Land and Water, Floreat Park, Western Australia 6014, Australia; [3]Faculty of Environmental Sciences, Griffith University, Nathan, Queensland 4111, Australia; [4]School of Mathematical Sciences, The Australian National University, Canberra,

A framework is proposed here to model catchment erosion, suspended sediment and phosphorus transport. It consists of two major components: an upland catchment model and an in-stream model. The upland model simulates streamflow (*Q*), suspended sediment (*SS*) and associated phosphorus (*P*) using rainfall data, and is calibrated to daily streamflow time series under historic conditions. Empirical relations are used to derive *SS* from *P* and *Q*. The in-stream model routes *Q*, *SS* and *P* from the outlet of upland catchments or in-stream nodes to nodes downstream. The upland conceptual rainfall–runoff model is parametrically efficient and very effective in predicting streamflow. The in-stream transport model can infer sources (resuspension and bank erosion) and sinks (settling) within a reach. Aerial photographs are used to assess the on-site effects of climate and land cover/use on erosion and the drainage network. Changes in land cover/use and the effects on the drainage network are related to the parameters in the rainfall–runoff model so that associated effects on *Q* (and hence *SS* and *P*) can be assessed. This modelling framework is being prototyped on the Namoi Basin in northern New South Wales, Australia and in the Ping Basin, Northern Thailand.

Introduction

A framework, which facilitates an appreciation of the local and cascading effects of land and water use options in a catchment system, may provide an

 Soil Erosion at Multiple Scales (eds F.W.T. Penning de Vries, F. Agus and J. Kerr)

essential contribution to improving resource management in being more sustainable, equitable and efficient. In this chapter the development of such a modelling framework is illustrated in terms of the physical effects on erosion and downstream water supply and quality. Extension to consider biological, social and economic effects is presented in Jakeman *et al.* (1997).

Definition of the physical problem

The problem being addressed here can be stated as follows:

- to infer the on-site and integral off-site effects of variations in climate, landscape attributes and land use/management in terms of erosion, water supply and water quality (e.g. suspended sediment and nutrient concentrations);
- to separate the effects of climatic forcing from the effects of landscape and land use/management; and
- to identify the relative contribution of different land (and water) areas to downstream effects, both historical and potential as these will vary with climate and land-use activities.

The scales associated with the above modelling will vary and comprise those that can be justified by the information content in the data available, yet are still useful for management purposes. Spatial data on erosion and the size of land units allowing satisfactory analysis dictate the scale of erosion effects. The stream gauge network dictates the scale of prediction of upland water quality effects.

The difficulty of the problem

The environmental modelling problem defined above is characterized by the following:

1. *Natural complexity:* Flows of water, sediment and nutrients, between and within media; each possess their own dimensions, time constants, spatial scales, throughputs and thresholds.
2. *Spatial heterogeneity:* The media (the atmosphere, land surface, subsurface, streams, etc.) are not uniform, and some transport processes may be non-conservative (e.g. involving deposition, resuspension).
3. *Sparse measurements:* System information will be disparate, spatially unrepresentative and contain pointwise measurement error; there will be limited observations of many internal system processes.

In the spirit of Clark (1986), we do not model such a problem in order to come up with a very detailed set of predictions. The combination of system complexity with data and model uncertainty makes such prediction unfeasible. Instead, we seek to build a framework and models to characterize the physical system, providing a basis for assessing how to intervene and influence desired land and water quality outcomes more to society's benefit.

Properties of the modelling framework

A modelling framework to treat this issue should possess the following properties:

- comprise a set of generic models, one for each subsystem type (e.g. rainfall–discharge, discharge–concentration, and in-stream flow), which allows investigation of the spatio-temporal system effects of climate forcing, landscape and land use;
- incorporate knowledge of the key processes in the catchment system;
- be compatible in conceptual complexity and discretization interval with the information content in the databases available; and
- link spatially distributed generic model components for simulation of cumulative effects at desired locations in the catchment network.

A central issue determining the precise prescription of the modelling framework is the database. In order to sharpen some of the foregoing discussion, the relevant database that is available for modelling runoff, erosion, sediment and nutrient concentrations in the Namoi Basin, New South Wales, Australia, is described in the next section.

Namoi Basin

Climate and land use of the Namoi Basin

The Namoi Basin (42,000 km^2) is located within the Murray–Darling Drainage Division in northern New South Wales (Fig. 21.1). Mean annual rainfall varies from 470 mm at Walgett in the west to 1100 mm at the top of the Great Dividing Range in the south east. As with many parts of South-East Asia, rainfall is summer dominant, characterized by high intensity, short duration thunderstorms which can result in flooding and significant erosion.

The Namoi Basin is a productive agricultural area supporting beef and sheep grazing, and both dryland and irrigated cropping. Dominant winter crops include wheat and barley, and the dominant summer crops are sorghum, sunflowers and cotton. Intensive land use includes dairying, cattle feedlots, piggeries and poultry farming. Approximately half of the catchment supports grazing, and one quarter of the catchment is used for cropping.

Climate and stream network

Precipitation and streamflow databases, which are necessary for inferring water availability and quality, are only available over scales of > 100 km^2 and vary considerably in representativeness and accuracy. Figure 21.2 shows the precipitation and discharge measurement network for the Cox's Creek, Mooki, Peel and Cockburn River tributaries of the Namoi River. The first two flow from the Liverpool Rangės and Plains. There are only two catchments under 200 km^2 with measured streamflow, and their record length varies from 11 to 21 years. The density of concurrent rain-gauge stations in these smallest stream-gauged catchments ranges from 93 to 150 km^2 per gauge.

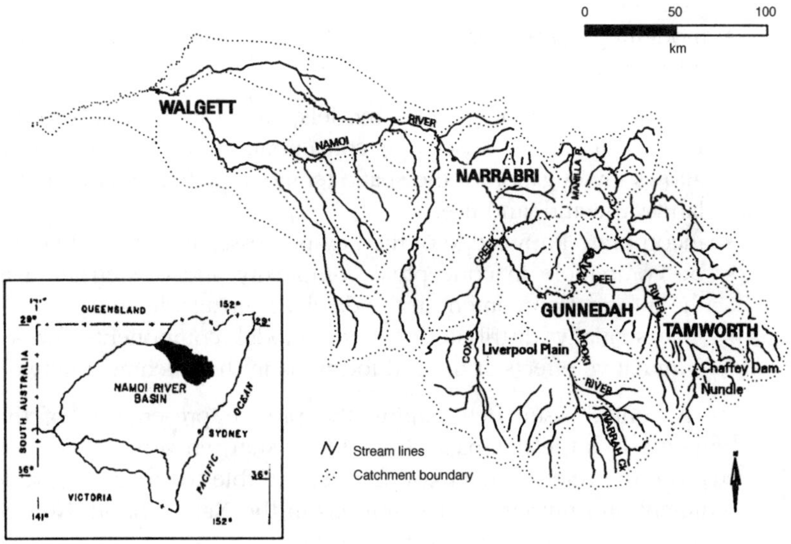

Fig. 21.1. Location of the Namoi Basin.

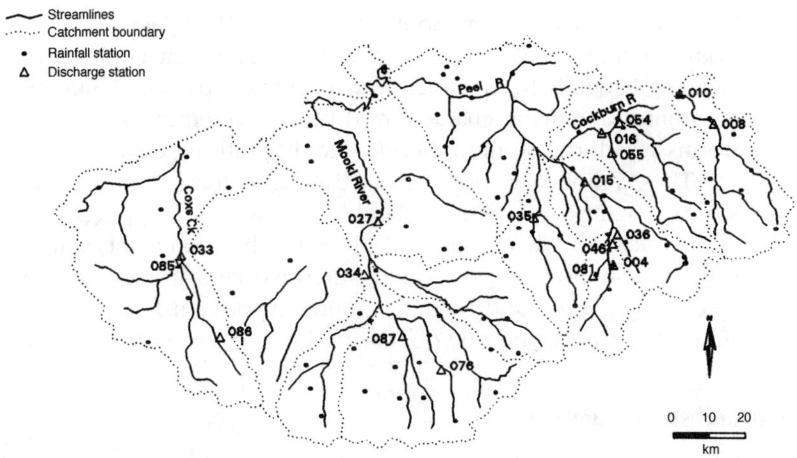

Fig. 21.2. The precipitation–discharge measurement network for the southern uplands of the Namoi Basin.

Hansen *et al.* (1996) have studied eight catchments in the Clarence Basin to the north-east to examine the predictive accuracy of a daily rainfall–runoff model. The stream-gauge rating and rain-gauge density were principal factors in the level of accuracy achieved. Duncan *et al.* (1993) found that rain-gauge density has a very strong effect on the accuracy of hydrograph parameters, with the standard error falling off as a power law with increasing gauge density.

Stream water quality measurements

The area shown in Fig. 21.2 contains the highest frequency water quality measurements in the basin. There are nine stations sampling water continuously and six of these have been in place since 1995. In addition to the above, there are five sites at which monthly grab samples are taken and six monitoring sites where weekly to fortnightly samples are taken. The authors have also instigated event sampling (~ 8 hourly) by landholders at five locations.

Landscape spatial data

Since the 1940s aerial photographs have been taken at approximately decadal frequency. These photographs have been used (Beavis *et al.*, 1997) to map land cover, streamlines, gullies and rills in the Cox's Creek catchment above Tambar Springs (gauge 033 in Fig. 21.2), in the Warrah Creek catchment (gauge 076) and in the Peel River catchment above Chaffey Dam (gauge 004). The maps have been digitized to estimate changes in gully and rill erosion through time and space in relation to land cover, climatic influences, soil and geological substrate.

Erosion and land cover data in the basin have been digitized for the 1980s. Land cover data, produced by satellite imagery, have recently been constructed by the New South Wales Department of Land and Water Conservation in the Liverpool Plains catchments for the summer of 1996/97 and the winter of 1997.

Sediment tracer measurements

Field observations of soils and sediments and analysis of natural environmental tracers in soil and water samples are being used to infer the genesis and, in some cases, the age of material in the basin network. For example, caesium and lead isotopes help differentiate between topsoil and subsoil contributions. Optical scanning luminescence techniques can be used to date the deposition of phosphorus in a sediment profile (Olley *et al.*, 1996).

Guidelines for Modelling Framework

In constructing a modelling framework to address the problem defined above, the following guidelines were used.

- Identify those system states and outputs which can be predicted with existing data, information and knowledge (and their spatio-temporal scales).

- Construct models of subsystem components wherein mean values of parameters and other moments to be inferred reflect subsystem response properties/characteristics.
- Relate characteristics to measurable subsystem properties (this may require a systematic data collection and causal analysis to find such relations in other locations where the requisite data can be acquired).
- Use empirical information for determining those outputs or states which cannot be modelled conceptually or physically (e.g. the outputs of the most complex processes involving chemical transformations).

The Upland Modelling Component

Given the above guidelines and the databases and model types available, the upland component of our modelling framework for predicting the on-site and spatially cumulative off-site effects of sediment and nutrient loss is presented below. Figure 21.3 summarizes the framework.

Rainfall–runoff model

The conceptual rainfall–runoff model IHACRES (Identification of Hydrographs and Components from Rainfall, Evaporation and Streamflow data) described in Jakeman *et al.* (1990), and updated in Jakeman and Hornberger (1993), is used to generate daily or higher frequency streamflow from rainfall and temperature inputs. Like most conceptual rainfall–runoff

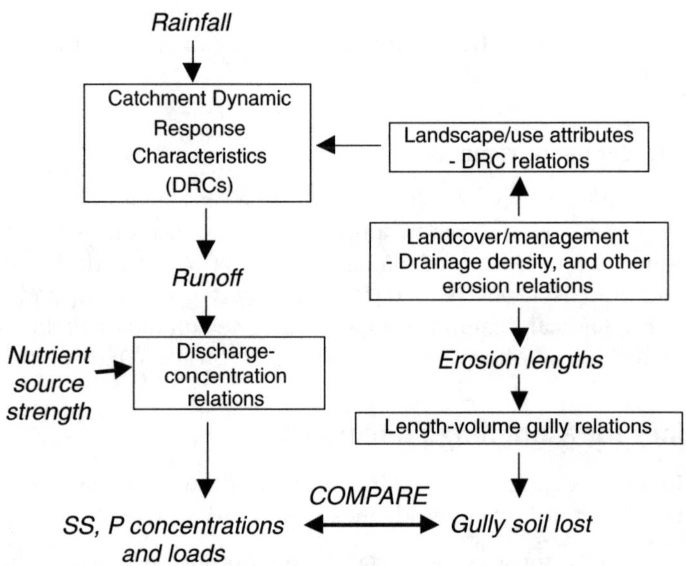

Fig. 21.3. Modelling framework for upland catchments.

models, it can predict streamflow with catchment-scale observations of precipitation and streamflow only. It has several advantages, most of which are not shared with other conceptual models, namely:

1. It predicts runoff well across a wide range of hydroclimatologies (see the example in Fig. 21.4), especially in its most recent forms with updated evaporative loss modules (Ye *et al.*, 1997).
2. It is highly efficient (five to seven parameters) and therefore its parameters can be estimated from relatively short periods of record (a few years of daily precipitation and discharge).
3. The parameterization encapsulates the response dynamics of catchment-scale behaviour allowing a characterization of similarities and differences in response between catchments (Jakeman and Hornberger, 1993).
4. The encapsulated model dynamics, derived easily from the model parameters, have been called dynamic response characteristics (DRCs) because their values are largely independent of climate in different model estimation/calibration periods (Jakeman *et al.*, 1993).
5. DRCs can be related to landscape attributes, e.g. quickflow recession rate corresponds to drainage density (Post and Jakeman, 1996), potential catchment moisture storage is influenced strongly by vegetation cover (Post, 1996).
6. The DRC known as the quick flow hydrograph peak (the quickflow response to a standardized pulse of rain) has been illustrated to relate well to long-term mean and 90-percentile turbidity (Post, 1996).

Discharge–concentration relationships

Relationships that allow material concentrations like suspended sediment and phosphorus to be derived from discharge are determined from the data.

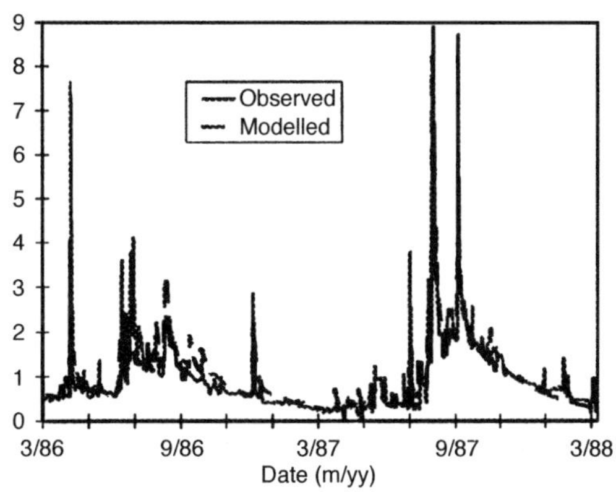

Fig. 21.4. Fit of the rainfall–runoff model to daily discharge in the Nam Mae Ngat at Ban Teen That catchment.

Various options exist to derive these relationships. One is to use a long-term relation between quickflow hydrograph peak and turbidity or suspended sediment routinely on an event basis. A second option is to relate cumulative material load to cumulative discharge through a power law. This has worked well (Barnes *et al.,* 1997; Crapper and Barnes, 1997) and has the advantage that the exponent in the power law relates to position in the landscape so that it becomes possible to infer an exponent value at locations between concentration measurement sites, especially where discharge is recorded.

Spatio-temporal landscape analysis

The aim of this analysis (see Beavis *et al.*, 1997) from digitization of aerial photographs is to assess the effects of climate, and categories of land cover and land management on lengths and volumes of rill and gully erosion and on the drainage network. The analysis is being used to produce a set of rules which state how climate, land cover and management affect on-site erosion and the drainage network. The predictive relationships between landscape attributes (e.g. drainage density or land cover) and dynamic response characteristics (e.g. quickflow hydrograph peak or runoff ratio), estimated by the rainfall–runoff modelling exercises, can then be used to modify the IHACRES model parameters.

Comparison of digitized photographs through time in the same subcatchment controls for geology, slope and sometimes land cover and management, thereby allowing the erosion effect of climate preceding the photographic period to be examined. Once the effect of climate has been determined, temporal comparisons of effects can be made within individual subcatchments identified to have changes in land management (such as contour banking) and then land cover (such as native pasture/woodland versus intensive pasture or cropping).

Comparison across catchments between the same time slices controls for climate to some extent. So, for example, changes in erosion over the same 10- and 20-year periods have been quantified for subcatchments with similar geology, soils and slopes. Once it is established from these analyses how to modify the IHACRES parameters for land cover and management type, the outputs of runoff and material concentration can then be simulated at all upland catchment sites for historic and hypothetical climate and land use. These are then inputs to the in-stream process model.

In-stream Modelling Component

The in-stream modelling component has a generic model linked to a landscape (stream morphology and riverine vegetation environment) classification and analysis, in this case to identify sources and sinks of sediment in reaches (Fig. 21.5).

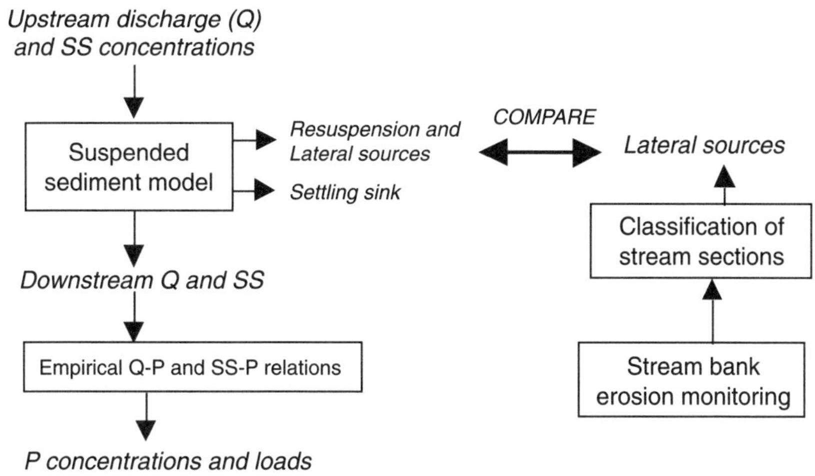

Fig. 21.5. Modelling framework for in-stream component.

In-stream model

The model STARS (Solute Transport with Advection, Resuspension and Settling) described in Dietrich and Jakeman (1997) has been developed to aid identification of the sources and sinks of sediment within reaches (Green *et al.*, 1997), and to route material through the reaches of a stream network. Such sources, which STARS quantifies, may derive from tributary loads, bank and bed erosion, and other effects including erosion from roads. The major features are mass conservation, advection, and settling and resuspension mechanisms. It is a conceptual model which makes certain necessary assumptions so that the parameters (four at present) associated with mechanisms can be estimated more reliably from upstream–downstream concentration measurements. For example, for sediment, there is a resuspension threshold velocity function which is a power law when flow rate is above a critical value and zero otherwise. Figure 21.6 illustrates the performance of the STARS model.

Stream bank erosion monitoring

Bank erosion sites are being monitored for a range of parent materials, stream size and power, and channel morphology. Each site was selected where bank recession was active on the outside of a bend, a relatively straight but widening channel section, or upstream portion of the inside of a bend. Changes in stream-reach and bank morphology over decadal times are being assessed using aerial photo analyses. Bank erosion sites similar to those being monitored are interpreted, and within each stream reach a spatial frequency distribution is assigned to each type of bank erosion monitored.

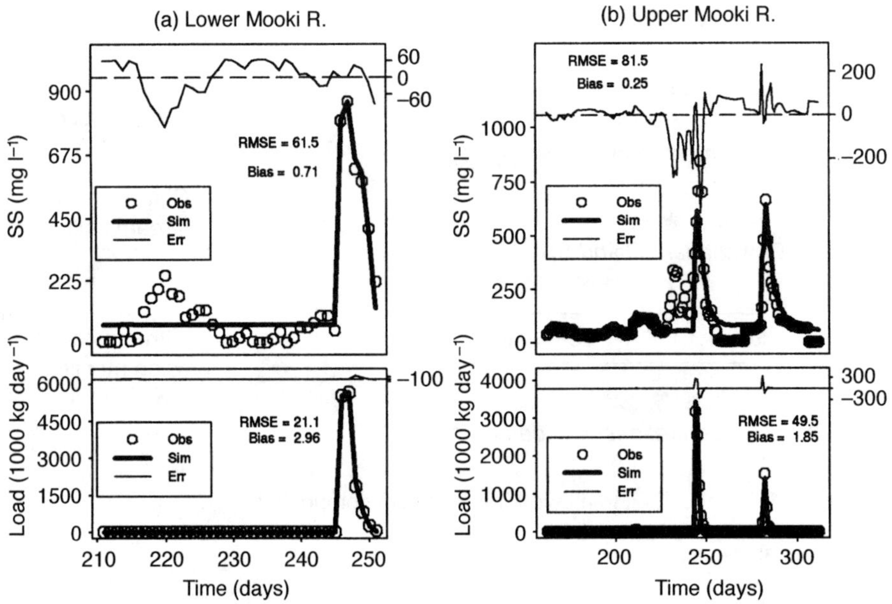

Fig. 21.6. STARS calibration to downstream suspended sediment concentration in Namoi Basin showing the direct model fit to concentration data (top) and the resulting estimates of sediment load (bottom) for (a) Lower Mooki River from Caroona to Ruvigne and (b) Upper Mooki by combining Warrah and Big Jacks Creeks upstream with flow adjustment to the Mooki River at Caroona. Times for both simulations are the day of the year, 1996, with the largest recorded event occurring around 7 September 1996.

Thus, an upper bound is estimated for sources of sediment from banks for given seasons and flow regimes (local redeposition is neglected).

For more precise quantification of the large-scale average bank erosion, the difference in suspended solid loads between upstream and downstream stations is related to a source term in the calibrated transport model, which in turn is related to bank erosion. The transport model makes it possible to quantify a mass of material from bank erosion, thereby allowing us to compare this with estimates derived from our spatial and temporal functional relationships between stream morphology (from aerial photo analyses) and measured bank erosion (Green *et al.*, 1997).

River Network Simulation Package

A computer program is being developed to simulate the effects of hypothesized climate and land use/management on stream discharge (*Q*), suspended sediment (*SS*) and nutrient concentration loads (*P* only at present) at selected points in a catchment or basin network. It is expected to be available by the end of 1998. It has a simple graphic interface to facilitate use by as wide a range of users as possible. It employs the framework and models referred to in this chapter. The package's screens and capabilities include:

1. A diagram of the total basin/catchment network and nodes available for simulation of *Q*, *SS* or nutrients (*P*).
2. Capability to select specified subcatchments or the total network for simulation of *Q* only, *Q* and *SS*, or *Q*, *SS* and *P* (concentrations and loads).
3. Selection of historical period for climatic inputs or importation of new climatic time series.
4. Specified parameter values for subsystem models (IHACRES rainfall–runoff model, *SS–Q* and *P–Q* relations, STARS in-stream model) for each subcatchment.
5. Guidelines specified regarding parameter changes to be expected with changes in land use/management and other landscape attributes.
6. Facility to modify parameter values for changes in landscape attributes as desired or according to guidelines in 5.
7. Tabulations and figures for system inputs and outputs.

Strengths and Limitations of the Framework

The framework presented here is largely a conceptual one, dosed appropriately with empiricism. It is sufficiently management oriented to be able to provide useful indications of the on-site and off-site effects of climate and land use, but not so empirical as to limit its ability unduly to evaluate management options.

The advantages of this approach are that it:

- is temporally continuous and the time step can be selected short enough for the model to be event responsive;
- attempts to separate climate, landscape and land-use effects;
- deals with gully and stream-bank erosion as sources of sediment, but can be combined with sheet and rill erosion models;
- assesses on-site (erosion) and off-site (water quality) effects;
- is based at the catchment scale for validation, but allows subscale variations in land cover/use to affect parameters at the catchment scale, and uses routing to obtain discharge, *SS* and *P* concentrations at larger scales (avoiding the scaling-up problem of a plot-scale approach); and
- is anchored to field measurements.

The model structures in our approach are very transportable to other locations, and do not demand an unreasonable monitoring investment. The whole methodology is data driven, mainly by rainfall, discharge and water quality measurements, the perceived weakness being that it cannot be applied immediately to other cases. However, any method will, in practice, require data to infer model parameter values.

The modelling procedure is computationally inexpensive, especially in 'what-if' mode. That is, once the parameters are estimated, simulation of climate and land-use options is virtually instantaneous. The model components are easily updatable so that parameters and relationships can be refined as more information and data comes to hand. Improvements to the approach will come from better understanding of upland and in-stream nutrient source strengths.

References

Barnes, C.J., Short, D.L. and Bonnell, M. (1997) Modelling water, nutrient and sediment fluxes using catchment scale parameters. *Hydrogeochemistry, Proceedings of the Rabat Symposium, April. IAHS Publ. No. 244.*

Beavis, S.G., Jakeman, A.J., Zhang, L. and Gray, S.D. (1997) Erosional history of selected upland catchments in the Liverpool Plains, New South Wales. In: McDonald, A.D. and McAleer, M.J. (eds) *Proceedings of the International Congress on Modelling and Simulation, MODSIM97, University of Tasmania, Hobart, 8–11 December*, pp. 277–281.

Clark, W.C. (1986) Sustainable development of the biosphere: themes for a research program. In: Clark, W.C. and Munn, R.E. (eds) *Sustainable Development of the Biosphere.* Cambridge University Press, pp. 5–48.

Crapper, P.F. and Barnes, C.J. (1997) Modelling sediment fluxes in large catchments. In: *Proceedings of the XXVII IAHR Congress on Water for a Changing Global Community, San Francisco, August.*

Dietrich, C.R. and Jakeman, A.J. (1997) A model for stream sediment transport with application to Murray and Murrumbidgee River reaches. In: McDonald, A.D. and McAler, M.J. (eds) *Proceedings of the International Congress on Modelling and Simulation, MODSIM97, University of Tasmania, Hobart, 8–11 December*, pp. 329–334.

Duncan, M.R., Austin, B., Fabry, F. and Austin, G.L. (1993) The effect of gauge sampling density on the accuracy of streamflow prediction for rural catchments. *Journal of Hydrology*, 142, 444–476.

Green, T.R., Beavis, S.G., Dietrich, C.R. and Jakeman, A.J. (1997) Observed streambank movement in relation to simulated in-stream suspended sediment transport. In: McDonald, A.D. and McAleer, M.J. (eds) *Proceedings of the International Congress on Modelling and Simulation, MODSIM97, University of Tasmania, Hobart, 8–11 December*, pp. 346–353.

Hansen, D.P., Ye, W., Jakeman, A.J., Cooke, R. and Sharma, P. (1996) Analysis of the effect of rainfall and streamflow data quality and catchment dynamics on streamflow prediction using the rainfall–runoff model IHACRES. *Environmental Software* 11, 193–202.

Jakeman, A.J. and Hornberger, G.M. (1993) How much complexity is warranted in a rainfall–runoff model? *Water Resources Research* 29(8), 2637–2649.

Jakeman, A.J., Littlewood, I.G. and Whitehead, P.G. (1990) Computation of the instantaneous unit hydrograph and identifiable component flows with application to two small upland catchments. *Journal of Hydrology*, 117, 275–300.

Jakeman, A.J., Chen, T.-H., Post, D.A., Hornberger, G.M., Littlewood, I.G. and Whitehead, P.G. (1993) Assessing uncertainties in hydrological response to climate at large scale. In: Wilkinson, W.B. (ed.) *IAHS Publication No. 214.* Institute of Hydrology, Wallingford, UK, pp. 37–47.

Jakeman, A.J., Ross, A.H., Wong, F. and Mackey, B.M. (1997) *Proceedings of the 2nd National Workshop on Integrated Catchment Management, Advancing Integrated Resource Management: Processes and Policies, The Australian National University, Canberra, 29 September–1 October.*

Olley, J., Caitcheon, G., Hancock and Wallbrink, P.J. (1996) *The Sources of Phosphorus and Sediment in the Namoi River Basin. CSIRO Division of Water Resources Progress Report, July.*

Post, D.A. (1996) Identification of relationships between catchment-scale hydrologic response and landscape attributes. PhD thesis, The Australian National University, Canberra, 301 pp.

Post, D.A. and Jakeman, A.J. (1996) Relationships between physical attributes and hydrological response characteristics in small Australian mountain ash catchments. *Hydrological Processes* 10, 877–892.

Ye, W., Bates, B., Viney, N.R., Sivapalan, M. and Jakeman, A.J. (1997) Performance of conceptual rainfall–runoff models in low-yielding ephemeral catchments. *Water Resources Research* 33, 153–166.

22

Designing and Implementing Outcome-oriented Soil Conservation Research

P. Lal, K. Coughlan and K. Menz
ACIAR, GPO Box 1571, Canberra, ACT 2601, Australia

Introduction

Chapters presented in this volume have mainly considered technical aspects of soil conservation, including measuring, modelling and predicting soil and nutrient losses, and addressed socio-economic analyses. In this chapter, the focus is on the development and management of outcome-oriented research. In our experience, poor planning, and poor project design and monitoring are more common reasons for projects not producing real outcomes than deficiencies in technical competency. A second contributing factor is the adoption of the traditional reductionist approach to research aimed at producing scientific results, without due attention being given to the extension and adoption of results, that has dominated our training. Whilst this approach has been critical in our understanding of the underlying principles and processes, it has not generated land management technologies that can be universally adopted by farmers and other decision-makers.

Over the last 15 years, the Australian Centre for International Agricultural Research (ACIAR)[1] has moved away from its emphasis on discipline-based biophysical and economic research at plot or farm level. ACIAR research has moved in three distinct directions:

1. Towards increasing involvement of relevant stakeholders and giving due recognition to identifying adoption pathways in the design and management of research.
2. Towards integration of disciplines to enhance the possibility of adoption of soil-conserving technologies at the farm level.

[1] ACIAR, established in 1982, is an Australian Government organization which funds collaborative research between Australian research providers and their counterparts in developing countries. Bringing together research groups collaboratively to address agricultural or natural resource problems which produce mutual benefits to Australia and to our partner countries is thus the central focus of ACIAR's activities.

3. Towards upscaling and integrating biophysical and economic research. This will allow bioeconomic assessments of on-farm and off-site effects at a catchment/regional or economy-wide level, for policy-makers to design appropriate government policies and farmer strategies that reflect both on-site and off-site impacts of soil loss.

This chapter will discuss six lessons learned by ACIAR in managing its portfolio of soil conservation research in South-East Asia.

Stakeholder Involvement

Stakeholder groups are often referred to either as 'change agents', such as government extension agencies or policy-makers who initiate the desired change in farming practices, or as 'target groups', who are receivers of information. Such groups are not mutually exclusive, and farmers can act as both target groups and change agents. Similarly, research results could be targeted at government agencies as the agencies of change. They can influence government policies and/or infrastructure support that can provide adequate incentives which empower farmers to act.

It is critical to ensure that the correct groups are involved in the research process, from the identification of the problem to design and implementation of the research project. In some projects, such as the on-farm research in the Philippines and Australia (see Swete Kelly and Gomez, Chapter 9, this volume), direct stakeholder involvement has been critical in ensuring the research remained relevant and appropriate for the needs of the farmers, and the results are readily adoptable. In Australia, the farmers were directly involved in the choice and monitoring of soil conservation technologies. This has resulted in the adoption of the minimum tillage technology by at least 60% of winter bean growers in the research study area.

However, it is not always evident who are the principal stakeholders. For example, where research is primarily targeted at government agencies, it is often not clear which agencies to involve, particularly since many government institutions are organized along sectoral lines and jurisdictional responsibilities do not reflect ecological boundaries. There are many examples that we can use. The one we use here is typical and no criticism of the groups concerned is intended. In the Philippines, the Department of Agriculture (DA) has the charter for agricultural extension, but only for 'alienable' lands of slope < 18%. However, in some remote developing areas, such as southern Mindanao where development of steep slopes is occurring, the DA is not active, and land is managed through the Department of Agrarian Reform (DAR). The DAR has no charter for either agriculture or extension, so the sustainability of research after its completion is doubtful. Therefore, unless both the DAR and the DA were involved in the design and the implementation of the research projects, the projects are unlikely to have the desired impact.

Where more than one agency has jurisdiction over management of soil conservation or there are likely to be many different stakeholder groups (such as farmers, government agencies, etc), the use of research advisory committees (RAC) could be relevant. Such committees could ensure relevance of research outcomes for their respective needs. For example in our bioeconomic project in Sri Lanka (see Menz and Grist, Chaper 3, this volume), off-site effects of soil loss from tea plantations affected the capacity of the downstream hydroelectricity plants. The involvement of the Ministry of Energy in the RAC ensured that these off-site effects were considered in the economy-wide analysis.

Ideally, stakeholder analysis should be an integral part of any R&D project, as is the case in our catchment study in Thailand (see Jakeman *et al.*, Chapter 21, this volume). In this project, direct involvement of the community in assessing their needs, aspirations and identifying alternative land-use options for detailed economic analysis is a critical component. Key government agencies with jurisdictional mandate in the upper Chao Phraya region will also be involved in various sub-projects in the project.

A consideration of stakeholders, and their needs, will also allow us to choose an appropriate scale for a research activity

Appropriate Scale of Research Activity

The research issue at hand and the management problem being addressed will no doubt dictate the choice of the scale of research. To illustrate this, we can look at two scenarios. In the first, the focus is mainly on-site loss of productivity due to soil erosion. The major stakeholder is the farmer, who bears the costs of loss in fertility and soil loss. She/he has the ability, with advice from research and extension workers, to remedy the situation. Field-scale research is appropriate in this situation. For example, in the work described in this volume by Ciesiolka and Rose (Chapter 18), Swete Kelly and Gomez (Chapter 9), and Cramb and Nelson (Chapter 7), farm-level studies are sufficient to address adequately the management issue.

In the second scenario, off-site effects on flooding and water quality are significant, and a wide range of stakeholders is involved. In this case, regional/catchment-scale research such as that reported by Jakeman *et al.* (Chapter 21) is necessary, to define both the on-site and off-site costs of the problem and the benefits accruing to different stakeholder groups from its solution. Often, large-scale soil erosion can have flow-on effects on other economic activities, such as was the case in Sri Lanka (see Menz and Grist, Chapter 3). In Sri Lanka, the government's ban on importing potatoes led to rapid expansion of potato farming on sloping lands, causing serious soil erosion and siltation of hydroelectric dams. In such circumstances, both cause and effect of erosion are at an economy-wide scale. Therefore, analysis at this level is appropriate, although biophysical and economic information at the farm and regional level may still be critical.

However, regardless of the scale of activity, scientific rigour must be maintained although, no doubt, the level of detail required will vary with the level of analysis. At the same time, a wider range of information, particularly social and economic data, needs to be considered as one moves away from plot/field level of analysis to farm/household, regional and economy wide.

Social and Economic Considerations

Social and economic factors, such as availability of labour, capital, and security of tenure, often determine a farmer's response to technological solutions presented by researchers and extension workers. For example, despite the availability of an appropriate hedgerow system, farmers surveyed in the project reported by Cramb and Nelson (Chapter 7) had not adopted the hedgerow-based system. One of the main reasons for this was the high labour cost in establishing and maintaining hedgerows. Lack of availability of labour during certain times of the year also was found to be a constraint in the adoption of hedgerow technology.

The results of this project also suggest that the opportunity cost of the farmer planting hedgerows is very high. When planting hedgerows, farmers are forgoing valuable land for what is seen to be unproductive use (at least in the short term). For farmers with limited capital, it may also mean keeping the children out of school or borrowing capital at high interest rates. In such circumstances, short-term costs of soil conservation can be large compared with the returns farmers may expect in the medium to long term.

Recognizing the social and economic constraints that the farmers are under and their short-term goals, researchers could also encourage conservation farming systems that meet their immediate needs. Farmers more readily accept conservation systems that incorporate cropping options that are profitable in the short term to balance any short-term costs of the system, such as establishing hedgerows. Examples in the project reported by Swete Kelly and Gomez (Chapter 9) have been vegetable cropping and the establishment of high-value fruit trees for longer-term income. An important part of these components is the incorporation of more sophisticated, integrated pest management systems. This illustrates the point that effective conservation farming systems are often equivalent to good farming practice in terms of crop agronomy and protection.

Therefore, although biophysical research may identify very effective technical solutions to the problem of soil erosion, it alone may not be sufficient to encourage farmers to adopt the technology (Cramb and Nelson, Chapter 7). One of the lessons that can be drawn is therefore that research focusing on technological solutions may not provide all the information needed to convince farmers to adopt soil conservation practices. Unless due recognition of social and financial characteristics of the farmers is taken, and the financial costs and benefits of the technologies are assessed, biophysical information is a necessary but not sufficient condition for farmers to adopt soil conservation techniques, even if it makes good scientific sense.

In some cases no amount of information would change the farmer's practice, particularly when the short-term costs are high and farmers are faced with the choice of feeding their families or conserving soil. In such cases, direct government assistance or change in government policy, such as a change in land tenure, may be required to change the farmer incentive, as suggested by Cramb and Nelson (Chapter 7).

Role of Researchers

It is better for researchers to adopt a structured but open mind about suitability of the alternative technologies rather than holding preconceived ideas about the appropriate (scientifically) technologies. As evident from our experiences in the Philippines, it is not wise to be dogmatic about optimal technological packages. For example, from the point of view of improving soil fertility, legume hedgerows could serve not only to provide a barrier but also improve the fertility of the soils due to their nitrogen-fixing capacity. Use of legume hedgerows has been encouraged by scientists throughout the world. However, grass hedgerows are more acceptable to farmers because they are easier to maintain, are less demanding on labour, and in fact are more effective as a barrier to sediment movement.

A further observation in our Philippines research was that where scientists adopted a facilitator role rather than be seen as providing all the solutions, farmers were more likely to take ownership of the research results and adapt technologies to be more suitable to their local conditions and their own circumstances.

Link Between Research and Adoption of Research Results

The assumption that the uptake of good technical solutions is automatic needs reconsidering, particularly in the case of land management practices where the short-term costs are significant. The results of ACIAR projects emphasize the importance of explicitly designing projects that recognize the continuum between research, extending the results and their adoption. By recognizing this, at the start relevant stakeholders can be involved in the design and implementation, either in an advisory capacity or as an active player in the research activity. By doing this researchers could ensure the research is relevant, that the outputs are targeted and appropriate for particular user groups, and that the user groups take ownership of the research and the results. For example, Australian farmers were actively involved in the project from the early stages of its design and this ensured their ownership of the soil conservation technology and that it developed and was assessed (Swete Kelly and Gomez, Chapter 9). The use of research advisory committees involving key policy-makers helped rapid uptake of research results by policy-makers in Sri Lanka (Menz and Grist, Chapter 3). The results reported to RAC, of which the Ministry of Environment was a member, led to a policy decision to minimize the conversion of less-erodible perennial crop lands to highly erodible seasonal/annual crops via provisions in the Central Environmental Authority Act.

Similarly, by explicitly considering the continuum between research and uptake of research results, the researchers can also ensure that the results are reproduced in a form suitable for the targeted group. Thus for example, in the case of methodological outputs targeted at other researchers, academic publications are the most appropriate means of disseminating the information to facilitate adoption. However, if the target group is the government policy-makers or the extension staff, then written outputs produced in a form more relevant to their needs, such as manuals, handbooks and policy briefs, are required to have significant impact.

If a research project focuses on, for example, producing scientific results, additional education strategies may be required for farmers to understand the consequences of soil erosion, and the short- and long-term benefits and costs of reducing soil loss, before adoption of soil conservation technologies could occur. In addition, as part of the adoption of any system, it may be important that farmers, and extension staff, are able to monitor the success of the change. Simple indicators for use by farmers and extension workers, such as soil colour, texture (consistency), and topsoil depth will give them a much stronger ownership of the process.

The impact of any research project will no doubt depend on the weakest link in the R&D chain. Therefore, when designing an applied research project, it is important to at least consider the key links in the chain and either include them in the project's design or else identify what other strategies would be needed before the impacts could be realized. Farmers may be totally convinced about the merits of a particular technology adoption but are constrained by either access to capital or something as simple as planting material for the hedgerow. Adoption of new practices may therefore require some degree of support from the government or other agencies such as NGOs or aid agencies.

Mode of International Collaboration

Perhaps the major lesson to be learnt from conducting and managing international projects (and particularly research in soil conservation) is not to try to be too ambitious. The tyranny of distance and the difficulty of maintaining good communications (particularly between a number of countries that have different languages and cultures) make relatively simple projects difficult. More often than not, the objectives in the project document are too ambitious and inadequate interaction between the Australian researchers and their counterparts in partner countries is often the single main reason for some of the original objectives not to be fully addressed. There is the real danger that communication breakdown will occur between collaborators. Often good communication occurs between the research leader in the ACIAR-commissioned organization in Australia and the individual collaborators, but communication between collaborators across countries may be limited to annual meetings. Greater use of modern information technology and availability of e-mail services in developing countries could help overcome some of the communications difficulties. However, such modern technologies will not in the short run help address the cultural sensitivities required in cross-cultural collaborative research.

Also, some of the discipline skills required for soil conservation research, such as hydrology, economics and social sciences, are not traditionally strong in many of our partner countries. In fact, many scientists well trained in agricultural areas such as soil science and agronomy are very poorly trained for carrying out soil erosion research. Expertise in economic research also varies from almost non-existent to marginal. In the case of economies in transition, it is almost non-existent because many of the economists are trained in Marxist economics and find market-based economic analysis a new field of study. Even in the presence of a strong expertise base, researchers from different disciplines are likely to have difficulties in communicating with each other, unless there is a greater degree of appreciation of the potential role individual disciplines could play in addressing the problem at hand.

Therefore, project leaders with a strong cultural sensitivity and collegiate approach to outcome-oriented research make better collaborative researchers than those who are singly focused on academic outputs within their individual discipline.

Reports of Working Groups

The symposium in Bogor was meant to exchange new insights, methodologies, and developments in thinking about soil erosion control. For two days, participants listened critically to presentations. The texts of most of those presentations are presented in these Proceedings. For another two days, participants discussed important and cross-cutting issues intensively in eight working groups. The groups were invited to be as explicit as possible, citing clear examples and giving strong recommendations on erosion control in general, and to IBSRAM specifically. Most groups did so remarkably well.

The reports of six groups are presented below. They are quite different in style due to the subjects, personal styles of the reporters and time pressure. We did not change this in order to preserve the group spirit. We are convinced, however, that you will enjoy the integrated insights and information contained in these working group reports.

Frits Penning de Vries, Fahmuddin Agus and John Kerr

Report of Working Group 1: Successful Technology Adoption Processes

Chair: John Kerr
Reporter: David Swete Kelly

Defining principles

Successful technology adoption is defined for two situations:

- Where there are no subsidies or material support associated with the technology, then success is characterized by spontaneous adoption and diffusion.
- Where material support or subsidies are provided, then success is determined by sustained adoption after the period of outside intervention.

A number of issues are related to this definition and need to be answered. These answers will be context specific and hence will vary from project to project. The key issues are:

- How much adoption are we talking about? Assessment of the magnitude of overall cost and benefits. The analysis may have to incorporate both on-farm and social impacts and can be performed *ex ante* or *ex post.*
- How is technology adaptation to be handled? Generally, adaptation or modification of the technology by the farmer is included as adoption provided the overall concept and erosion outcomes are maintained.
- What time frame are we considering? Generally, overall success needs to be evaluated in the short to medium term after the life of the project. However, evaluation during the life of the project should be an essential part of project management.
- Adoption itself should be determined by evidence that the farmer is investing in the technology, i.e. that material, labour or capital is being invested to maintain or improve the technology. This will obviously include evidence of spread to other areas of the farm. The fact that, for example, legume hedgerows are still on the property at the time of evaluation does not necessarily reflect adoption unless the farmer is maintaining and/or expanding the area.
- Negative indicators are also very powerful. Just as spontaneous adoption is important, spontaneous disadoption does occur.

The definition also raises the issue of subsidies in relation to sustainability. Project evaluation should take this aspect into account. Adoption that can be sustained when subsidies are in place cannot be considered as successful technology adoption if the subsidies themselves are not sustainable.

Elements to achieve successful adoption of conservation technologies

Successful adoption is related to a wide range of issues, some of which relate to the technology and some to the environment within which adoption is expected to occur. Adoption of conservation technologies need not require on-farm work. It can be facilitated at the broader scale as well. Some of the important issues are:

- Institutional mechanisms. Institutions or community groupings at the local, regional and national scale often provide an environment which impacts both positively and negatively on adoption. Specific institutions which assist adoption must be identified and mechanisms to establish or enhance these must be in place.
- Policies. Again, the policy environment will impact greatly. National policies related to price, land tenure and trade, for example, have impacted greatly. Local policies related to investment or other issues should also be considered.
- Development and extension process. At the local level, the way in which the technology is to be developed and extended will obviously play a major role in influencing adoption. Much work has been done on these processes. The topic is so important that the working group recommends that:

IBSRAM, through its Management of Soil Erosion Consortium, coordinates a working group to review the issue of successful adoption processes with the view to publishing methodological guidelines. This group can initially be facilitated via email. Several non-MSEC participants expressed interest in taking part as well.

Case studies: features and implications of successful adoption

Case studies are of two types:

- approaches to resource conservation; and
- farming systems or practices which conserve the soil and are very widely adopted.

Successful adoption case studies are not as rare as many at the meeting believe. There are cases of spontaneous adoption over very large areas, but often these cases do not fit out traditional concept of erosion control practices. It is contended that this is a fault of our concepts. Erosion control should not be seen as a defeatist issue. Simple and indigenous techniques occur widely and often over large areas. Techniques such as contour ploughing, for example, are often overlooked because they do not fit the research paradigm of a sophisticated developed technology.

The following are the case studies we have chosen to illustrate this point. For each case study we have identified the distinguishing features that facilitate successful adoption. We have then listed the implications these have for new work on increasing adoption of erosion control technologies. The cases are: (i) The Landcare movement; (ii) Participatory technology development and extension; (iii) Rice terraces; (iv) Agroforestry, specifically for industrial plantations; (v) Reduced, minimum and zero tillage; (vi) Vegetative contour barriers (including field boundary vegetation); and (vii) Cross-slope ploughing.

(i) Landcare
Features
- Bottom-up approach
- Farmers are the solution not the problem
- People are not told what to do: the group defines its agenda
- Great public reputation and esteem

Implications for adoption of conservation technologies
- Rural people have little political clout in developing countries
- Top-down approach is often used
- Oriented toward transferring full packages in the past
- Need increased public awareness of importance of rural issue

(ii) Participatory technology development
Features
- Based on local communities
- Community is empowered
- People work together on common problems
- Small at present but approach looks good

Implications
- Collective action is needed and must come from a local institution
- Should not be prescriptive as to the format of the groups. Local institutions set and implement priorities
- It reorients approach to research
- Working together is a prerequisite to success
- The approach is applicable to a wide range of situation and problem areas (if industrial crops)

(iii) Rice terraces
Features
- 2000 years old
- Allow water management of paddy rice
- Technology is familiar and physical environment was selected because of elevation
- Common objective and action
- Production, not conservation, is the objective
- Rice and good rice production is the basis of the whole religion and culture
- Tenure must be secure
- Role of off-farm income – ambiguous effect
- Deep soil resource needed

(iv) Industrial plantations in Malaysia (rubber, coconut, oil palm, coffee, cocoa, tea)
Features
- Originally on large farms
- Commercial orientation
- Area is increasing

- Export and organized marketing system
- Many enabling policies
- Need to intensify production (mature agronomic inputs and technology or components for conservation)
- Intensive research and development extension support
- Tenure

Implications
- Infrastructure
- Collective action to take advantage of economics of scale
- Technical/agroclimatic constraints – limited area/variety of application.
- The approach does not work everywhere

(v) Minimum tillage
Features
- Broad management benefit, not just conservation
- Reduced cost
- More flexibility in planting time
- Saves labour

(vi) Vegetation contour barriers (field handling vegetation)
Features
- Only minimal cost versions are attractive to farmers (establishment, maintenance)
- Benefits must be immediate
- Agronomic conditions matter
- Elements of this approach are widespread
- Does not work everywhere

(vii) Cross-slope ploughing
Features
- Transection from up and down in last decade
- No cost technology
- Easy
- Immediate gains: seed, fertilizer, infiltration

Report of Working Group 2: Status of Soil Erosion in South-east Asia

Chair: Keith Syers
Reporter: Djoko Santoso

Definition

The group decided that the soil erosion problem for the humid regions was essentially limited to water erosion. Tillage erosion can involve significant soil movement, but its economic impact is much less.

Existing data

There is a considerable amount of data and information available within NARS and other international organizations. **It is important that existing information be collated systematically, synthesized, and interpreted at the national level. It is recommended that IBSRAM facilitates this activity in conjunction with NARS and international organizations.**

Upscaling

Scaling is a critical issue and it requires an interdisciplinary strategy. Linking plot to catchment and then up to the national scale requires information on delivery ratio and the composition of sediment. Such information is not available for the region. Lack of knowledge or limited information about processes at the catchment scale restricts the use of available models. The problems at the country level are different from the sum of the problems at the plot and catchment level because of the interactive effects of processes at a larger scale. Furthermore, some decisions, which influence soil erosion at the national level, do not require detailed information at catchment and plot levels.

Scaling at the national level can be facilitated by adopting the following four approaches: (i) soil scape (ii) farming/cropping system, (iii) ecoregion, and (iv) interaction between soil, climate, and land cover. Application of GIS can facilitate assimilation of spatial information.

Nutrient balance

Soil erosion is an important process governing nutrient loss. Other processes of nutrient loss include removal in harvested crop products, non-return of residues, leaching, and gaseous loss (for nitrogen). Nutrient balance is an important approach for putting nutrient loss by erosion in context, and can also assist in developing proper soil management practices. Modelling can be useful in partitioning nutrient loss. Research is needed on enrichment ratios and on the impact of nutrient depletion on soil erosion.

NARS perspectives

Soil erosion assessment and its impact on agronomic productivity and economic loss have high priority with all national governments of the region. There are on-going programmes in the region, but countries use different methodologies. It is recommended that NARS use quantitative methods of assessment which will facilitate sharing the information. Despite the considerable amount of work being done at the plot scale, little quantitative information exists on historic soil loss.

NARS would appreciate assistance with technical support from within and from outside the region. Examples of the current projects in the region include:

- Laos: (i) National soil inventory, (ii) Sloping land project in five provinces, (iii) Soil survey and land allocation at country scale.
- Thailand: (i) DLD land management programme over the last 25 years in collaboration with FAO, ACIAR, USAID, IBSRAM, etc.
- Indonesia: (i) CSAR programmes with USAID, World Bank, ADB, ISRIC, IBSRAM and CIRAD.
- Vietnam: (i) IBSRAM, (ii) Swedish Agricultural University project, (iii) Programme with CIAT, (iv) National soil conservation project, (v) Collaboration with AGCD (Belgium).

Impact assessment

It is probably more logical to consider on-site and off-site impacts rather than agronomic and economic impacts. On-site agronomic and economic impacts are interrelated. Off-site economic impacts are complex, there are few experimental data, and these impacts can be positive or negative. The positive impacts are mostly land related but can have a time lag because the deposited soil is not immediately productive.

Off-site effects include those on water quality, aquaculture, sedimentation in waterways and reservoirs, and health hazards. Quantification of the effect of erosion on agronomic productivity at different levels of severity (slight, moderate, severe, and deposition) is needed. This experiment, on a natural landscape to assess the effect of past erosion, should be complemented by experiments to study the impact of present erosion and predict that of the future erosion rate. Erosion severity impact should be quantified using geomorphology and ^{137}Cs techniques. The data from this project can be used for model validation. The new experiments proposed should preferably be done in conjunction with IBSRAM-*ASIALAND* and ACIAR sites in the region.

Although the need for on-going and priority future research cannot be overemphasized, the overriding need is to collate, synthesize, and interpret existing information.

Report of Working Group 3: Magnitude of Catchment-level Erosion

Chair: Guy Trébuil
Reporter: Bofu Yu

We define, for the purpose of this report, catchment-level erosion as soil erosion from those areas of the landscape without significant gully and stream bank erosion. These catchments would broadly correspond to the headwater region of the first-order streams, and their lower boundary is delineated by the emergence of permanent stream channels. These catchments are henceforth called microcatchments because they are usually small in size, probably ranging from 1 to 10 ha depending on the slope.

Study of hydrology and soil erosion at the scale of the microcatchment is required for the following:

- To isolate the effect of the existing or changing crop and land management practices. In larger catchments, sediment movement will be related to other processes such as stream bank erosion. In addition, changes in land management at the microcatchment level can result in a change in the proportion of surface and subsurface flows, with significant consequences for the hydrology of larger catchments.
- To identify, by survey and measurement, areas of erosion hazard in the microcatchment, where urgent conservation practices may be required.

Monitoring, describing, analysing and modelling hydrological and erosion processes at the microcatchment scale represents a significant advance in our attempt to address the important issue of sustainable land management, and poses new challenges to the scientific community. In contrast to a typical erosion plot of the order of 100 m^2 (with an artificial plot boundary, a simple runoff mechanism, and little net deposition), microcatchments have a natural boundary and complicated topography. Both Hortonian overland flow and saturation overland flow can occur within various parts of these microcatchments. In addition, there can be significant zones of sediment deposition and temporary sediment storage can occur over a range of time scales.

In this report, we present a range of methods that can be used at the microcatchment scale, and for each method we have considered a range of relevant issues as outlined in the terms of reference (see Table WG3.1). We highlight research and other activities that will follow from this discussion. In addition, we have included a section on erosion ecology, interpreted to mean spatial interaction at the catchment scale.

Table WG3.1. Appropriate method for soil erosion studies at microcatchment scale.

	Method				
Terms of reference	Catchment survey (10^4–10^5 m^2)	Paired catchment (10^4–10^5 m^2)	Plot studies (10^2 m^2)	Modelling including GIS	Laboratory based
Objectives	To identify erosion hazards, processes and causes, to assist plot site selection and conservation planning	To evaluate land-use practices and to quantify the magnitude of the impacts of land-use change (with nested catchments and plots within them)	To quantify individual erosion processes and causes in relation to land-use practices	To describe, simulate, and predict the magnitude of erosion and its spatial and temporal variation at catchment scale	To understand the fundamental erosion processes of relevance to catchment scale
For whom	Planners, farmers and researchers	Planners, researchers	Farmers, NARES, researcher	NARES, consultants, planners	Modellers, scientists
Bottlenecks	Costs/GPS	Costs calibration, duration tenure/access, social issues	Data management, upscaling, extrapolation, long project duration	Distributed parameter models addressing spatial variability	Equipment availability and costs
Existing data sets	Malaysia (Morgan), Senegal (P. Perez), N. Thailand (F. Turkelboom)	Springvale in central Queensland, Nepal, FRIM (Malaysia), Latin America (ORSTOM), FRIM (Malaysia, Baharuddin Kasram)	Many, but only some, such as Springvale, Goomboorian, Mauritius (C. Ciesiolka), and Nepal (Ins. Hydrol.), are nested within larger monitored catchments	Kemaman, Malaysia (Ghulam, MARDI), Springvale (C. Ciesiolka)	Many, for example from Griffith tilting flume

Continued

Table WG3.1. *Continued*

Terms of reference	Catchment survey (10^4–10^5 m^2)	Paired catchment (10^4–10^5 m^2)	Plot studies (10^2 m^2)	Modelling including GIS	Laboratory based
Action on existing data sets/further analysis	Review the effectiveness of the survey approach	Desk-top review of tropical case studies by MSEC. Forestry and Water resources Departments are sources of case studies	Derivation of erodibility for extrapolation to microcatchments	B. Yu R. Collins P. Perez	C.W. Rose B. Yu W. Hogarth
Knowledge gaps	Need for a standardization of methodology	Questionable assumptions at times, lack of understanding of processes	Upscaling method in variable landscape	as Bottlenecks above	Understanding the role of grass strips and slope change
Action on knowledge gaps	*Field Guide for Erosion Hazard Assessment,* F. Turkelboom (in preparation)	Is appropriate model available? If no, MSEC to work on it	Griffith University	*Surface hydrology, sediment transport dynamics, nutrients balance and flux at the Kemaman site in Malaysia* (B. Yu, Ghulam M. Hashim, C. Rose)*	Further experimentation
Quality assurance	By comparison with *in situ* measurement of soil loss and indigenous knowledge	Is calibration OK?	Site, equipment and data management	Choice of model relevant to scale and assumptions	Publication in international journals
Measurement	Rill volumes/density, remote sensing, deposition of radio tracers	Flume and pumping samplers	Weather station, soil properties, tipping buckets, etc.	Not applicable. Measurements are inputs to models	

*Further action required.

A framework for investigation

For a specified catchment, determine if erosion is a problem.

NARES/NGOS collaborate to identify the scale/scope of the problem. Tools might include:

- GIS/remote sensing
- PRA
- Technical surveys (soil, land use, etc.)
- Literature review

If 'no' problem was identified, then 'forget' as a low priority issue.

If 'yes', produce 'definitive' outputs, e.g.:

- Maps – erosion hazard, sediment sources and sinks
- Stakeholder perceptions/needs/support acceptance
- Potential sites

What is the appropriate scale?

- Catchments >10 ha
- Microcatchments 1–10 ha
- Plots 100–500 m^2
- Microplots 1 m^2

Each scale has specific requirements and will yield specific outcomes.

Erosion ecology

Erosion plot ecology has been interpreted to mean spatial interaction at the catchment scale. Scale can be considered in both a temporal and areal dimension. Thus choose the scale in relation to the study objectives. The difficulties associated with the siting of a flume or another measuring device in a channel are best solved at the first step, i.e. before choosing the plot size.

The first important considerations for siting a flume are the following:

- Choose the stream segment where sediment moves freely, i.e. not a depositional zone
- Choose a flume type that will pass all sediment without deposition in the structure

These important features are best considered by accounting for the following:

- Bed slope
- Channel geometry and plan form
- Channel sediment
- Baseflow
- Contributing area – this might be as large as 20 ha
- Catchment land uses
- Costs
- In linear flow elements: channel flow can be routed by standard equations through the linear elements such as roads, gullies and irrigation channels

Second, identify the land-use mosaic by a land-use survey, noting hillslope geomorphology. A suggested procedure is:

1. Identify the land use interfaces (e.g. fence line, hedges, secondary forest, grass) and linear flow lines such as paths, tracks, channels, and ditches.
2. Carry out an on-farm erosion survey to check the performance of these linear flow lines and interface elements.
3. Identify the crucial interfaces and elements and check by using some simple measuring devices, e.g. troughs and ponds.

These can be followed by an experiment phase where two sets of experiment are at the interfaces of the mosaic combinations; select treatment combinations characteristic of the mosaic found on major land forms. Use this for plots. The number of plots is related to the number of mosaic combinations, e.g. upland rice, paddy, cabbage, maize, beans and orchards.

Report of Working Group 4: Assessing the Off-site Impacts of Land-use Changes

Chair: Amado Maglinao
Reporters: Anthony Jakeman and Thomas Enters

Introduction and rationale

Underestimation of the value of soil and water and the impacts of economic activities, both on-site and off-site, is a major factor contributing to widespread soil, water and land degradation. The continuous depletion of soil resources and the increasing scarcity of high quality water resources threaten prospects for sustainable economic growth. These processes have socio-economic implications whose magnitudes, especially off-site, are currently poorly understood and frequently misperceived. This is particularly so because natural scientists have inadequately communicated their knowledge to economists and decision-makers and neglected in their research designs the interconnectedness within and between ecosystems. In addition, economists have frequently relied on plot study results and assumptions that could not withstand any close scrutiny.

Decision-makers require information of immediate and future costs of land-use changes in order to prioritize problems and to design appropriate incentive systems for rehabilitating landscapes and reducing soil erosion. Multidisciplinary research on the off-site impact of land-use changes will develop new tools and provide solutions for land and water management problems, and support the development of technologies and more effective political instruments. Catchment studies of various hierarchical scales will address the environment–economy interactions and complex upstream–downstream interrelationships. Built on cost–benefit analysis and a scenario-based approach, the research will explore and assess in economic terms the interface between environmental supply and societal demands for goods and services derived from land and water resources.

Identification of potential environmental effects

The first step entails the identification of potential effects and their economic impacts. Its result provides the framework for further detailed studies and takes the form of a matrix or checklist of environmental services and related problems (Table WG4.1). In addition to the effects of land-use change and soil erosion, soil movement and the redistribution of soil over shorter distances, i.e. not necessarily leaving a particular watershed, needs attention. The potential impact varies among catchments. It is therefore necessary to base further research on case studies.

Quantification of impacts

The previous step is a prerequisite for the quantification of biophysical variables and provides answers to the question of why to measure any

Table WG4.1. Off-site biophysical impacts and affected socio-economic sectors.

Water quantity	Water quality	Both
Environmental services and problems		
Changes in water yields and regimes	Changes in sediment and bed-loads, nutrients and pollutants (organic and inorganic)	
Affected sectors		
Energy	Health	Agriculture Fisheries Navigation and transport Industrial, residential areas Biodiversity Climate (carbon sequestration) Sociocultural effects

environmental attribute. The 'why' provides the link between upstream and downstream and focuses on the supply of environmental goods and services and their demand by society. The methods to be applied address this interconnectedness between upstream supply and downstream demand through research to be conducted, expressed in simple terms, in both locations.

Valuation of environmental impacts

During the valuation, quantified supply and demand are translated into monetary terms based on cost–benefit analysis. There are three broad approaches that can be used:

- Change of productivity
- Replacement cost
- Preventive expenditure

The first approach does not rely on proxy measures, is logical and straightforward, although it provides meaningful results only if the impacts are properly quantified in the previous step. It requires not only information on upland changes and their potential impacts, but also a careful investigation of actual effects in the lowlands, i.e. off-site. Social surveys of stratified population samples provide detailed accounts of:

- Population structure
- Livelihood strategies and wealth of downstream residents
- Sociocultural aspects
- Economic activities
- Prices for various products and services
- Demand for goods and services
- Future trends for the above four categories

Based on social surveys, social impact assessments provide insights into those aspects that cannot be quantified and are required for contingent valuation (CV) during which participants of lowland communities are asked about their willingness to pay for upstream soil and water conservation. CV is a useful tool for assessing economic impacts where environment–economy interactions are difficult to predict. Cost–benefit analysis permits collective action that can impose harm on individuals. Therefore, the distributional and equity effects of downstream impacts are assessed in an additional step based on information concerning population distribution, livelihood strategies and wealth.

Scenario development

The forward looking assessment requires in the final step that values are assessed according to different scenarios of future developments. It is similar to a sensitivity analysis that is part of the cost–benefit analysis, but provides broader insights into expected impacts and cumulative effects. In the scenario development, the analyst focuses on the supply of goods and services that the upstream environment supplies to the lowlands and on the demand that the lowland communities have for upstream supplies. Both environmental supply and societal demand change in response to external and internal forces. For policy-makers it is important to balance decreases in supply with increases in demand, which may be brought about by factors such as population increase, economic growth or institutional changes amongst others.

Three scenarios based on five criteria are proposed at this point (Table WG4.2):

- The pessimistic, in which in general things are getting worse
- The neutral, in which the future is like the past
- The optimistic, in which developments are positive and things are getting better

Each scenario has a different effect on supply and demand of goods and services such as clean water. For example, it can be hypothesized that in the 'neutral' case, the demand for clean water would continue to grow slowly, while its supply would change due to continued deforestation of upland areas. Competition for water resources would further increase because of an increase in cultivated and irrigated areas as well as an increased demand for

Table WG4.2. Scenarios.

	Pessimistic	Neutral	Optimistic
Economic growth	+/–	+	++
Population growth	++	+	+/–
Environmental policies	+/–	+/–	+
Ecological stability	–	+/–	+
Climatic variability	–	+/–	+

clean water from the non-agricultural sector. According to this development, prices would rise slowly which would ultimately lead to ever-increasing costs if supply and demand cannot be balanced.

Finally, the scenario-building assesses the impacts of extreme events, that is when certain thresholds are exceeded. This includes, for example, the requirements for minimum streamflow and the need for occasional flooding of swamp habitats. No methods are proposed for this step as it needs to be assessed in individual case studies, depending on whether and what type of thresholds exist and whether they can be exceeded by land-use changes.

Off-site impacts of land-use changes on biophysical and economic aspects have been elaborated. Data (indicators) required for assessing such impacts are summarized in Table WG4.3.

Case studies

The following reports have recently reviewed methodological approaches to assessing economic impacts of soil and water conservation:

- Clark, R. (1996) *Methodologies for the Economic Analysis of Soil Erosion and Conservation. CSERGE Working Paper GEC 96-13.* University of East Anglia, Norwich.
- Enters, T. (1998) *Methods for the Economic Assessment of Soil Erosion and Soil Conservation. Issues in Sustainable Land Management No 2.* International Board for Soil Research and Management, Bangkok, Thailand, 102 pp.

Both provide insights into the advantages and drawbacks of using particular methodologies and serve as a basis for designing case studies of the valuation of off-site impacts. Their scope is limited to valuation aspects and methods for quantifying environmental variables are excluded.

A large number of case studies have been executed. Regarding the biophysical research component, the following bottlenecks and gaps still exist:

Table WG4.3. Methods for assessing the off-site impacts of land-use changes.

Indicators	Methods
Water yield	Water balance of the entire watershed. The measurement should be supplemented with regular (5 to 10 yearly) mapping of land use, and soil and topographic maps/data
Water quality	Regular monitoring of bedload and suspended sediment. Determination of nutrient and biocide concentration leaving the catchment; COD and BOD determinations
Change of off-site productivity	Economic analysis of change in productivity in off-site area, by taking into account other developments such as fertilizer and pesticide use

- Methods for scaling-up from plot to catchment
- Sparse monitoring, lack of historic data for some variables
- Expensive equipment and high labour costs
- Comparability of data and mega-data
- Efficient and robust models with low data requirements
- Wider application of known geochemical techniques for tracing and sourcing sediments, nutrients and biocides
- Wider application of physical field techniques for sediment movement on smaller scales, particularly soil movement that causes negative externalities without entering any water bodies
- Protection of equipment from vandalism
- Scientific innovation and training
- Well-trained human resources

Regarding the socio-economic research component, the following bottlenecks and gaps are recognized:

- Real integration of biophysical and socio-economic disciplines
- Limited knowledge of the interconnectedness within and between ecosystems
- Limited knowledge of the economy–environment relationships
- Limited experience with some valuation methods such as contingent valuation
- Historical data
- Limited understanding of the relationship between poverty, population growth and land-use changes
- Active involvement in the valuation process of all stakeholders

Conclusions

Previous studies have suffered from the following main deficiencies:

- A monodisciplinary approach
- Focus on plot studies
- Focus on upstream activities
- Lack of scenario development

The information and prescription above address these four drawbacks of earlier studies, give directions for future studies and provide broad guidelines for quantifying and valuing impacts during interdisciplinary case studies. They take the research activities from the plot to the catchment and from the uplands to the lowlands. In addition, impacts are assessed according to different scenarios. The proposed approach will improve our knowledge of what happens to the lowlands when changes in the uplands occur. Investigating the off-site impacts in a holistic approach will result in new knowledge and improve the basis for decision-making, although it is not possible to translate every aspect considered into monetary terms.

Report of Working Group 5: A Framework for Promoting the Soil Erosion Issue with Policy-makers

Chair Rattan Lal
Reporter: Kep Coughlan

Any framework for action should be activity based. The group considered that inappropriate communication of information, rather than lack of data, was the key issue. There are data shortages and data are not always provided in the most useful form, but presentation and communication have often been weak. Thus there is an urgent need to collate, synthesize, interpret and package information for use by policy-makers.

Two main thrusts are envisaged: economic and environmental. The impacts on productivity are hard to cost, but the emphasis on costs at the national level, rather than loss of income at the farm level, should be recognized. Dramatic examples of the environmental consequences of soil erosion (massive flooding), as well as the costs resulting from off-site sediment effects, should be given prominence. The impacts on carbon mineralization and biodiversity should not be ignored.

Future soil erosion experiments should produce data that are useful for economic impact assessment.

Recommendation: **IBSRAM should provide a framework for use by NARES for the collation, synthesis, interpretation and packaging of soil erosion information for use by policy-makers. Areas where erosion is not a problem should be communicated, as well as areas where it is or could become a problem.**

Report of Working Group 6: Participatory Approaches for Soil and Water Conservation

Chair: Hans-Dieter Bechsted
Reporter: Guy Trébuil

The discussion focused on the objectives and bottlenecks of participatory approaches, and on actions needed.

Objectives

Participatory approaches for soil and water conservation aim:

On the technological level:
- To improve sharing of knowledge between farmers and 'outsiders' (researchers, extension agents, policy-makers).
- To improve the relevance of the research agenda.
- To increase the impact and efficiency of soil conservation efforts.

On the socio-economic level:
- To assist in managing on-farm diversity (of farming systems, physiography).
- To contribute to empowerment, community development and equity.

Bottlenecks

Some bottlenecks of participatory approaches were detected, related to methodology and implementation:

Methodology:
- The information gathered is site specific and cannot always be extrapolated.
- Participatory exercises often pay too little attention to the differentiation in farmer groups.

Implementation:
- Communication skills of the implementers are sometimes too weak.
- Involving all stakeholders and managing potential conflicts during a participatory process can be a complicated matter. For example: what to do if the outcome of a participatory process is in conflict with current policy?
- Lack of staff to conduct a full participatory process.
- A participatory process can lead to overexpectations in the target group; or when many participatory exercises have already been conducted, some farmers start to give 'pleasing answers' (the effect of 'overgrazing').
- 'Participation' is becoming a 'buzz-word' and covers a lot of activities which have nothing to do with the original idea of participation.

Gaps in the participatory methodology

The following gaps were identified:

- Proper methods for the evaluation of a participatory process.

- Difficulties with knowledge integration: how to integrate the information of different stakeholders and disciplines.
- A lack of methods for data processing of qualitative data.
- New models, such as 'dynamic modelling simulation' and multi-agent systems'.

Recommended actions

Action related to training:
- Improve mediation and negotiation skills of practitioners (e.g. by 'role plays').
- Training in farming system analysis in the function of soil water conservation (e.g. identification of different household strategies and household groups).
- Design of participatory approaches, especially for soil conservation applications (e.g. recognition of 'indigenous soil conservation strategies').
- Training in data processing of qualitative and quantitative data.

Actions related to methodology:
- Development of a multi-agent, development decision, support system for soil and water.
- Conservation (e.g. IRRI–ORSTOM–CIRAD project in northern Vietnam).
- Simulation models to evaluate the impact of policies.

Actions related to policy:
- Policies for institutional collaboration (e.g. Eco-regional initiatives).

Case studies

Some examples of projects where participatory approaches were applied in the framework of soil conservation development:

- Aga Khan project (Pakistan)
- Myrada (India)
- Consortium for sustainable agriculture on hillsides (CIPASLA)
- SANREM project (the Philippines)
- IRRI–ICRAF experience (the Philippines)
- Farmer participatory research (CIAT)
- Soil fertility research (Mae Jo University, northern Thailand)

See also for case studies: RRA/PLA notes of IIED, UK.

Index